IMMOBILIZED AFFINITY LIGAND TECHNIQUES

IMMOBILIZED AFFINITY LIGAND TECHNIQUES

Greg T. Hermanson
A. Krishna Mallia
and
Paul K. Smith

ACADEMIC PRESS, INC.
Harcourt Brace Jovanovich, Publishers

San Diego   New York   Boston   London   Sydney   Tokyo   Toronto

This book is printed on acid-free paper. ∞

Copyright © 1992 by ACADEMIC PRESS, INC.
All Rights Reserved.
No part of this publication may be reproduced or transmitted in any form or by any means, electronic or mechanical, including photocopy, recording, or any information storage and retrieval system, without permission in writing from the publisher.

Academic Press, Inc.
1250 Sixth Avenue, San Diego, California 92101-4311

United Kingdom Edition published by
Academic Press Limited
24-28 Oval Road, London NW1 7DX

Library of Congress Cataloging-in-Publication Data

Hermanson, Greg T.
    Immobilized affinity ligand techniques / Greg T. Hermanson, A. Krishna Mallia, and Paul K. Smith.
        p.   cm.
    Includes bibliographical references and index.
    ISBN 0-12-342330-9
    1. Ligand binding.   2. Immobilized ligands (Biochemistry)
    3. Immunoadsorption.   4. Affinity chromatography.   I. Mallia, A. Krishna.   II. Smith, Paul K. (Paul Keith)   III. Title.
    QP517.L54H47   1992
    574.19'28--dc20   92-6321
                                                                                 CIP

PRINTED IN THE UNITED STATES OF AMERICA
92 93 94 95 96 97   EB   9 8 7 6 5 4 3 2 1

This book is dedicated to all those who labor in the arena of Life Science.

It is not the critic who counts; not the man who points out how the strong man stumbled or how the doer of deeds could have done them better. The credit belongs to the man who is actually in the arena, whose face is marred by dust and sweat and blood, who strives valiantly; who errs and comes short again and again; who knows the great enthusiasms, the great devotions, who spends himself in a worthy cause, who at best, knows in the end of the triumph of high achievement, and who, at worst, at least fails while daring greatly, so that his place shall never be with those timid souls who know neither victory or defeat.

Theodore Roosevelt

CONTENTS

Preface   xvii
Introduction   xxi

1. THE MATRIX

1.1	General Considerations	1
1.2	Natural Supports	6
	1.2.1 Agarose	6
	1.2.2 Cellulose	9
	1.2.3 Controlled Pore Glass and Silica	11
1.3	Synthetic Supports	15
	1.3.1 Acrylamide Derivatives	16
	1.3.1.1 Polyacrylamide Beads	16
	1.3.1.2 Trisacryl	19
	1.3.1.3 Sephacryl	22
	1.3.1.4 Ultrogel AcA	25
	1.3.1.5 Azlactone Beads	28
	1.3.2 Methacrylate Derivatives	31
	1.3.2.1 TSK-Gel Toyopearl HW	31
	1.3.2.2 HEMA	34
	1.3.2.3 Eupergit	37
	1.3.3 Polystyrene and Its Derivatives	39
	1.3.3.1 Poros	39
	1.3.3.2 Polystyrene Balls, Plates, and Devices	41
	1.3.4 Membranes	45

2. ACTIVATION METHODS

2.1	General Considerations		51
2.2	Procedures		53
	2.2.1	Amine Reactive Chemistries	53
		2.2.1.1 Cyanogen Bromide (CNBr)	53
		2.2.1.2 *N*-Hydroxy Succinimide Esters	57
		2.2.1.3 Carbonyl Diimidazole	64
		2.2.1.4 Reductive Amination	69
		2.2.1.5 FMP Activation	79
		2.2.1.6 EDC-Mediated Amide Bond Formation	81
		2.2.1.7 Organic Sulfonyl Chlorides: Tosyl Chloride and Tresyl Chloride	85
		2.2.1.8 Divinylsulfone	88
		2.2.1.9 Azlactone	90
		2.2.1.10 Cyanuric Chloride (Trichloro-s-Triazine)	95
	2.2.2	Sulfhydryl Reactive Chemistries	98
		2.2.2.1 Iodoacetyl and Bromoacetyl Activation Methods	98
		2.2.2.2 Maleimide	100
		2.2.2.3 Pyridyl Disulfide	103
		2.2.2.4 Divinylsulfone	107
		2.2.2.5 Epoxy or Bisoxirane Activation	107
		2.2.2.6 TNB-Thiol	107
	2.2.3	Carbonyl Reactive Chemistries	110
		2.2.3.1 Hydrazide	110
		2.2.3.2 Reductive Amination	115
	2.2.4	Hydroxyl Reactive Chemistries	118
		2.2.4.1 Epoxy (Bisoxirane) Activation Method	118
		2.2.4.2 Divinylsulfone	119
		2.2.4.3 Cyanuric Chloride	119
	2.2.5	Active Hydrogen Reactive Chemistries	120
		2.2.5.1 Diazonium	120
		2.2.5.2 Mannich Condensation	126
	2.2.6	Photoreactive Cross-Linkers	132

3. IMMOBILIZATION OF LIGANDS

3.1	Small Ligands		137
	3.1.1	Spacer Arms	137
		3.1.1.1 Diaminodipropylamine	141

		3.1.1.2	1,6-Diaminohexane	144
		3.1.1.3	6-Aminocaproic Acid	146
		3.1.1.4	Ethylene Diamine	147
		3.1.1.5	Amino Acids as Spacers	150
		3.1.1.6	Aminated Epoxides	153
		3.1.1.7	Use of Anhydrides to Extend Spacer Length	153
		3.1.1.8	Blocking or Capping Agents	156
	3.1.2	Sugars and Glycosaminoglycans		158
		3.1.2.1	Preparation of Melibiose–//–Polyacrylamide by Reductive Amination	158
		3.1.2.2	Preparation of Lactose–//–Sepharose 4B by the Bisoxirane Method	161
		3.1.2.3	Preparation of N-Acetylglucosamine–//–Sepharose 4B by the Bisoxirane Method	161
		3.1.2.4	Preparation of D-Mannose–//–Sepharose 4B by the Divinylsulfone Activation Method	164
		3.1.2.5	Preparation of Heparin–//–Sepharose 4B	165
	3.1.3	Inhibitors and Various Biospecific Binders		167
		3.1.3.1	Preparation of Immobilized p-Aminobenzamidine	167
		3.1.3.2	Preparation of Immobilized D-Biotin	169
		3.1.3.3	Preparation of Immobilized 2-Iminobiotin	170
		3.1.3.4	Preparation of Immobilized Pepstatin A	172
	3.1.4	Immobilized Dyes		173
		3.1.4.1	Immobilized Cibacron Blue F3GA	174
		3.1.4.2	Preparation of Immobilized Procion Red HE-3B	176
		3.1.4.3	Phenol Red and Thymol Blue	176
	3.1.5	Metal Chelators		179
		3.1.5.1	Preparation of Immobilized Iminodiacetic Acid	180
		3.1.5.2	Preparation of Immobilized Tris(carboxymethyl)ethylenediamine	181
	3.1.6	Hydrophobic Ligands		183
		3.1.6.1	Preparation of Immobilized Octylamine	184
		3.1.6.2	Preparation of Immobilized Benzylamine	185
	3.1.7	Immobilization of Drugs and Receptor Ligands		185
		3.1.7.1	Preparation of Immobilized Methotrexate	186
		3.1.7.2	Preparation of Immobilized Alprenolol	187
		3.1.7.3	Preparation of Immobilized 17β-Estradiol	188
		3.1.7.4	Preparation of Immobilized Tetrahydrocannabinol	193
		3.1.7.5	Immobilization of Ethynyl Steroids	194

3.2 General Protein Immobilization 195

- 3.2.1 Preparation of Immobilized Serum Albumin — 195
  - 3.2.1.1 Coupling of Albumin Using CNBr Activation — 196
  - 3.2.1.2 Coupling of Albumin Using Periodate Activation — 196
- 3.2.2 Preparation of Immobilized Avidin, Streptavidin, and Monomeric Avidin — 197
  - 3.2.2.1 Coupling of Avidin or Streptavidin Using CNBr Activation — 199
  - 3.2.2.2 Coupling of Avidin or Streptavidin Using Periodate Activation — 200
  - 3.2.2.3 Preparation of Monomeric Avidin — 200

3.3 Immobilization of Enzymes 202

- 3.3.1 Preparation of Immobilized TPCK-Trypsin — 203
- 3.3.2 Preparation of Immobilized Pepsin — 205
- 3.3.3 Preparation of Immobilized Papain — 206
- 3.3.4 Preparation of Immobilized *Staphylococcus aureus* V8 Protease — 207
- 3.3.5 Preparation of Immobilized β-D-Galactosidase — 208

3.4 Peptide Antigens, Antibodies, and Immunoglobulin Binding Proteins 210

- 3.4.1 Immobilization of Peptide Antigens — 210
  - 3.4.1.1 Coupling Peptides via Amine or Carboxylate Groups — 211
  - 3.4.1.2 Coupling Peptides via Sulfhydryl Groups — 214
  - 3.4.1.3 Coupling Peptides via Active Hydrogens — 216
- 3.4.2 Immobilization of Antibody Molecules — 217
  - 3.4.2.1 Preparation of Immobilized Anti-Human Serum Albumin — 223
  - 3.4.2.2 Preparation of Immobilized Anti-α_2-Macroglobulin Using Immobilized Protein A Cross-Linked with DMP — 224
  - 3.4.2.3 Site-Directed Immobilization of Antibody — 226
  - 3.4.2.4 Immunoglobulin Coupling to Polystyrene — 231
- 3.4.3 Immunoglobulin Binding Proteins — 244
  - 3.4.3.1 Preparation of Immobilized Protein A — 244
  - 3.4.3.2 Preparation of Immobilized Protein G — 246
  - 3.4.3.3 Preparation of Immobilized Protein A/G — 247
  - 3.4.3.4 Preparation of Immobilized Mannan Binding Protein — 250
  - 3.4.3.5 Preparation of Immobilized Protein B — 251

3.5 Immobilization of Lectins — 252
- 3.5.1 Preparation of Immobilized Concanavalin A — 253
- 3.5.2 Preparation of Immobilized Jacalin — 254

3.6 Immobilized Nucleic Acids — 255
- 3.6.1 Adsorption onto Cellulose — 255
- 3.6.2 Ultraviolet Irradiation Technique — 257
- 3.6.3 Entrapment in Cellulose Acetate, Agar, or Polyacrylamide — 257
- 3.6.4 CNBr Activation — 258
  - 3.6.4.1 Preparation of Immobilized HeLa DNA Using CNBr Activation — 258
  - 3.6.4.2 Preparation of Immobilized Poly (A) Using CNBr Activation — 259
  - 3.6.4.3 Preparation of Immobilized Single-Stranded RNA Using CNBr Activation — 259
- 3.6.5 Diazonium Coupling — 259
  - 3.6.5.1 Preparation of Immobilized Single-Stranded DNA Using Diazonium Activation — 260
- 3.6.6 Carbodiimide Coupling — 263
  - 3.6.6.1 Preparation of DNA–//–Sephadex G-200 Using Carbodiimide Coupling — 263
  - 3.6.6.2 Preparation of Oligo(dT)–//–Cellulose Using Carbodiimide Coupling — 264
- 3.6.7 Cyanuric Chloride Activation — 265
  - 3.6.7.1 Preparation of Double-Stranded DNA-Cellulose Using Cyanuric Chloride Activation — 265
- 3.6.8 Coupling With Carboxymethyl-Cellulose — 267
  - 3.6.8.1 Preparation of Calf Thymus DNA-Carboxymethyl-Cellulose — 267
- 3.6.9 Epoxy Activation — 268
  - 3.6.9.1 Immobilization of Calf Thymus DNA Using Epoxy-Activated Cellulose — 268
  - 3.6.9.2 Coupling of Nucleotide Homopolymers to Epoxy-Activated Cellulose — 270
- 3.6.10 Periodate Oxidation of RNA — 270
  - 3.6.10.1 Preparation of Immobilized RNA by Periodate Oxidation — 270
- 3.6.11 Reversible Complex Formation with Immobilized Boronic Acid — 272

3.7 Thiphilic Adsorbents — 274
3.8 Immobilized Disulfide Reductants — 275

4. TECHNIQUES OF THE TRADE

4.1 Measurement of Activation Level and Immobilized Ligand Density   281
    4.1.1 Measurement of Activation Levels Using Ninhydrin   282
    4.1.2 Determination of Immobilized Protein Using Bicinchoninic Acid   284
    4.1.3 Qualitative Assay for Amine, Hydrazide, or Sulfhydryl Functionalities Using 2,4,6-Trinitrobenzenesulfonate   286
    4.1.4 Assay of Hydrazide Functionalities Using Bicinchoninic Acid   287
    4.1.5 Assay of Biotin Binding Sites on Immobilized Avidin or Streptavidin   288
    4.1.6 Qualitative Determination of Immobilized Protein Using Coomassie Dye   290
    4.1.7 Measurement of Ligand Coupling by Difference Analysis   291
    4.1.8 Direct Absorbance Scan of the Immobilized Ligand   291
    4.1.9 Measurement of Sulfhydryl Groups Using Ellman's Reagent   293

4.2 Affinity Techniques   294
    4.2.1 Column Packing and Use   295
        4.2.1.1 Disposable Plastic Minicolumns   295
        4.2.1.2 Larger Columns   300
    4.2.2 Buffer and Flow Rate Considerations   304
        4.2.2.1 Binding Buffers   304
        4.2.2.2 Nonspecific Interactions   306
        4.2.2.3 Methods of Elution   307
        4.2.2.4 Flow Rates   310
    4.2.3 The Sample   312
    4.2.4 Measuring Capacity or Activity of Immobilized Ligands   313

5. SELECTED APPLICATIONS

5.1 Purification   317
    5.1.1 Purification of Trypsin Using Immobilized p-Aminobenzamidine   317
    5.1.2 Purification of Avidin or Streptavidin Using Immobilized Iminobiotin   319
    5.1.3 Purification of Lectins   321

	5.1.3.1	Purification of Wheat Germ Lectin Using Immobilized N-Acetyl-D-Glucosamine	322
	5.1.3.2	Purification of Castor Bean Lectins	323
	5.1.3.3	Purification of Jacalin Using Immobilized Melibiose	325
	5.1.3.4	Purification of *Griffonia (Bandeiraea) simplicifolia* Lectin I Using an Immobilized Melibiose Column	326
5.1.4		Purification of Human Serum Albumin on an Immobilized Anti-HSA Immunoaffinity Column	328
5.1.5		Purification of α_2-Macroglobulin Using Immobilized Anti-α_2-Macroglobulin Immunoaffinity Column	329
5.1.6		Purification of Immunoglobulin G Using Protein A, Protein G, or Protein A/G	331
5.1.7		Gentle Immunoaffinity Chromatography	333
5.1.8		Isotype Elution Buffers for Purification of Mouse IgG Subclasses	334
5.1.9		Purification of Immunoglobulin M	336
	5.1.9.1	Purification of IgM Using Immobilized C1q	336
	5.1.9.2	Purification of IgM Using Immobilized Mannan-Binding Protein	337
5.1.10		Purification of Mouse IgG_1 from Mouse Ascites Fluid Using Thiophilic Gel Chromatography	340
5.1.11		Purification of Human Serum Albumin Using Immobilized Cibacron Blue F3GA	341
5.1.12		Purification of Phosphoamino Acids, Phosphopeptides, and Phosphoproteins Using Iron Chelate Affinity Chromatography	343
5.1.13		Purification of Human Plasma α_2-Macroglobulin Using Immobilized Metal Chelate Affinity Chromatography	344

5.2 Scavenging — 346

5.2.1		Removal of Detergents from Protein Solutions	347
	5.2.1.1	Removal of Detergents from Protein Solutions Using Extracti-Gel D	347
	5.2.1.2	Removal of Detergents From Protein Solutions Using Bio-Beads SM-2	350
5.2.2		Removal of Lipids from Protein Solutions	351
	5.2.2.1	Removal of Fatty Acids from Serum Albumins by Lipidex 1000 Chromatography	351
	5.2.2.2	Removal of Lipoproteins Using Cab-O-Sil	351
5.2.3		Removal of Endotoxins from Protein Solutions	352
	5.2.3.1	Removal of Endotoxin Using Immobilized Polymyxin B	354

	5.2.3.2	Removal of Endotoxin from Protein Solutions Using Immobilized Histidine	355
5.2.4	Removal of Proteases		355
	5.2.4.1	Immobilized Aprotinin	356
	5.2.4.2	Immobilized Soybean Trypsin Inhibitor	358
	5.2.4.3	Immobilized p-Aminobenzamidine	359
	5.2.4.4	α_2-Macroglobulin, Carrier-Fixed	359

5.3 Catalysis and Modification — 360

- 5.3.1 Immobilized Enzymes — 360
  - 5.3.1.1 Immobilized Trypsin — 361
  - 5.3.1.2 Immobilized Anhydrotrypsin — 365
  - 5.3.1.3 Immobilized Chymotrypsin — 369
  - 5.3.1.4 Immobilized Anhydrochymotrypsin — 370
  - 5.3.1.5 Immobilized Pepsin — 372
  - 5.3.1.6 Immobilized Papain — 377
  - 5.3.1.7 Immobilized Bromelain — 379
  - 5.3.1.8 Immobilized Carboxypeptidase Y — 380
  - 5.3.1.9 Immobilized *Staphylococcus aureus* V8 Protease — 381
- 5.3.2 Immobilized Disulfide Reductants — 382
  - 5.3.2.1 Reductions of Peptides Using Immobilized Reductants — 383
  - 5.3.2.2 Reduction of Proteins Using Immobilized Reductants — 383
  - 5.3.2.3 Reduction of Oxidized Glutathione Using Immobilized Reductants — 384
  - 5.3.2.4 Reduction of Insulin to Insulin A and B Chains Using Immobilized Reductants — 385
  - 5.3.2.5 Reduction of Lysozyme and Ribonuclease A Using Immobilized Reductants — 386

5.4 Analytical Applications of Affinity Ligands — 388

- 5.4.1 Use of Immobilized Aminophenyl Boronic Acid to Quantify Glycated Hemoglobin in Red Cell Hemolysates — 388
- 5.4.2 Use of HPLAC for Automated Affinity Assays — 394
- 5.4.3 Immunoassay Techniques Using Immobilized Affinity Ligands — 398
  - 5.4.3.1 Enzyme Immunoassays Using Flow-Through Affinity Chromatography Columns (Immunography) — 399

	5.4.3.2	Enzyme Immunoassays Using Flow-Through Affinity Membrane Systems	406
5.4.4		Affinity Electrodes and Biosensors	410

Appendix — 417

References — 421

Index — 431

PREFACE

When we started the outline for this manual, we had one goal: Create an updated text on affinity chromatography that would be useful to both the expert and novice. With a combined total of well over 35 years of laboratory experience making and using affinity matrices, we hoped to create a practical book of methods that really work. What resulted was far more than we originally envisioned and much more than just another manual on affinity chromatography.

Over the past decade, immobilized affinity ligand technology has grown so rapidly that the chromatographic aspect has become just one of many application areas. What was once a niche dedicated almost entirely to solving purification problems has now become a diverse field of broad utilization. Instead of addressing just affinity chromatography, the book naturally evolved into a resource on how to design, make, and use immobilized affinity ligands in a variety of systems.

We have divided the use of these affinity systems into four main application areas: (1) purification devices to isolate target molecules from complex solutions; (2) scavenging reagents to remove unwanted contaminants; (3) modification or catalysis to effect specific transformations; and (4) separation tools to produce analytical determinations. Readers interested in using affinity techniques in any of these application areas can find a wealth of practical information on how to design a successful system.

The flow of the book is arranged to take the reader through the entire process of designing custom affinity supports. Chapter 1 discusses the basis of matrix selection. Examples include porous beaded matrices, nonporous particles, membranes, and a variety of plastic devices, to name a few. Complete descriptions of each matrix material's chemical and physical properties are provided as well as tips on how to best handle and use them. Recent years have seen the introduction of many new support materials useful in each of the main affinity application areas. Updated information is provided here which, to our knowledge, has not been discussed in any other book on affinity techniques.

Chapter 2 outlines the choices of activation chemistries used for coupling affinity ligands. Every functional group suitable for immobilization techniques is considered for both the matrix and the ligand. In every case, easy-to-follow protocols are given for both the activation and coupling steps. Several novel chemistries described in this section have not been previously discussed in other affinity books. For convenience, the book is liberally illustrated with chemical reactions, graphs, and diagrams describing key principles.

Chapter 3 provides a compendium of ligand immobilization examples. Discussions and protocols involving the immobilization and use of spacers, small biospecific molecules, peptides, proteins, antibodies, enzymes, nucleic acids, and many other affinity ligands are included. The affinity ligand that can solve a particular need can most likely be found within this chapter.

This book provides extensive treatment of immobilized affinity ligands for use in immunochemical techniques. The text provides information on matrix design for heterogeneous immunoassay development and thoroughly describes the purification, immobilization, and fragmentation of antibody molecules using a number of specialized affinity supports.

Any immobilized affinity ligand should be carefully optimized for its intended application. In Chapter 4, we provide assay methods to test for the degree of ligand loading and total matrix capacity, logical guidelines for media or buffer optimization, and tips on how to best "fine-tune" performance of any affinity support. These are well-honed, practical methods that will result in the best possible affinity separations regardless of the particular application.

Finally, Chapter 5 presents the numerous applications of immobilized affinity ligands. The discussion follows the four main categories of purification, scavenging, modification and catalysis, and analytical determination.

Topics distinctive to this manual include: (1) The use of the Mannich condensation reaction for immobilization of active hydrogen-containing molecules—especially useful for some drugs, steroids, and dye molecules; (2) azlactone chemistry for coupling amine and sulfhydryl-containing ligands; (3) the immobilized mannan-binding protein for purification of IgM; (4) techniques for site-directed immobilization of antibody molecules and their use in immunoassays or immunoaffinity chromatography; (5) fragmentation of IgG and IgM using immobilized enzymes; (6) techniques for immobilization on plastic surfaces and membranes; (7) complete techniques for immobilization of RNA and DNA; (8) new methods for the analytical use of immobilized affinity ligands in flow-though immunoassay formats using minicolumns and membranes as well as the latest in biosensor design; and (9) use of immobilized affinity ligands to remove detergents, lipids, proteases, endotoxins, metals, and other unwanted contaminants.

By the examples given throughout this manual, it is our hope that the reader will began to appreciate the beautiful and powerful idea of affinity interactions—not just with respect to chromatographic manipulations but also as a fundamental glue of nature.

The authors wish to thank the many excellent scientists at Pierce Chemical Company who, over the years, labored with us in the field of immobilized affinity ligands. Without their dedicated work, parts of this book would not exist. We also thank Joan Stockburger and Tara Asche for their expert help in preparing several figures.

Greg T. Hermanson
A. Krishna Mallia
Paul K. Smith

INTRODUCTION

The term "support" or "media" (as in "affinity support" or "affinity media") is usually understood to refer to a combination of (1) a ligand (usually of some known molecular configuration), that is firmly attached (e.g., immobilized), often by covalent means, and (2) a matrix (usually a solid insoluble substance).

To use an immobilized ligand in a separation strategy, the user must first decide whether to purchase a commercially prepared support or make the support of choice. This manual is intended to aid in this decision by allowing the reader to evaluate commercial options more properly and by providing sufficient examples to accomplish in-house preparations.

There are many valid reasons to seek out commercially prepared supports; convenience and speed perhaps top the list of reasons to explore commercial alternatives. Additionally, the buyer is assured that the purchased support was prepared by an "expert." Perhaps the best reason to consider a commercial source for affinity media is that commercial sources, by and large, provide a product well-characterized in terms of ligand loading and/or capacity.

The option to seek a commercial source assumes that the desired configuration can be purchased. Ideally, the affinity support should contain the correct ligand, attached optimally to the best matrix available. A compromise almost always accompanies the decision to purchase such a support. Whether or not the compromise is tolerable depends on the separation problem.

The Trouble with Affinity-Based Separation Strategies

It has been our repeated experience that we can tolerate a number of suboptimal separations, as long as the process is reliably reproducible. The reproducibility of the process in an affinity-based system is affected by the purchase or preparation of the separation media. If the preparation of an immobilized ligand is indeed an exercise yielding a good but unique object, all hope for a reproducible separation process is doomed. Even with commercially available supports such as immobilized protein A, there are enormous differences in selectivity and capacity from one manufacturer to another, not to mention lot-to-lot variations from the same manufacturer. Without a doubt, making reproducible supports is a major problem inherent with affinity-based separations. No other separation technique is plagued with quite so many vagaries because no other separation technique offers so many options with respect to matrix selection, ligand selection, and method of ligand attachment.

Fortunately, the situation is not hopeless. Reproducible affinity supports *can* be made or purchased and used repeatedly. In support of this statement, we offer the example of boronate affinity chromatography described in greater detail in Chapter 5. Aminophenyl boronic acid can be consistently immobilized to a cross-linked agarose support and used to capture and quantify levels of a blood protein with a between-batch coefficient of variation of less than 5%.

An objective of this manual is to provide sufficient information by way of example to allow the reader to prepare or purchase an affinity support that not only will perform well but will produce reproducible results. We have structured this manual in the sequence of events one might follow to first construct and then use an affinity support. Chapter 1 deals with matrices and their selection, Chapter 2 with matrix activation, and Chapter 3 with ligand attachment. Chapter 4, "Techniques of the Trade," is truly the heart of the matter, since here we describe how to use the combination of components described in the previous chapters. Chapter 5, "Selected Applications," is an attempt to teach the material in Chapter 4 by example.

The order of the topics was determined with convenience in mind. There exists, however, the possibility that the chosen presentation is misleading on at least two counts. First, we do not want to imply by our "modular" presentation of affinity-based separations that the technique consists of simply "plugging" together matrix, activation, and ligand. Each component of the combination influences the other to a large extent, and the entire combination must function as a system in an environment of buffers and temperatures selected to bring out the best performance characteristics of the individual affinity support. On many occasions, in our experience, a change in matrix material or activation chemistry will result in a behavior of the same ligand radically different from that seen on a reference support. We call this a "matrix effect" or an "activation effect," which is another way of saying we do not fully understand the result. We do know the effect is real and often dramatic.

A second problem with our presentation of affinity-based separation systems is that, although the order of topics is logical from a synthetic standpoint, the design and planning of an affinity support is best done in the reverse order. In other words, the objectives of the separation should direct the design of the method. If an affinity technique is deemed the appropriate methodology, then the ligand should direct the choice of activation and matrix.

Affinity techniques are most useful and powerful when the objectives are:

1. *Isolation and purification* of target macromolecules when purity demands are high and concentration or purity of the target molecule is low in the initial sample. The inherent selectivity of affinity techniques offers (and delivers) high purities at high yields under the mildest of conditions.

2. *Capture and detection* of macromolecules for analytical purposes. Frequently, small synthetic affinity ligands have the power and selectivity of antibodies and can be obtained and used in a similar fashion. Affinity binding of a low-concentration analyte to an insoluble support represents concentration and enhancement of analytical sensitivity.

3. *Selective removal (scavenging)* of undesirable contaminants from process streams, buffers, or media. The selective nature of affinity techniques allows the removal of trace amounts of contaminants without otherwise altering the composition of the initial sample.

4. *Enzymatic catalysis and chemical modification* Immobilized enzymes or reagents efficiently modify target molecules economically and usually without contaminating the final product.

In the chapters of this manual we have included representative examples of each of the four utility areas just listed. We have done this not in an attempt to document what has been done but in an attempt to illustrate what is possible.

1

THE MATRIX

1.1 GENERAL CONSIDERATIONS

A "matrix" is any material to which a biospecific ligand may be covalently attached. Typically, the material to be used as an affinity matrix is insoluble in the system in which the target molecule is found. Usually, but not always, the insoluble matrix is a solid. Hundreds of substances have been described and employed as affinity matrices, from ground shrimp shells (chitin) to grass pollen to sea sand to volcanic pumice and, of course, to processed seaweed extract (agarose).

When designing an affinity support for any separation problem, perhaps the most important question to answer is whether a reliable commercial source exists for the desired matrix material in the quantities required. Fortunately, a large number of reliable suppliers offers a wide range of practical and efficient matrices. An appendix at the end of this manual lists vendors of various support materials that have been found to be very useful. Many vendors will describe their products as being the "best" available. Despite their tendency to exaggerate the virtues of their product, many vendors have active research programs that have been producing new and improved matrix materials.

The commercial availability of a wide range of high quality, high performance, and economical matrices represents a dramatic improvement over what was available before 1980. Still, at the time of this writing, no "perfect" or "best" matrix material for every application is available, in our opinion. Before choosing a matrix for a particular application, it is useful to ask the following questions.

1. *Is the matrix available from a reliable commercial source in convenient quantities?*

Commercial availability is an essential attribute, because companies that manufacture matrix materials do it better, more reproducibly, and more economically than individual investigators can. Manufacturers know procedures and pitfalls and have the proper analytical expertise and equipment to verify their results. A bonus in dealing with a reputable supplier is the wealth of technical help available from such a vendor. A well-equipped separations laboratory should have a variety of commercially available matrix materials at hand, as well as a full range of vendor catalogs and technical data sheets on file.

2. *Does the matrix have an abundance of easily derivatizable functional groups?*

The preferable functional group for subsequent activation and attachment of an affinity ligand is the primary hydroxyl group. Primary hydroxyls are amenable to a wide assortment of activation procedures, but do not contribute to nonspecific binding of nontarget molecules. The most popular and useful matrix materials have an abundance of either naturally occurring or synthetically incorporated primary hydroxyls. Beaded agarose has a natural abundance of primary and secondary alcohols. Agarose perhaps represents the ideal functionalized surface since a large population of secondary hydroxyls remains to preserve the hydrophilic nature of the matrix even after primary hydroxyls are consumed in activation or cross-linking reactions.

Supports with an abundance of primary amines (for example, polyethyleneimine) or carboxylic acids (such as carboxymethyl cellulose) are easily derivatizable, but resulting supports may contain residual nonspecific ion-exchange effects.

3. *Does the matrix have good mechanical and chemical stability?*

The mechanical and chemical stability of a matrix material is most severely challenged during the chemical activation and ligand coupling procedures. If a matrix cannot withstand solvents, oxidizing conditions, pH variations from 3 to 11, and moderate mixing operations, its utility is severely limited.

Mechanical stability usually pertains to the ability of a matrix to withstand high pressure and flow rates without disintegrating or becoming deformed. In this respect, rigid solid supports such as glass, silica, or highly cross-linked synthetics are presumed to be preferable to soft gel particles such as agarose. This is not always the case since another important aspect of mechanical stability is resistance to abrasion, which can be tested by rubbing a few moist particles between the thumb and forefinger to see if gentle rubbing causes the matrix to form a slime or paste.

4. *Does the matrix possess good flow characteristics?*

The flow characteristics of a matrix depend largely on particle size, particle size distribution, and rigidity. For laboratory columns of 1 L capacity or less, a particle size of 100 ± 40 µm is a good balance between flow rate and capacity (discussed subsequently).

A narrow particle size distribution is not as critical a requirement for successful separations based on affinity chromatography as it is for ion-exchange or reverse-phase techniques. Ion-exchange separations are dynamic reiterative sorption–desorption events, whereas most affinity interactions are once-on/once-off processes. Sepharose CL-6B, as commercially supplied, has a fairly wide particle size distribution of 45–165 µm, which is adequate for affinity chromatography but hardly qualifies as a high performance matrix.

Although particle size distribution is not critical to affinity chromatography, this parameter cannot be disregarded. A narrower particle size range will result in better and more efficient column capacity because of less frequent column channeling, greater concentration of final eluted product because of a reduced void volume, and greater flow rates because of the elimination of "fines." Fines are small breakdown fragments of matrix material or tiny beads that are significantly smaller than the average particle in the support. These small fragments or beads can severely restrict column flow.

From a practical point of view, the user must accept the particle size distribution supplied by the vendor, except with respect to fines. Taking a few minutes to decant fines, and taking care not to create new fines by abrasive mixing techniques prior to packing the column can save hours and prevent frustration.

5. *Does the matrix have good capacity for the target molecule?*

The word "capacity" is commonly used to mean how much of a target molecule can be selectively bound per unit volume of support material. Typically, for particulate supports, capacity is expressed as milligrams of target molecule per milliliter of swollen beaded support. Capacity of nonbeaded supports, such as membranes or other flat surfaces, is expressed in terms of mass of target molecule (usually in micrograms) per area of matrix surface (usually in square centimeters). The literature is inconsistent on what would appear to be a fairly simple issue. Some authors use terms such as "mass (of target molecule) per gram of damp gel" or "mass per gram of dry unswollen matrix." In this manual, we refer to the volume of the support as the volume in the column under running conditions (i.e., settled gel volume).

A more practical meaning of capacity includes some appreciation of time as a parameter, in addition to a mass per unit volume. A support that has the "high" capacity rating of 20 mg target molecule/ml support is of little practical significance if it takes 5 days to process one affinity-based purification cycle. Therefore, a convenient working definition of efficiency for a process based on a particular matrix, for example mg/ml/day or mg/column/cycle, would be useful. Membrane-based systems may offer considerably enhanced efficiencies because of shortened loading, washing, and elution cycle times if final product purities and concentrations are satisfactory to the user.

The capacity of a support matrix is a function of *accessible* matrix surface area and *effective* ligand density attached to that surface. Naturally, smaller particles with smaller pores will have the greatest surface area. Unfortunately, one cannot choose an affinity matrix solely on the basis of the surface-area data printed on the label of the matrix container. More important than raw surface-area data is how that surface interacts with the target molecule. Large targets and/or large ligands require a wide pore structure; preferably on the order of 50–100 nm. The configuration of the pores within the matrix is as important as the size of the pore in a matrix (i.e., is the pore a "dead end" or does the pore channel completely through the particle?). Beaded agarose and newer synthetic supports such as beaded azlactone polymer have been demonstrated to yield high capacity supports for large (>100,000 dalton) ligands or targets because of their wide perfusive pore network.

A similar argument can be made for the importance of ligand density. How the target molecule "sees" and effectively binds the ligand is as important, and often more important than, total ligand density.

Many investigators advocate the routine use of a "spacer" between the matrix and the ligand to enhance accessibility and binding (capacity). When comparing spacer and nonspacer options, we find situations in which spacers help and situations in which they do not. Ligand orientation is also an important consideration. For example, if an antibody is immobilized with its antigen binding site facing the matrix, capacities will be lower than if the antigen binding site were facing the target solution.

Choosing an affinity matrix to maximize capacity and efficiency does not have to be an empirical exercise. Rational choices of matrix can be made by gaining as much information as possible about the target molecule (size, shape, nature of binding with the ligand in solution) and about the matrix (surface area per unit volume, pore size, pore structure, ligand density potential).

6. *Is the matrix material free of unknown nonspecific absorption effects?*

Nonspecific ionic and hydrophobic binding to nontarget molecules normally can be reduced by proper choice of buffer and temperature *if the user is aware of the effect*. Certain ionic or hydrophobic characteristics of a matrix material may be exploited, for example, nonspecific binding of antibodies to polystyrene surfaces. Also, many nonspecific effects of an affinity support are a result of activation, spacer arm attachment, and ligand coupling rather than of the matrix itself.

While commercial vendors of matrix materials have taken considerable measures to remove spurious ion-exchange residues (chemical removal of sulfate groups from agarose) or obliterate them (e.g., various polymer coatings on silica or glass supports) certain synthetic polymer matrices (e.g., Trisacryl, Eupergit, TSK, and azlactone) that were designed to be free of nonspecific effects all have a subtle but discernible hydrophobic character.

Most commercial matrix materials are acceptable if their nonspecific behaviors are known. Unknown characteristics can compromise the purity of the final product.

7. Is the matrix material "user friendly"?

We tried to avoid this over-used platitude, but "user friendly" is a more accurate adjective than "convenient" or "appropriate" when describing acceptable matrices. A checklist of "friendly" (or "unfriendly") characteristics of matrix materials follows. Many of these characteristics are derived from the topics already discussed. A positive response to the following questions is indicative of a "user friendly" matrix.

1. Is it possible to get experimental quantities of the matrix (or must I buy a large quantity just to try it)?
2. Does the matrix swell or shrink insignificantly with changes in solvents or pH (or should I anticipate widely varying flow rates during the various steps of the affinity chromatography)?
3. Can I eliminate extensive de-fining prior to using this matrix (or must I spend hours decanting)?
4. Do I have a wide range of activation chemistry options (or must I modify the chemistry of the matrix prior to use)?
5. Can I activate and couple ligand to this matrix without using copious amounts of organic solvents (or must I set up facilities for properly handling and disposing of organic wastes)?
6. Is this a robust matrix that I can heat, stir, sterilize, solvent exchange, freeze, thaw, and derivatize without discernible changes in performance (or must I treat the matrix and derived supports with utmost care during all phases of preparation, use, and storage)?

7. If I have technical questions regarding the matrix, does the vendor have technical literature and/or a technical service group (or must I discover through trial and error how best to use or how not to misuse this matrix)?
8. If the process based on this matrix works well on a small scale, will I be able to scale up without difficulty using the same matrix material (or, in order to have a large scale separation based on this affinity interaction, will I have to begin again with a different matrix)?
9. Will the matrix result in an affinity support that will give high yields of adequately pure target protein with a single pass (or will additional purification steps be required)?

8. *Is the matrix economical?*

To compare the price of 1 L of matrix material to 1 L of another is easy, but not very meaningful. We have attempted to emphasize in the preceding questions that, if you cannot use the matrix in its available form, it is too expensive.

1.2 NATURAL SUPPORTS

Natural polysaccharide matrices such as agarose are nearly always adorned with highly ionic sulfate or carboxylate residues, requiring considerable human engineering and processing before they are suitable for use as affinity matrices. Silica, alumina, and glass, included in this section because of their natural abundance, must also be processed before they can be exploited as affinity supports.

All who practice affinity-based separation owe a considerable debt to pioneers such as Hjertén, Porath, Wilchek, Cuatrecasas, Anfinson, and Regnier, who first learned about and taught the rest of us how to use the matrix materials described in this section.

1.2.1 Agarose

Structure and Properties

As shown in Figure 1.1, the polysaccharide agarose has a primary structure consisting of alternating residues of D-galactose and 3-anhydrogalactose. The sugar building blocks of processed agarose provide an uncharged hydrophilic matrix with an abundance of primary and secondary alcohols, which can be used for activation and attachment.

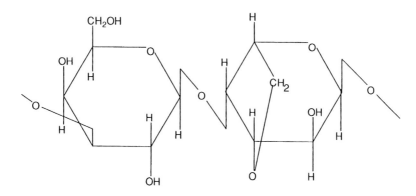

FIGURE 1.1
Partial structure of agarose.

Agarose is more than its simple content of sugar residues. The secondary and tertiary structures of agarose can be described as single fibers spun into a yarn of multiple fibers that is in turn "knitted" into a remarkable fabric with large accessible pores. The knitted porous structure of agarose is knotted at the juncture of the pores with strong hydrogen bonds that can be disrupted (causing the beads to dissolve) with heating or in the presence of strong denaturants such as guanidine hydrochloride. Agarose beads that have been chemically modified with covalent cross-links introduced with reagents such as epichlorhydrin or divinyl sulfone (DVS) will not undergo liquification under denaturing conditions.

Because of its enhanced structural stability, cross-linked agarose (e.g., CL-6B; Pharmacia) usually is preferred over the non-cross-linked variety for most affinity applications that require harsh activation or use conditions. Unfortunately, the added chemical and mechanical stability of agarose gained by cross-linking results in some loss (approximately 30–50%) of activatable hydroxyl groups that are consumed in the chemistry of cross-linking.

The large accessible pore structure of beaded agarose yields affinity supports with superior capacity (in terms of weight of target molecule per volume of support) when the target molecule (or ligand) is a large protein or polysaccharide. The addition of chemical cross-links to stabilize beaded agarose does not reduce porosity significantly, since both cross-linked and non-cross-linked 6% beaded agarose (e.g., Sepharose CL-6B, Sepharose 6B; Pharmacia) will accommodate protein or polysaccharide targets (or ligands) of molecular weight of at least 1 million. Beads with a lower density or content of agarose (e.g., Sepharose 2B, 4B; Pharmacia) have even greater porosity and will accommodate macromolecular targets or ligands in excess of 10 million daltons.

Immobilization Methods

Cross-linked (CL) beaded agarose is suitable to use with any activation method described in Chapter 2. The ability of CL beaded agarose to withstand a wide range of solvents, temperatures, and redox conditions makes it quite versatile.

On the other hand, non-cross-linked agarose beads are limited to cyanogen bromide CNBr) activation (see Chapter 2, Section 2.1.1). Although a few additional methods of activation are possible with beaded non-cross-linked agarose (e.g., DVS, bisepoxy DVS) the use of these methods will result in a cross-linked agarose bead. If a cross-linked agarose bead is desired, it is preferable to obtain a cross-linked matrix from a commercial source, to insure reproducibility.

Handling

Beaded agarose matrix is available commercially (Table 1.1; Pharmacia; BioRad) and is supplied preswollen as a thick paste in water containing a preservative to prevent microbial growth. The beads are sufficiently mechanically stable to be slurried and stirred without fear of grinding or producing fines. The thick paste is best handled by diluting it in water or buffer to produce a suspendable slurry. The slurry percentage should be estimated by centrifugation or by settling in a graduated cylinder. Wash the matrix thoroughly on a Buchner funnel (equipped with a glass fiber filter pad) or a fritted glass filter with water or activation buffer to remove the preservative use.

Avoid freezing beaded agarose supports. Freezing causes irreversible damage to the bead structure that results in smaller pores and poor chromatographic behavior. Under normal circumstances, therefore, one should not attempt to lyophilize beaded agarose matrix or derived supports. An exception can be made if 15% lactose is added to the gel slurry to stabilize the beads prior to freezing. The lactose stabilizes the three-dimensional structure of the agarose polymers and permits freeze-drying of the matrix. CNBr- and tresyl-activated agarose supports are supplied commercially as freeze-dried preparations using this lactose stabilization technique.

Never heat non-cross-linked beaded agarose to a temperature greater than 40°C or the particles will dissolve. Cross-linked agarose beads, how-

1.2 Natural Supports

TABLE 1.1
Commercially Available Agarose Supports

Company	Trade name
Pharmacia LKB	Sepharose 2B, 4B, 6B
	Sepharose CL-2B, CL4B, CL-6B
	Superose
	Sepharose Fast Flow
BioRad	Bio-Gel A
IBF	Ultrogel

ever, may be heated to 110–120°C at pH 7.0 for 11 hr without significant changes in porosity or flow rate.

Cross-linked beaded agarose will tolerate a pH range of 3–14 for extended periods without suffering measurable hydrolysis. Non-cross-linked agarose should not be exposed to pHs outside a range of 4–9 for extended periods.

Cross-linked beaded agarose will tolerate a wide range of water-miscible solvents such as acetone, dioxane, alcohols, tetrahydrofuran (THF), dimethylformamide (DMF), dimethylsulfoxide (DMSO), or pyridine if the exchange into these solvents is done in a stepwise gradual manner. Cross-linked beaded agarose will tolerate denaturants and mild oxidizing conditions (e.g., periodic acid). Strictly avoid the use of solvents, denaturants, and oxidants with non-cross-linked agarose beads.

Beaded agarose will compress under pressure causing flows in columns to slow drastically if too high an inlet pressure or hydrostatic pressure is applied to the column. Non-cross-linked, low agarose content beads (e.g., Sepharose 2B) are more compressible than cross-linked, high agarose content beads (e.g., Sepharose CL-6B). Beaded agarose columns work best if they are less than 0.5 m in height.

De-fining commercial beaded agarose supports is not necessary prior to use.

Storage of unsubstituted agarose beads is best accomplished at room temperature as a 0.02% sodium azide aqueous slurry. Never allow the beads to dry.

1.2.2 Cellulose

Because cellulose-based matrices are available and economical, the greatest potential of this material is in industrial processes. The full potential of cellulose as an affinity support has not yet been fully developed. For instance, newer beaded celluloses from "regenerated" cellulose are macroporous, hydrophilic, and show great versatility and capacity. Membrane formats are popular cellulose matrices. Recently Memtek Corporation has introduced hydrophilic affinity microporous membranes composed of reactive reinforced cellulosic polymers. This affinity membrane provides reactive aldehyde sites for covalent coupling to the amino groups of proteins and other ligands. Reinforced cellulosic polymers are suitable for pleated and specially cut membranes. The cracking and breaking common to other cellulosic membranes are eliminated due to their proprietary reinforced structure. Fibrous cellulose particles are extremely important as matrices for DNA-bearing supports (see Chapter 3, Section 3.6).

Structure and Properties

Cellulose consists of a linear polymer of 1,4 β-D-glucose (Fig. 1.2). Cellulose does not possess the sophisticated secondary and tertiary structure of agarose, and can be imagined as a tangled wad of string. As a result, unsubstituted cellulose fiber has no effective porosity or extended surface area. Although this effect does not allow enhanced capacities for ligand and target molecules, this "surface-only" matrix might be appropriate or even advantageous for certain detection applications in which washing away nontarget molecules is important, since this is not easily done on porous supports.

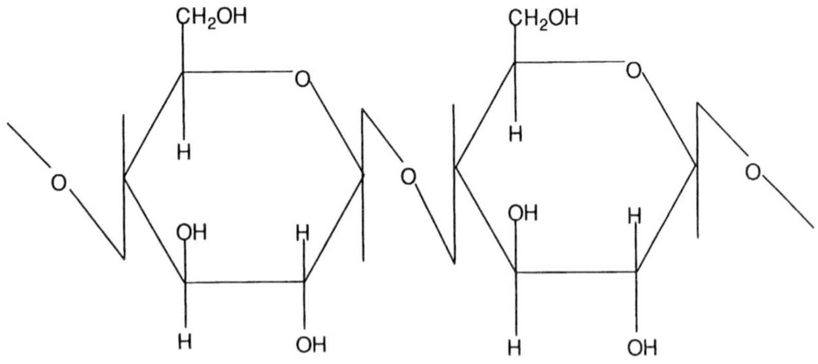

FIGURE 1.2
Partial structure of cellulose.

Immobilization Methods

Fibrous cellulose can withstand a wide range of solvent and denaturing conditions and mild oxidation conditions. Thus, fibrous cellulose can undergo any of the activation chemistries described in Chapter 2. We have had less experience in our laboratories with beaded regenerated celluloses. Oxidation of beaded cellulose with periodate in our hands, led to severe particle damage and is not recommended. Activation methods that form cross-links, for example, carbonyl diimidazole (CDI) and DVS, appear to alter the porosity of beaded cellulose.

Cellulose has better acid stability than alkaline stability. Working pH should be maintained in a range of 3–10. At pH 7, cellulose may be autoclaved at 115°C.

Handling

Fibrous cellulose is supplied as a dry powder by a variety of sources (Whatman, BioRad, Schleicher and Schuell). Material as supplied is often contaminated with fines that must be removed by a series of slurry formation settling and decanting procedures from water prior to use. Although cellulose fiber is inexpensive, this extra processing step may represent a significant hidden cost. Flow rates of support matrix fashioned from fibrous cellulose are typically half of those normally seen with beaded support material; capacities for average-sized protein targets are about one tenth those of macroporous supports.

Slurries of fibrous cellulose can be agitated with overhead stir paddles without fear of particle-to-particle grinding. Bottom-mounted stir bars should be avoided since they may grind fibers against the surface of the reaction vessel, creating new fines.

Fibrous cellulose supports may be dried or maintained as a 50% aqueous slurry in 0.02% sodium azide for storage.

1.2.3 Controlled Pore Glass and Silica

Porous glass, silica, alumina, and certain zeolites have the inherent advantage of rigidity over soft gels of agarose or beaded cellulose. The remarks that follow address the structure, activation, and handling of controlled pore glass (CPG), but are equally applicable to the other inorganic matrices just mentioned, as well as to nonporous glass surfaces.

Structure and Properties

Controlled pore glass is manufactured by heat treating borosilicate glasses to approximately 800°C, which causes the glass to separate into distinct silicate- and borate-rich phases. Rapid cooling followed by acid etching will leach out the borate-rich phases and leave behind a very regular network of pores. The remaining structure has strong ion-exchange character contributed by silanol (-Si-OH) groups. The same silanol functions that contribute so strongly to nonspecific interactions with proteins can fortunately be exploited as anchor points for coating chemistries that will largely obliterate this undesirable characteristic.

CPG is extremely brittle. Particles of CPG are best swirled gently in reaction medium, rather than stirred or mixed, to avoid particle grinding. Using a rotary evaporator unit for performing chemistry on glass is desirable for two reasons. (1) Air trapped in the porous matrix must be removed and replaced with reaction solvent. This is easily done by pulling and breaking a vacuum several times while the particles and solvent are in a rotary flask. (2) Most rotary evaporator units are equipped with good controls for the rate of turning the flask. A very slow rate of turning will insure good mixing without tumbling and grinding of the particles. New reactants may be introduced while maintaining a slight vacuum on the slowly turning flask without taking the apparatus apart.

CPG is very resistant to acid but will rapidly degrade at pH greater than 8.0. Glass will also dissolve at an appreciable rate in deionized water. Working buffers should be neutral and contain at least 0.05 M salt or, preferably, be slightly acidic.

CPG can withstand enormous pressures without compression or collapse, and can withstand any organic solvent as well as temperatures in excess of 300°C.

Immobilization Methods

Plain CPG glass must be surface modified before it can have any utility as an affinity support. The following protocol for attaching a glycerol-type coating is a modification or combination of methods described by Weetal (1969) and Regnier *et al.* (1976) (Fig. 1.3).

COVALENT COATING OF CPG TO FORM A NEW DERIVATIZABLE SURFACE

1. Gently add the desired quantity of glass particles to a rotary evaporator flask and cover the matrix with 0.05 M HCl. Attach the flask to the rotary evaporator unit and, while turning slowly, intermittently pull and break a water aspirator vacuum.

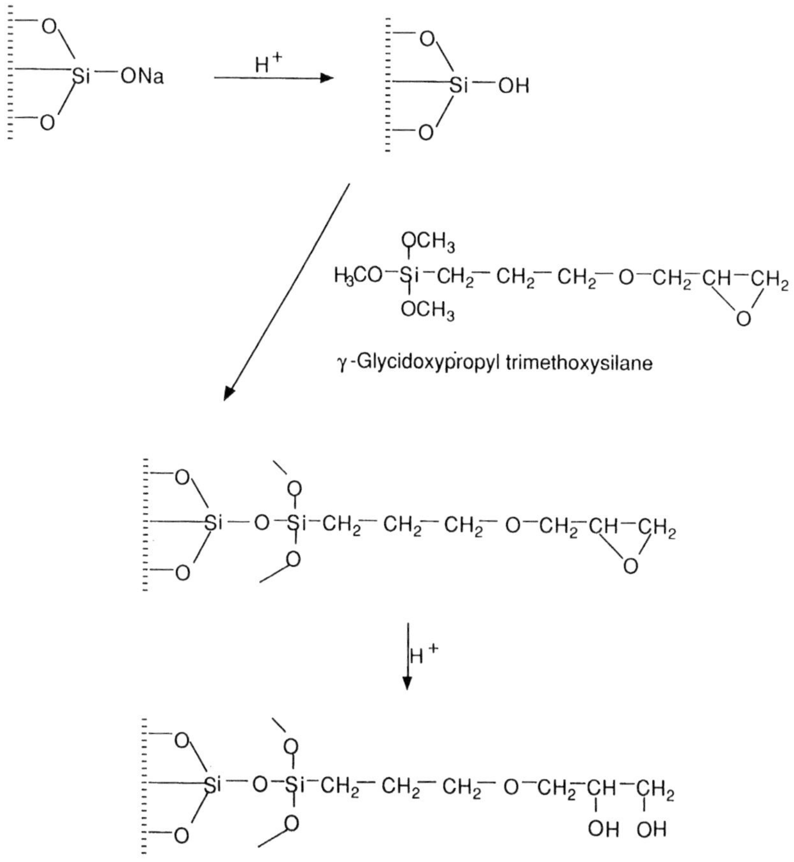

FIGURE 1.3
Reaction sequence for the preparation of glycerol-coated glass.

NOTE: The objective here is to replace sodium ion with hydrogen (via ion exchange) from the surface of the glass and to remove air from within the pores of the glass.

Continue the acid conditioning for about 1 hr at room temperature.

2. Filter the acid-washed particles on a Buchner funnel equipped with a glass fiber filter pad. Wash briefly with deionized water followed by acetone. Allow to air dry on the filter pad.
3. Return the acid-treated glass to a dry rotary evaporator flask and cover the particles with toluene. Remove trapped air as described in Step 1.

With slow turning, externally heat the flask in the bath of the unit to 40–60°C and continue application of vacuum until the toluene begins distilling from the flask. Continue distilling and collecting toluene under vacuum until water is no longer seen azeotropically co-distilling with the toluene.

NOTE: We find this drying step to be essential to achieve a thin monomeric coating of glass by coupling with silane in the next step. Failure to remove water will result in silicone polymer coatings that will plug up the pore structure of the matrix.

4. Add 10% γ-glycidoxypropyl trimethoxysilane (Aldrich) in toluene to the dried glass particles until all particles are covered. Remove any entrapped air and rotate overnight (10–12 hr) at slow speed and room temperature.
5. Filter the silane-coated particles on a Buchner funnel equipped with a glass fiber filter and wash with acetone. Transfer the acetone-washed particles to a clean rotary evaporator flask and again cover with 0.05 M HCl. Remove trapped air and, with slow turning, heat to 50–60°C for 1–2 hr.

NOTE: This step converts the newly attached epoxy function to a *cis*-glycol function. From this point, a wide range of options are available for subsequent activation and ligand attachment chemistries (see Chapter 2).

6. Finally, wash the glass particles with acetone on a Buchner funnel and air dry.

Properly coated CPG prepared by this procedure will show negligible nonspecific binding of protein. To test coating efficiency, we usually challenge a small column of the coated glass particles (preequilibrated with 0.05 M phosphate, 0.15 M NaCl, pH 7.0) with a solution approximately 5 mg/ml hemoglobin (human red cell hemolysate is satisfactory) in the same buffer. Recovery of hemoglobin after chromatography and washing with five column volumes of buffer should exceed 99% (measured as O.D. at 414 nm) when challenging the coated matrix with 5 mg hemoglobin per gm of glass.

Other functional silanes such as aminopropyl triethoxysilane may be bonded to glass, silica, or alumina matrix materials using the approach just described.

1.3 SYNTHETIC SUPPORTS

Synthetic supports are produced by polymerization of functional monomers to give matrices suitable for affinity-based separations. In some cases, copolymer derivatives have been produced from the combination of a naturally occurring matrix material (such as agarose) and synthetic monomers.

The number of commercially available synthetic materials for affinity-based separations is increasing. These supports typically have superior physical and chemical durability and can withstand the rigors of process-scale purifications better than the natural soft gels. Conventional natural gels compress easily and cannot tolerate the higher pressures created by great column height or high flow rates. Most polymeric matrices can withstand pressures of at least 2–3 bar without collapse, and the best can approach HPLC-type pressure gradients without failure. The physical strength of the synthetic supports often allows for increased linear flow rates and, thus, greater throughput, dramatically reducing purification times. These advantages are the primary reasons for the current growing popularity of polymeric supports in the biotechnology industries.

The newest polymeric supports are also more tolerant of changes in ionic strength or buffer composition. Many can withstand organic solvents without swelling or shrinking, and some can tolerate extremes in pH without matrix decomposition. These features allow more flexibility in washing and regenerating large columns and make sodium hydroxide sterilization feasible. In addition, because of their synthetic nature, the base matrices are usually microbe resistant.

In a few cases, even nonporous synthetic polymers have found use in affinity separations. At one extreme, very small solid beads ($\leq$5-µm diameter) have been used successfully in HPLC-type situations or in immunoaffinity agglutination assays. Surface derivatization of these particles provides enough ligand density for many affinity-based interactions. At the other end of the spectrum, large solid macrobeads or balls ($\geq$3 mm diameter) have been used for the immobilization of ligands in immunoassay applications. Although these large balls are not useful for classical chromatographic separations, their potential use in immunoassay determinations has made them quite popular for various ligand immobilizations. Other polymeric support forms such as tubes, multiwelled plates, or sticks, are also suitable for the immobilization of ligands. Specific applications for these unique affinity supports range from immunochemistry to cell culture and enzymology.

The incorporation of monomers with suitable functional groups can provide activation sites on these supports for ligand immobilization. Usually synthetic supports are made with monomers that contain primary or secondary hydroxyl groups that not only lend hydrophilicity to the resultant matrix, but also allow compatibility with most coupling chemistries. This section discusses synthetic supports most suitable for affinity chromatography or immunoaffinity applications. In some cases, the same base matrix is also useful for gel filtration applications or is further derivatized to provide ion-exchange properties. However, the primary emphasis here will be on their relative usefulness in affinity systems.

1.3.1 Acrylamide Derivatives

1.3.1.1 Polyacrylamide Beads

Structure and Properties

Porous polyacrylamide beads were, perhaps, the first synthetic polymer chromatography supports to be described in the chemical literature and gain popular acceptance by biochemists. The support is made from a copolymerization of acrylamide and the crosslinking agent N,N'-methylenebis(acrylamide), as described by Hjerten and Mosbach (1962). Figure 1.4 shows the reactions involved in the preparation of polyacrylamide beads and the structure of the final cross-linked product. Although this type of support was originally designed primarily for gel filtration applications, some of the larger pore size varieties have found use in affinity separations.

BioRad Laboratories (Richmond, California) is a principle supplier of these supports and markets them under the trade name Bio-Gel P. BioRad currently offers nine different products with pore sizes ranging from an exclusion limit of 1800 daltons to a maximum of 400,000 daltons. Their P-200 gel is especially well suited for most affinity applications involving protein purification, since it has a fractionation range of 30,000 to 200,000 daltons.

The main advantages of polyacrylamide-based matrices are their resistance to microbial attack, relatively good pH stability (pH 2–10), excellent chemical stability (due to a polyethylene backbone), and low nonspecific binding characteristics (due to the lack of a biopolymer matrix based on sugar residues). The low nonspecific binding makes polyacrylamide beads with immobilized carbohydrate ligands suitable for the purification of carbohydrate-binding proteins, such as lectins.

Unfortunately, because of some of their unfavorable properties (to be discussed), polyacrylamide-based supports have never become very popular for use in affinity separations. The primary difficulties that limit its use are poor mechanical stability and slow flow rates. Considering only gels with reasonable porosity for proteins, the maximum linear velocity at

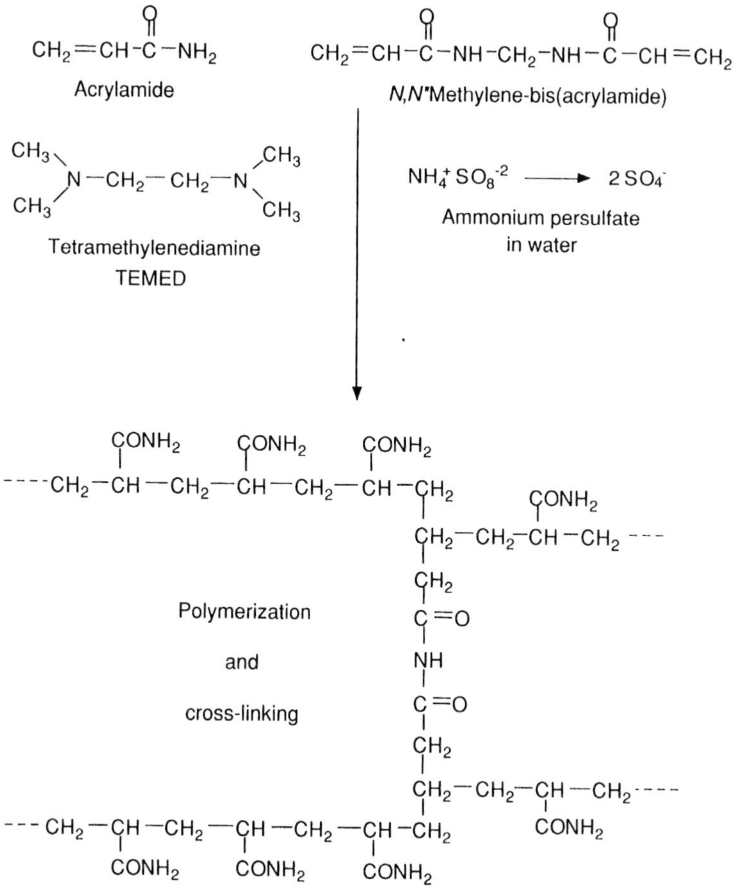

1.3 Synthetic Supports

$$CH_2=CH-\overset{\overset{\displaystyle O}{\|}}{C}-NH_2$$
Acrylamide

$$CH_2=CH-\overset{\overset{\displaystyle O}{\|}}{C}-NH-CH_2-NH-\overset{\overset{\displaystyle O}{\|}}{C}-CH=CH_2$$
N,N'Methylene-bis(acrylamide)

$$\underset{CH_3}{\overset{CH_3}{\diagdown}}N-CH_2-CH_2-N\underset{\diagdown CH_3}{\overset{\diagup CH_3}{}}$$
Tetramethylenediamine
TEMED

$$NH_4^+ \, SO_8^{-2} \longrightarrow 2\,SO_4^-$$
Ammonium persulfate in water

↓

Polymerization
and
cross-linking

$$\begin{array}{c}
\quad\quad\;\;CONH_2\quad\;\;CONH_2\quad\;\;CONH_2\\
\quad\quad\;\;|\quad\quad\quad\;\;|\quad\quad\quad\;\;|\\
---CH_2-CH-CH_2-CH-CH_2-CH-CH_2\\
\quad\quad\quad\quad\quad\quad\quad\quad\quad\quad\quad\quad\quad\quad|\quad\quad\;\;CONH_2\\
\quad\quad\quad\quad\quad\quad\quad\quad\quad\quad\quad\quad\quad\;\;CH_2-CH_2-CH-CH_2---\\
\quad\quad\quad\quad\quad\quad\quad\quad\quad\quad\quad\quad\quad\;\;CH_2\\
\quad\quad\quad\quad\quad\quad\quad\quad\quad\quad\quad\quad\quad\;\;C=O\\
\quad\quad\quad\quad\quad\quad\quad\quad\quad\quad\quad\quad\quad\;\;NH\\
\quad\quad\quad\quad\quad\quad\quad\quad\quad\quad\quad\quad\quad\;\;C=O\\
\quad\quad\quad\quad\quad\quad\quad\quad\quad\quad\quad\quad\quad\;\;CH_2\\
\quad\quad\quad\quad\quad\quad\quad\quad\quad\quad\quad\quad\;\;CH_2-CH_2-CH-CH_2---\\
---CH_2-CH-CH_2-CH-CH_2-CH-CH_2\quad\quad CONH_2\\
\quad\quad\;\;|\quad\quad\quad\;\;|\quad\quad\quad\;\;|\\
\quad\quad\;\;CONH_2\quad\;\;CONH_2\quad\;\;CONH_2
\end{array}$$

FIGURE 1.4

Polyacrylamide matrices are formed by the polymerization of acrylamide and the water soluble cross-linking agent N,N'-methylene-bis(acrylamide). Free radical initiation is usually performed by ammonium persulfate and catalyzed by the addition of TEMED.

---

which buffer can be pumped through a column of polyacrylamide beads is unacceptably slow when compared with other chromatography support materials. For instance, Bio-Gel P-200 will give linear flow rates of only 10–15 cm/hr, even when using very large beads 150–300 μm in diameter. If the bead size is reduced to a "normal" range of 80–150 μm, the typical flow rate drops to only 3–6 cm/hr. These figures should be compared with even the standard cross-linked agarose supports, which are capable of delivering flow rates in the range of 30–100 cm/hr. The better polymeric matrices can achieve flow rates of at least several hundred cm/hr without failure.

Poor mechanical stability is apparent even when handling or manipulating polyacrylamide beads. Although the gel does not fracture, it is easily compressed. This characteristic directly relates to the height of a column that can be packed without matrix compression at the bottom of the gel bed. Once a gel starts to collapse under pressure, either due to its own weight or because of flow rate-induced pressure drops, the chromatographic properties of the support are greatly compromised. Gel beds of polyacrylamide exceeding about 50 cm in height give unacceptably slow flows, probably due to matrix compression.

Another serious deficiency of polyacrylamide is its tendency to shrink or swell in various solvents or buffer compositions. Often, a change in salt strength or merely the addition of a counter ligand for elution will cause drastic changes in the bed volume of the column. This phenomenon can easily cause changes in porosity or gel compression in isolated regions of the column and result in altered or poor chromatographic behavior.

Immobilization Methods

Derivatives of polyacrylamide gels can be made easily by transamidination. In this reaction, the indigenous carboxamide linkages in the matrix are broken and replaced by another amide bond resulting from the linkage of primary amine groups on the ligand. In this way, aminoethylamide derivatives can be prepared using an excess of ethylene diamine while stirring at 90°C. Similarly, hydrazide modifications can be made reacting with an excess of aqueous hydrazine at 50°C. [Complete protocols for preparing these modified supports are given in Chapter 2 (Section 2.2.1) and mentioned in Chapter 3 (Section 3.1.1).] Some of these modified forms of polyacrylamide beads are now available commercially under the tradename Enzacryl® from Koch–Light, Ltd. (Colnbrook, England) or BioRad.

Another method of activation for polyacrylamide gels, using glutaraldehyde, provides a multivalent aldehyde-containing intermediate useful for coupling amines via reductive amination. This method is also fully described in Chapter 2 (Section 2.2.1.4).

Handling

Most polyacrylamide gels are supplied dry. To hydrate them, slowly add beads to buffer while stirring vigorously. The volume of buffer used should be equal to about twice the expected volume of the wetted gel. Slow addition and vigorous stirring prevents clumping. Agitation can be done by a number of methods, such as using a paddle or rocking, but it is good practice to avoid stirring bars since they have a tendency to damage most gels by their grinding action. Obtaining fully hydrated beads will take about 4 hr of stirring at room temperature or 1 hr at 90°C.

After hydration, the gel should be washed with buffer, presumably the buffer that will be used in any subsequent coupling chemistries. Although many gels can be conveniently and quickly washed by gentle suction filtration using a glass funnel with an embedded fritted glass filter, it is best to use a Buchner type funnel with a glass fiber filter pad with polyacrylamide gels. The reason for this is the tendency of polyacrylamide to quickly clog the fritted glass filters and stop or severely slow the filtration process. The filtration process may be accelerated by using a vacuum produced by a water aspirator or a vacuum pump. High vacuum is not required to obtain good flow rates. Periodic mixing of the gel before reapplying vacuum will assure uniform washing.

1.3.1.2 Trisacryl

Structure and Properties

The Trisacryl line of beaded polymer matrices made by IBF Biotechnics (Columbia, Maryland; manufacturing facilities in France) represents novel synthetic supports based on a derivative of an acrylamide monomer. The tradename for the product comes from its basic building block (*N*-acryloyl-2-amino-2-hydroxymethyl-1,3-propane diol), a compound composed of three hydroxymethyl groups attached to the carboxamide of acrylamide. Figure 1.5 shows the chemical structure of this unique monomer as well as the polymeric structure of the matrix. The final beaded support is actually a copolymer of the tris-hydroxymethyl monomer and another hydrophilic divalent monomer (*N*,*N*'-diallyltartradiamide) added as a cross-linking agent (French patent no. 7,702,391).

The hydroxymethyl triads attached as side groups to every other carbon of the resulting polyethylene backbone surround the polymer chain and lend considerable hydrophilicity to the support. The result is a chromatographic matrix with very low nonspecific binding for biological molecules. Careful regulation of monomer ratios and concentrations has produced gels of discrete porosities and exclusion limits, some of which are well suited for affinity applications.

Trisacryl GF-2000 is produced with a particle size distribution range of 40–80 μm and has an exclusion limit of greater than 10 million daltons,

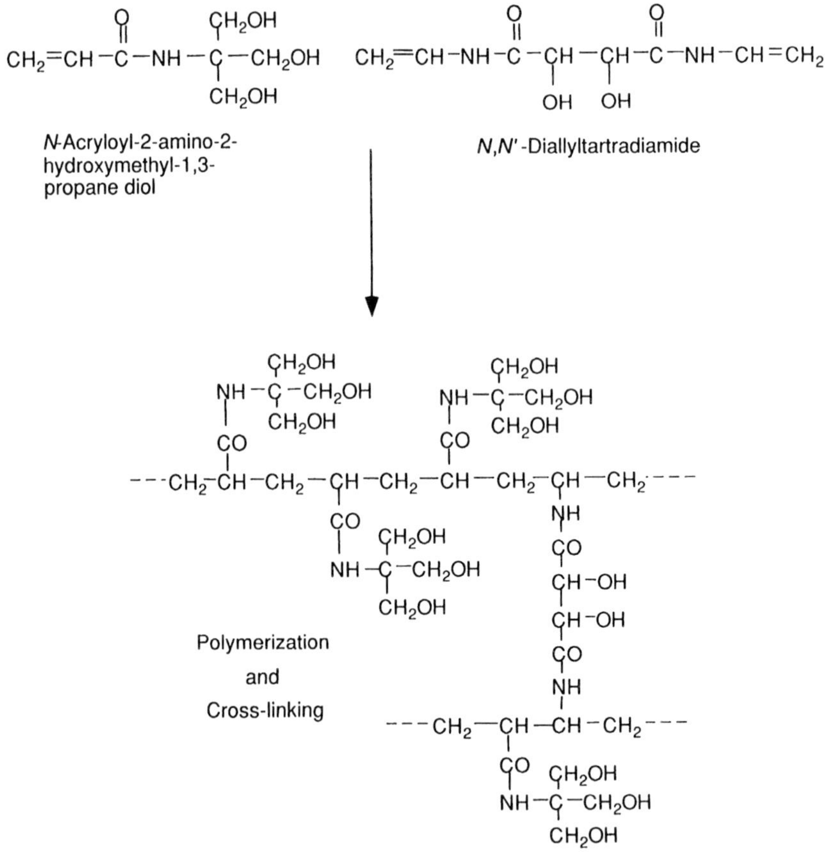

FIGURE 1.5

Trisacryl supports are made by the copolymerization of a unique trihydroxylic derivative of acrylamide and the cross-linking agent N,N'-diallyltartradiamide.

---

making it the most appropriate choice for protein separations. The GF-LS form of this gel has identical porosity, but has a bead size in the range of 80–160 μm for use in the higher flow rate applications often required in process-scale purifications.

Unlike the basic polyacrylamide beads described earlier, the Trisacryl polymer derivatives provide much better mechanical stability and flow rate properties. The beads are more rigid in nature and do not collapse as easily as the softer gelatinous supports. The matrix can withstand up to 2–3 bars of pressure, making it suitable for most low to intermediate pressure operations. A pressure drop of 1–2 bar will generally provide a linear flow rate of 100–200 cm/hr, making the support excellent for larger scale purifications. Additionally, since it is a synthetic polymeric support, Trisacryl has excellent resistance to microbial attack and can also tolerate extremes of pH (from 1 to 11) without decomposition. More basic pH environments slowly begin to break down the amide bonds and will degrade the matrix. For this reason, although dilute NaOH sterilization is possible (0.2 N), the gel should not be left in contact with a highly alkaline solution for long periods. The support is also said to be stable from −20° to 121°C and compatible with all commonly used denaturants, organic solvents, and detergents.

Immobilization Methods

The activation and coupling protocols available for the immobilization of affinity ligands on Trisacryl gels are many because of its preponderance of primary hydroxyl groups. As a general guideline, any activation procedure that is suitable for hydroxyl-containing supports will work well. Appropriate direct activation methods include CDI, CNBr, tresyl, tosyl, DVS, and epoxy (see Chapter 2). Most small ligands can be immobilized at high densities by many of these methods with good resultant capacities for affinity purifications. Secondary activation methods can then be built on these initial immobilizations to couple other specific functionalities. However, one problem is often observed with rigid polymer matrices: protein immobilization yields are frequently lower than those normally obtainable with agarose-like soft gels. This is a recurring problem with most matrices that are mechanically rigid, perhaps because of some restrictions in pore diameter near the core of the beads.

To overcome this deficiency, we have developed a modified periodate oxidation and coupling procedure that allows immobilizing protein at higher densities than traditional activation protocols. Although the method of direct periodate oxidation to aldehydes is not possible since the Trisacryl hydroxyl groups are not on adjacent carbons, there is a two-step alternative that makes this chemistry feasible. For superior protein immobilization yields, prior modification of the support with glycidol followed by subsequent periodate treatment will provide a support with extremely high coupling capacity via reductive amination (see Chapter 2, Section 2.2.1.4). This coupling procedure can be used to immobilize proteins at even greater yields than those obtained using the best chemistries on agarose. IBF does not supply a preactivated form of this gel.

Handling

Trisacryl supports come preswollen as a thick slurry of gel, containing 1 M NaCl and 0.02% sodium azide as preservative. The gel is easily dispensed by shaking to suspend the settled beads or by scooping the heavy gel cake out with a spatula. Dilution of the beads with water or buffer containing a preservative can be done to produce a more suspendable mixture. To accurately measure a quantity of any preswollen gel it is best to use settled volume rather than mass. Although some protocols specify a number of grams of wet gel cake, differences in what constitutes "wet" can drastically alter the gel to water ratio actually present in a sample. To insure reproducibility in immobilization reactions, it is necessary to reliably measure the portion of gel present in the slurry. This measurement can be performed using several simple methods. A well-mixed aliquot of the gel slurry can be centrifuged (2000 rpm, 5–10 min) in a graduated tube and the quantity of settled gel and total slurry volume determined. Alternatively, a portion of the slurry can be allowed to settle by gravity in a graduated cylinder or tube (minimum settling time, 24 hr) and the same volumes determined. These values can then be used to calculate the slurry percentage (the fraction of the total volume that is swollen gel) so the original slurry can be sampled by volume containing a known volume of gel.

Trisacryl is a robust mechanically stable gel that can be routinely handled without fear of matrix damage. The support can be washed using either a suction filter funnel with an embedded fritted glass filter or a Buchner funnel with a glass fiber filter pad. Filtration may be accelerated by vacuum created by a water aspirator or a vacuum pump. High vacuum is not required to obtain good flow rates for washing. Mixing the gel after the addition of fresh buffer and before reapplying vacuum will insure uniform washing. The gel should be washed free of the preservative solutions before activation or coupling chemistries are performed. Washing with 4–5 bed volumes of activation or coupling buffer should be sufficient to prepare the gel for subsequent operations.

Over time, the fritted filter may become slow due to embedded particles, but clogged filters can be easily cleaned using concentrated sodium hydroxide (50%) for brief periods (1 hr) to dissolve the support material. Alternatively, backflushing with water may also remove the gel particles.

1.3.1.3 Sephacryl

Structure and Properties

Sephacryl, developed by Pharmacia (Piscataway, New Jersey), is a unique copolymer produced by combining a derivative of a naturally occurring matrix material and a synthetic monomer. In this support, the excellent

1.3 Synthetic Supports

gel filtration properties of beaded dextran gels (such as the Sephadex line from Pharmacia) have been enhanced by polymerizing allyl dextran with the cross-linking monomer N,N'-methylene-bis(acrylamide). The resulting polymer is believed to contain linear chains of polymeric glucose molecules held together by bis(acrylamide)-induced intramolecular cross-links as well as linear portions of polymerized bis(acrylamide). The polymerization reactions and the theoretical structure of the final beaded support are shown in Figure 1.6.

Although Sephacryl was originally designed for gel filtration separations, it has been used to immobilize ligands in very few cases. However, appropriate pore size varieties of Sephacryl should be able to be used for affinity chromatography. A choice of Sephacryl S-300 or S-400, which have exclusion limits of 1.5 million and 8 million daltons, respectively, should give the necessary spatial accommodation for affinity-based interactions. The relatively new Sephacryl HR series has been optimized further in its degree of cross-linking and particle size to give a support with higher flow rate potential than the original SF series. This gel has a particle size distribution of 25–75 μm and will allow linear flow rates of up to 100 cm/hr in process scale.

The unique polymeric structure of the Sephacryl support provides a significant increase in stability over Sephadex-type cross-linked beads. The gel can withstand all commonly used chromatography solvents without shrinking or swelling. It can also be autoclaved or cleaned in place with 0.5 N NaOH.

Immobilization Methods

Dextran-based gels have abundant hydroxyl groups for activation and coupling of affinity ligands. Most of the basic methods useful for the activation of hydroxyl groups will also work well with Sephacryl. However, since the presence of primary hydroxyls is extremely limited (because the terminal dextran chain C-6 hydroxyls are probably tied up in cross-links), the reactivity of the remaining secondary hydroxyls may be somewhat lower than that of other matrices that contain mainly primary hydroxyls. The initial activation methods (including CDI, CNBr, tresyl, tosyl, DVS, and epoxy) used to attach ligands or spacers to matrices should work well. Once the appropriate spacer or intermediate is immobilized using such primary coupling chemistries, secondary activation methods can be employed to immobilize other specific functional groups. (See Chapter 2 for further details.) Pharmacia does not supply a preactivated form of this gel for ligand immobilization.

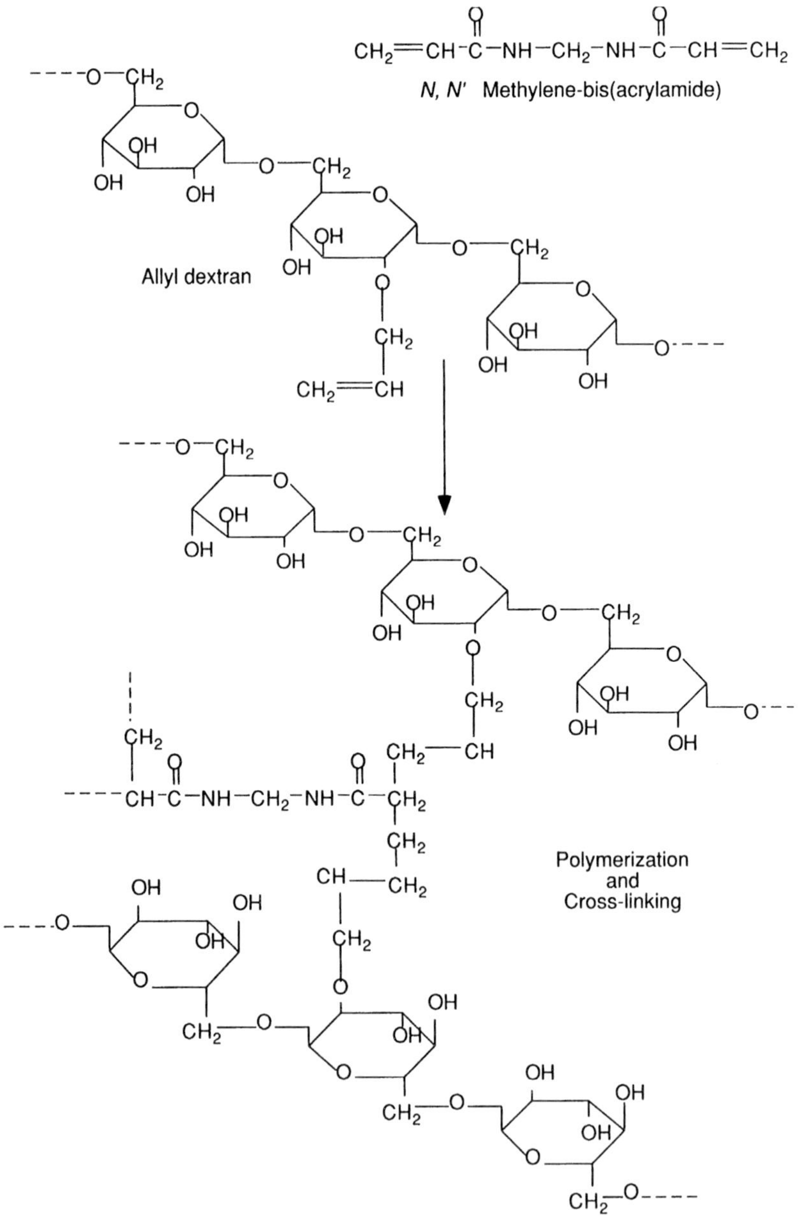

FIGURE 1.6
Sephacryl matrices consist of allyl dextran polymerized with the cross-linking agent N,N'-methylene-bis(acrylamide).

Handling

All Sephacryl media are supplied preswollen in an aqueous solution containing a preservative. The gel is mechanically stable, so the thick slurry can be shaken or stirred to suspend the particles. Dilution with water or buffer containing a preservative can be done to produce a more suspendable mixture. The slurry percentage should be measured by centrifugation or by settling in a graduated cylinder to accurately determine the amount of gel sampled (see Section 1.3.1.3). Wash the required portion of gel using a Buchner funnel with a glass fiber filter pad or a funnel with a fine grade fritted glass filter. Since the particle size diameters can be as small as 25 µm, it may be best to use the Buchner funnel with a filter pad to prevent clogging. The washing steps may be accelerated by use of a vacuum produced by a water aspirator or vacuum pump. High vacuum is not required to obtain good flow rates. Periodic mixing of the gel before reapplying vacuum will insure uniform washing. The gel should be washed thoroughly with activation or coupling buffer to remove preservatives prior to use.

1.3.1.4 Ultrogel AcA

Structure and Properties

The standard Ultrogel line of products from IBF is a group of agarose-based supports similar to the Pharmacia Sepharose varieties. The Ultrogel AcA gels are custom-designed mixed matrices produced from a blend of agarose and polyacrylamide. In this case, the high resolution potential of polyacrylamide is combined with the greater mechanical stability of agarose to give a synergistic support somewhat superior to supports formed from either type alone. Five different mixed matrices are available from IBF, all with a particle size distribution of 60–140 µm. The final percentage of polyacrylamide versus agarose governs the fractionation range and exclusion limit of the resultant support material. The greater the percentage of polyacrylamide in the gel, the lower the exclusion limit of the support. The gels most suitable for protein chromatography include Ultrogel AcA 34 and Ultrogel 22, with exclusion limits of 750,000 and 3 million daltons, respectively. The "34" type is comprised of 3% polyacrylamide and 4% agarose, whereas the "22" type has 2% of each polymer component. The chemical makeup of the individual polymers present in Ultrogel AcA and a schematic of the theoretical polymer matrix structure in the beads are given in Figure 1.7.

Another variation of this composite support is the addition of iron oxide (Fe_3O_4) particles in the core of the bead, giving magnetic character to the

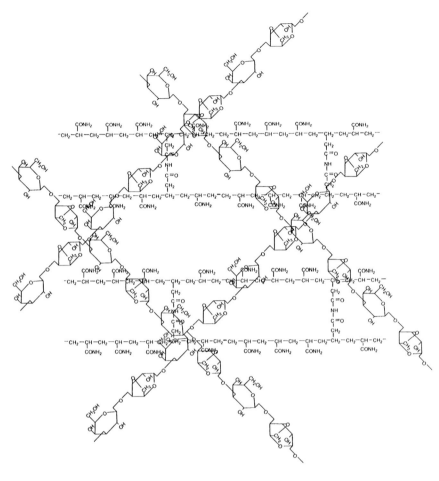

FIGURE 1.7

Ultrogel matrices are complex mixtures of polyacrylamide and agarose polymers knitted together in one support material.

support. This matrix is available under the name Magnogel AcA 44, with an exclusion limit of 200,000 daltons.

Although many supports originally designed with gel filtration properties in mind can be used easily for affinity chromatography, the Ultrogel AcA series may not be the best choice for immobilizing ligands. The primary advantage of polyacrylamide beads in affinity techniques was the lack of carbohydrate interference in the matrix itself. Adding agarose to the gel does enhance the mechanical stability of the matrix, but defeats the purpose of using polyacrylamide in affinity separations. The agarose component is also a non-cross-linked polymer that has limitations in its solvent compatibility (some organic solvents may damage the gel) and temperature stability (due to the potential for agarose melting). A high concentration of denaturants (urea or guanidine hydrochloride) or detergents (e.g., SDS) should also be avoided since they could disrupt the matrix integrity. In addition, the sensitivity of both non-cross-linked agarose and polyacrylamide to extremes in pH is reflected in its rather limited working range of pH 3–10.

Immobilization Methods

Since the recommended working temperature range for Ultrogel AcA is 2° to 36°C, some of the modification chemistries (transamidination in particular) designed specifically for the polyacrylamide component (needing 50–90°C) are not possible. However, the polyacrylamide portion of the matrix still can be activated by glutaraldehyde to allow amine-containing ligands to be coupled (see Chapter 2). This active aldehyde-containing matrix is commercially available from IBF using the Ultrogel AcA 22 support. The non-cross-linked agarose component also can be initially activated using conventional means for modifying a hydroxyl-containing support. These methods include CNBr, DVS, and epoxy. Other methods that require solvent exchange into a nonaqueous environment may damage or shrink the gel and should be avoided.

Handling

Ultrogel AcA is supplied as a preswollen aqueous slurry containing 2 mM sodium citrate, pH 6.0, and 0.02% sodium azide as preservative. The gel can be sampled by creating a uniform slurry (dilution of the supplied gel with water or buffer may be necessary) and measuring the gel percentage according to the methods described for Trisacryl. Prior to activation, the gel should be washed with 4–5 bed volumes of the appropriate buffer. This can be done using either a suction filter funnel with an embedded fritted glass filter (coarse) or a Buchner funnel with a replaceable glass fiber filter pad. Filtration may be accelerated by vacuum (produced by a water aspirator or a vacuum pump). High vacuum is not required to obtain good flow rates for washing. Periodic mixing of the gel before reapplying vacuum will insure uniform washing.

Over time, the fritted filters may clog with embedded particles. Clogged filters can be cleaned easily by briefly incubating (1 hr, room temperature) and washing with concentrated sodium hydroxide (50%), washing with water, and then incubating with concentrated sulfuric acid, followed by another water wash. This protocol will dissolve any Ultrogel AcA support material that has become trapped in the filter.

1.3.1.5 Azlactone Beads

Structure and Properties

3M Corporation has recently developed a new rigid polymeric matrix that has great potential for use in affinity separations. The support combines excellent chromatographic properties with a high level of mechanical strength to provide superior linear flow rates and high binding capacities for process scale-up operations. The base support comes preactivated with a novel azlactone coupling chemistry that results in outstanding yields for protein immobilization (U.S. patent no. 4,737,560 and 4,871,824). The highly cross-linked, hydrophilic beads are produced by a copolymerization of the monomers vinyldimethyl azlactone (oxazolone) and *N,N'*-methylene-bis(acrylamide) (Fig. 1.8). By controlling the azlactone/bis monomer ratio, supports with 1 meq/gm (100 µmmol/ml) to 3 meq/gm (300 µmol/ml) of azlactone functionality have been produced.

The azlactone group is highly reactive toward nucleophiles such as amines, thiols, and, to a lesser extent, alcohols. The ring opening reaction will couple ligands from pH 4–9 to give stable amide linkages with an indigenous spacer molecule. Although the azlactone group is labile in an aqueous environment, the hydrolysis rate is very slow compared with the rate of ligand coupling. Proteins and other small ligands are coupled in extraordinarily high yields in less than 1 hr. The maximum immobilization density reportedly achieved for protein A is over 30 mg/ml gel. Even with this very high loading level, the support maintained a molar binding ratio for IgG of 1:1. Human IgG itself has been immobilized to a density of over 20 mg/ml. These values should be compared with those of other supports, normally maximal at 5–10 mg/ml gel using traditional coupling chemistries. (For further details on activation and coupling, see Section 2.2.1.8 in Chapter 2.)

The internal motif of the azlactone polymeric matrix provides mechanically strong beads that give the base support excellent physical and chromatographic properties. The polymer is resistant to degradation in extreme pH environments, it is reported to be stable at pH 1–14. The dried matrix initially will swell in aqueous buffer to a density of about 10 ml/gm and give an average particle size of 54 µm. Once fully swollen, the matrix does not shrink or swell with changes in organic or ionic media. The porosity of

1.3 Synthetic Supports

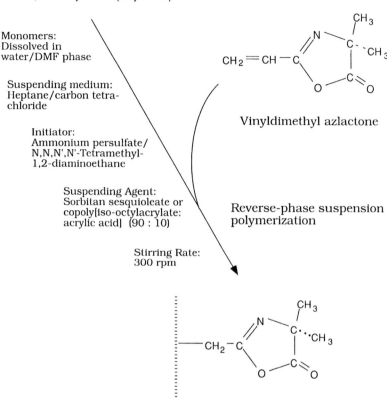

FIGURE 1.8
Azlactone beads are formed by the copolymerization of the monomer vinyldimethyl azlactone and the cross-linking agent N,N'-methylene-bis(acrylamide).

the hydrated support provides ample space for protein–protein interactions (50-nm pores); the matrix also has a high level of surface area available for immobilizing large biomolecules (250 m^2/gm).

The strength inherent in the bead structure of this matrix allows high pressure operation (up to 1000 psi), which immediately translates into the ability to pack large columns without fear of support collapse. Very high linear flow rates are also possible (up to 4500 cm/hr), which lends itself to process-scale applications requiring high throughput. Even under gravity operations at laboratory scale, the support will exceed the flow rates of cross-linked agarose by 160%.

Many chromatography supports that can physically withstand high flow rates cannot practically be used at these levels due to drastically reduced capacity. The dependence of capacity on flow rates is often the result of the internal pore structure of the gel. As the linear flow is increased, the mobile phase moves along the paths of least resistance, avoiding or missing most of the smallest or constrained pores. Thus, an affinity gel that is able to bind 10 mg of a protein at gravity flow rates may be able to bind only 2 mg at 5 times that flow. Azlactone beads have the curious property of being able to maintain surprisingly high capacity even at significantly elevated flows.

This characteristic, along with the excellent immobilization yields, may make azlactone polymers excellent candidates for process-scale purification applications. At the time of this writing, azlactone matrices have not been made commercially available so this support has yet to be proven in the field.

Immobilization Methods

Unlike most chromatography supports, azlactone beads contain an excellent general purpose primary functionality for coupling ligands that is indigenous to the matrix itself. The azlactone ring chemistry may be used to immobilize primary amine- or sulfhydryl-containing ligands with high efficiency. This relatively new technique of ligand coupling may soon become a method of choice, especially for protein immobilization. The reported coupling yields and density of ligand attachment with this chemistry deserve careful consideration. Subsequent secondary activation methods may also be used (after attaching the appropriate intermediate compound) to facilitate the coupling of other specific groups. A complete discussion of azlactone chemistry is given in Chapter 2.

Handling

Azlactone beads are supplied dry because of the instability of the ring under aqueous conditions. Since the support in its active form is susceptible to long term hydrolysis, careful sampling should be done to prevent the introduction of moisture. Resealing the bottle under nitrogen is a good practice. Storage of properly sealed containers at room temperature is acceptable.

The beads should be accurately weighed to determine the final volume the sample will occupy after hydration (0.1 gm will swell to 1 ml gel). Since the support as supplied contains no preservatives or other added compounds, it is not necessary to wash the gel prior to the addition of ligand. The dry beads, therefore, can be added directly to a ligand-containing solution. To aid in the hydration process, application of a mild vacuum will eliminate trapped air that can cause significant frothing. (This technique should be used for an organic solvent-based coupling reaction as well.)

The beads are best washed using a Buchner funnel with a removable glass fiber filter pad, since the support is not easily dissolved so cleaning an embedded fritted glass filter is difficult. Once the beads have been hydrated in a buffer containing a dilute detergent, they can be washed free of detergent without further frothing caused by air bubbles.

1.3.2 Methacrylate Derivatives

1.3.2.1 TSK-Gel Toyopearl HW

Structure and Properties

TSK-Gel Toyopearl resins are manufactured by Tosoh Corporation (Japan) and distributed worldwide by TosoHaas, a joint venture company of Tosoh and Rohm & Haas. Through most of the 1980s, these resins were distributed under the tradename Fractogel TSK by EM Science, a division of E. Merck.

The Toyopearl polymer is a porous semirigid spherical gel synthesized from methacrylate-based monomers. Copolymerization of glycidyl methacrylate, pentaerythritol dimethacrylate, and polyethylene glycol gives a complex resin structure that is rich in hydroxyl groups and ether bonds (Fig. 1.9). The surface environment formed by this type of polymer network is only mildly hydrophilic and may even display some hydrophobic tendencies under certain conditions. The internal matrix is composed of thick interwoven polymer agglomerates that lend considerable mechanical stability to the gel.

Toyopearl supports can withstand pressures up to 100 psi without collapse, making them very suitable for scale-up operations. At such pressures, the gel will easily exceed linear flow rates of 200 cm/hr. In addition, the unusual chemical strength of the ester bonds in methacrylate-based polymers gives the support excellent resistance to extremes in pH, being stable in a range of about pH 2–12. Concentrated solutions of urea,

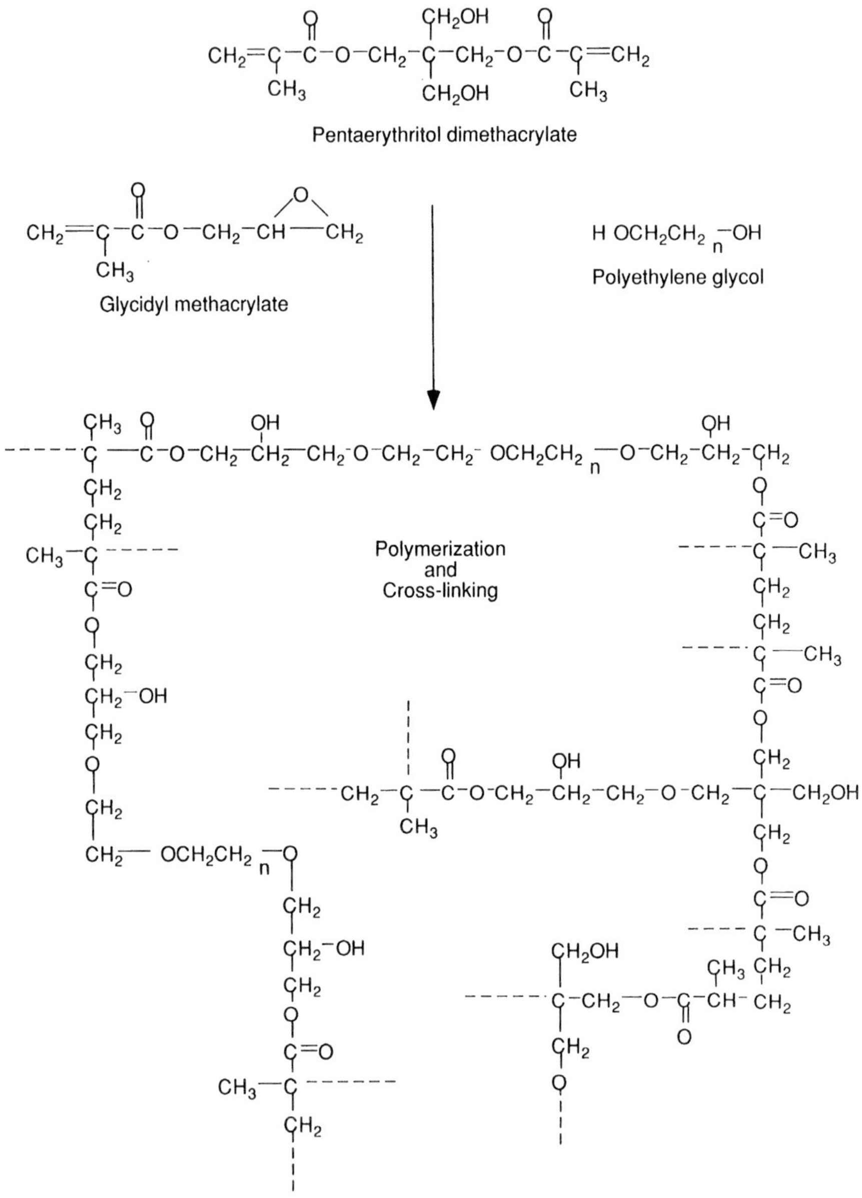

FIGURE 1.9

Toyopearl supports are methacrylate polymers made by the copolymerization of glycidyl methacrylate, polyethylene glycol, and the cross-linking agent pentaerythritol dimethacrylate.

1.3 Synthetic Supports

guanidine hydrochloride, detergents, or organic solvents will not damage the matrix; the gel will not support microbial growth.

TSK-Gels originally were designed for gel filtration applications. Their superior mechanical and chemical stability takes them a step beyond the traditional soft gel dextran-based supports. By carefully controlling the polymerization conditions, TSK varieties ranging from small particle HPLC-type gels to larger particle process-scale supports have been created. Most of the process-scale supports are available in several particle sizes, designated "S" (superfine; 20–40 µm), "F" (fine; 30–60 µm), and "C" (coarse; 50–100 µm). In general, the smaller particles will give higher resolutions whereas the larger particles will give greater flow rates at a given pressure. In all particle sizes, from HPLC to process, a variety of large pore size supports are available for protein purification purposes. The ones that accommodate the widest range of proteins are usually the best choices for use in affinity chromatography. In particular, we have found that the Toyopearl HW-65F resin (particle size, 30–60 µm) with a fractionation range of 50,000–5 million daltons for proteins gives adequate access for affinity interactions to take place within the matrix.

Immobilization Methods

Toyopearl gels contain a mixture of primary and secondary hydroxyl groups within their matrix structure. The support is easily activated by the chemistries most suited to hydroxyl-containing gels, including CNBr, CDI, tresyl, tosyl, epoxy, and DVS. Decreases in immobilization yields (compared with soft gels such as agarose) may be observed when coupling some proteins because of the rigidity and inaccessibility of the smallest pores within the matrix. Modification of Toyopearl with glycidol followed by periodate oxidation (similar to the Trisacryl procedure described previously) produces a matrix with superior coupling capacity for amine-containing ligands, particularly proteins. Under alkaline conditions, glycidol will modify Toyopearl's hydroxyl groups through its epoxy end. The other end of the molecule contains two hydroxyls on adjacent carbons that can be periodate-cleaved to give an aldehyde. The net probable result is the replacement of each original alcohol group of the matrix with a reactive formyl group. Subsequent immobilization by reductive amination will give excellent coupling yields with proteins (see Chapter 2).

Handling

Toyopearl supports come preswollen as a thick aqueous slurry containing a preservative. A known volume of gel should be taken (volume can be determined by the protocol described in Section 1.3.1.1) and washed thoroughly with water. Washing operations should be done in a Buchner funnel containing a glass fiber filter pad. Filter funnels containing an attached fritted glass filter should be avoided, because they will quickly clog and any embedded Toyopearl particles cannot be easily dissolved away for cleaning. With bulk quantities of gel (>1 L) a degree of "fines" (very small particles) may be noticed actually coming through the filter pad. Continue washing until these particles are completely removed. Alternatively, fines may be removed by decantation. Any stirring operations done for activation or coupling protocols should be accomplished with a paddle stirrer and not a stir bar which will damage the matrix. Other than these few precautions, Toyopearl supports are very robust and will not be easily damaged in handling.

1.3.2.2 HEMA

Structure and Properties

HEMA is another synthetic support made by the copolymerization of methacrylate based derivatives. The development of this type of support material originated in Czechoslovakia in the early 1970s. The gel was formerly marketed under the tradenames Separon and Spheron and more recently acquired by Alltech Associates (Deerfield, Illinois). The spherical matrix consists of a heterogeneous suspension of 2-hydroxyethyl methacrylate and ethylene dimethacrylate, copolymerized to form a highly cross-linked hydrophilic bead. The structure of the polymer (Fig. 1.10) is dominated by the presence of primary hydroxyl groups sticking out from a polyvinyl backbone. The chains are held together and stabilized by periodic cross-links formed by the ethylene dimethacrylate monomer. The three-dimensional structure of the matrix actually consists of smaller very dense polymeric microspheres that are cross-linked together to form a porous mesh, creating the larger macrospheric beads. This configuration produces both macropores, formed between aggregates of microspheres, and micropores, formed within the polymeric structure of the microspheres themselves. The resultant physical stability of this configuration is reflected in the ability of HEMA to withstand pressures exceeding 1000 psi (maximum of 4400 psi for some varieties) at maximal linear flows of over 2000 cm/hr.

The chemical structure of HEMA forms numerous tertiary α-carbonyl esters of pivalic acid, one of the most stable and least hydrolyzable esters known. The support is stable between pH 2 and pH 13, but will withstand brief washings with 2 N NaOH and 2 N HCl for cleaning and sterilizing purposes. The matrix is very resistant to chaotropic agents, detergents, or-

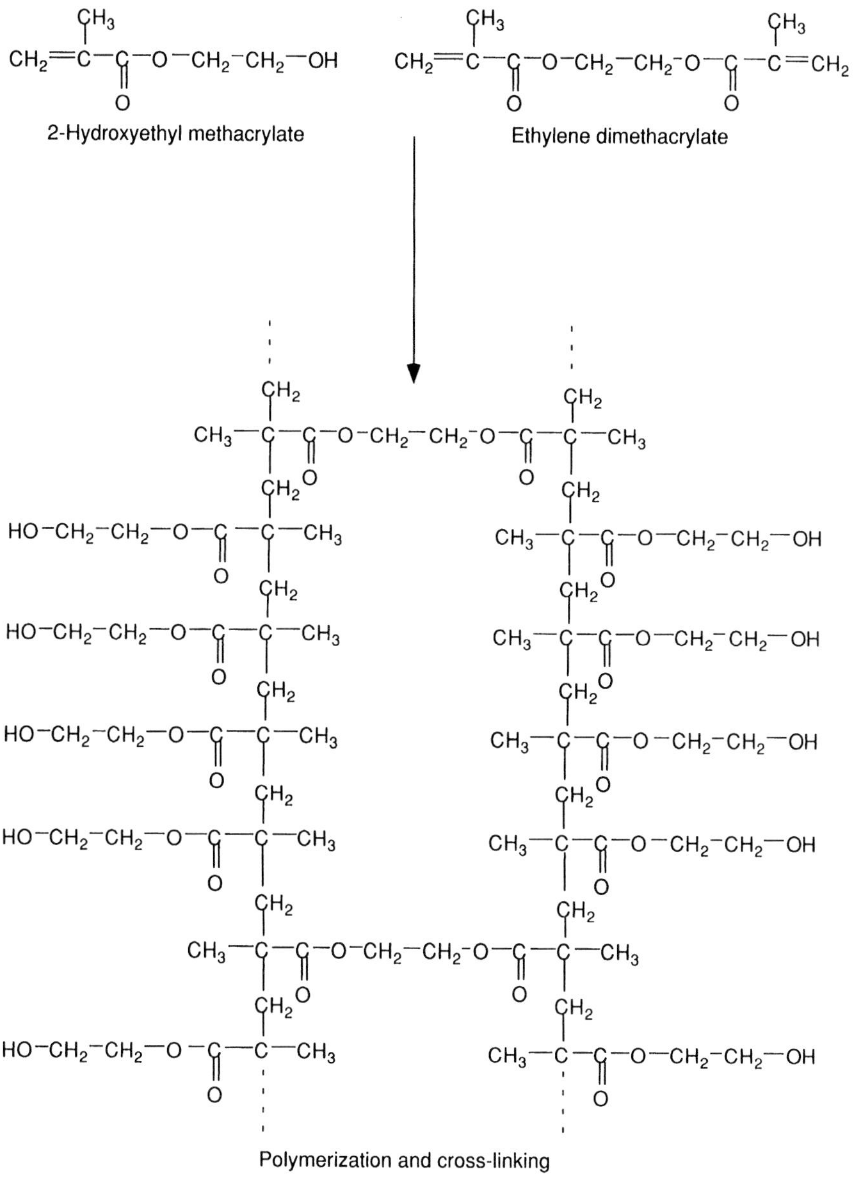

FIGURE 1.10

HEMA is produced by the copolymerization of 2-hydroxyethyl methacrylate and the cross-linking agent ethylene dimethacrylate.

ganic solvents, and common buffer salts. The support is also stable to autoclaving and will not decompose at temperatures up to 170°C.

The standard line of HEMA matrices is primarily designed for the HPLC market, with both 5-μm and 10-μm versions of the support available in prepacked columns. For scale-up operations, a 60-μm particle support has become available with virtually the same chromatographic properties as the HPLC versions. Having such a combination of analytical- and process-scale supports allows a quick transition from the research level to process-scale purification. The analytical version of the matrix allows for easy development of automated methods using an HPLC unit, whereas the larger particle supports are usually used in large-scale production operations.

Bulk quantities of the gel are supplied dry. Hydration of the matrix should be done in a high salt buffer with stirring. One unique aspect of this polymer is that the dried form of the gel maintains the physical characteristics of the hydrated support quite well. The swelling properties of the matrix are thus minimal, resulting in about 2.4 ml of hydrated gel per gram of dry beads.

Immobilization Methods

HEMA-AFC BIO is a form of the gel (distributed by Alltech) that has been developed and preactivated for use in affinity chromatography. Two activation chemistries are commercially available: one uses epoxy coupling and the other couples via DVS. Several activation levels are also provided, ranging from a low of 5–10 μeq/gm to over 1400 μeq/gm. Since the base matrix is rich in hydroxyls, other coupling protocols such as CNBr, CDI, tresyl, and tosyl can be used with success (see Chapter 2). Although we have not tried this approach, the glycidol/periodate method of generating aldehydes and subsequently coupling via reductive amination should also work well for HEMA (see previous section on Trisacryl).

Handling

The HEMA supports that are geared toward HPLC separations usually will come as prepacked columns for which all activation and coupling procedures are done in a chromatographic mode. Bulk handling of HPLC matrices can be tedious, since the small particle sizes (typically less than 10 μm) do not filter or settle well. In contrast, the process-scale support is much easier to handle. Since this support has a larger particle size distribution (average 60 μm), it flows well under gravity and may be manipulated for activation and coupling in a manner similar to that used for other affinity matrices.

1.3 Synthetic Supports

For the process-scale support, filtering and washing operations should be done using a Buchner funnel with a glass fiber filter pad. Gentle suction initiated by an aspirator or vacuum pump provides good flow rates for washing. The use of a fritted glass filter should be avoided, because cleaning the pad may be difficult due to the high chemical stability of the matrix.

1.3.2.3 Eupergit

Structure and Properties

Eupergit (pronounced "Oy-per-git") is a beaded polymeric support produced by Rohm Pharma (Darmstadt, Germany) and available in the United States through Richard Scientific (Novato, California) and Spectrum Medical Industries (Houston, Texas; sold under the tradename Spectra/Cryl). The base matrix is made by a copolymerization of methacrylamide, N,N'-methylene-bis(methacrylamide), and a component containing a reactive oxirane group (glycidyl methacrylate or allyl glycidyl ether). A hypothetical structure of the polymerized product is shown in Figure 1.11.

Eupergit C, the base-activated matrix, is available in four forms with average particle size distributions of 250 μm, 150 μm, 30 μm, and a nonporous variety at 1 μm. Analysis by electron microscopy of the porous beads has revealed the matrix morphology to contain large channels and cavities ranging in diameter from about 0.1 μm to 2.5 μm (100–2500 nm). These macropores are constructed from a network of small microbeads strung together throughout the Eupergit bead structure. However, the exclusion limit of the small pores within the microbeads is only about 200,000 daltons, indicating a very tightly packed polymer mesh. This morphology provides excellent physical and chemical stability. The support is capable of withstanding up to 4000 psi and remaining unaffected by organic solvents, buffer salts, detergents, and chaotropic agents. The matrix will swell on hydration to give about 2.5 ml gel per gram of polymer (for the 150-μm beads). Once fully swollen, further changes in buffer composition have only minimal effects on gel volume.

Immobilization Methods

Because of the incorporation of oxirane monomers in the manufacture of Eupergit, the matrix comes preactivated and ready to couple ligands through its epoxy groups. The direct immobilization of primary amine-, sulfhydryl-, or hydroxyl-containing ligands is possible using this chemistry (see Chapter 2 for a complete description of epoxy coupling procedures). An unusual aspect of the Eupergit coupling protocol is the use of 1 M po-

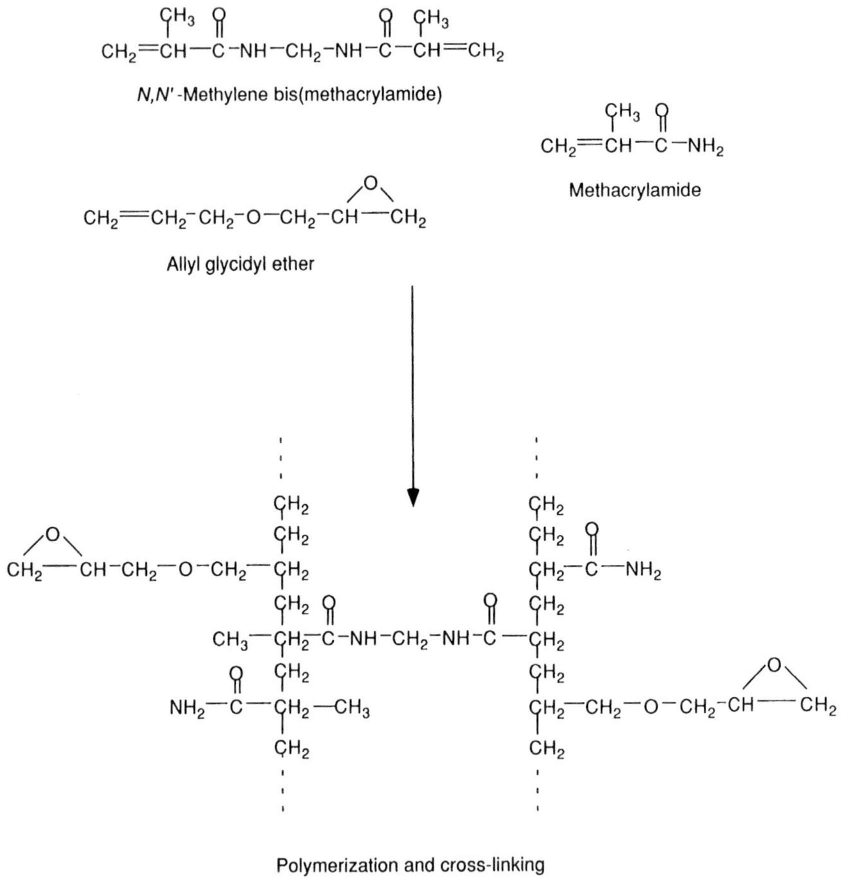

$$CH_2{=}CH-\underset{CH_3}{\overset{}{C}}-\overset{O}{\overset{\|}{C}}-NH-CH_2-NH-\overset{O}{\overset{\|}{C}}-\underset{CH_3}{\overset{}{C}}-CH{=}CH_2$$

N,N'-Methylene bis(methacrylamide)

$$CH_2{=}CH_2{-}CH_2{-}O{-}CH_2{-}\overset{\overset{O}{\diagup\diagdown}}{CH\quad CH_2}$$

Allyl glycidyl ether

$$CH_2{=}\underset{CH_3}{\overset{}{CH}}-\overset{O}{\overset{\|}{C}}-NH_2$$

Methacrylamide

↓

Polymerization and cross-linking

FIGURE 1.11

Eupergit is made by the copolymerization of methacrylamide, N,N'-methylene bis(methacrylamide), and a component containing a reactive oxirane group (glycidyl methacrylate and/or allyl glycidyl ether).

---

tassium phosphate, pH 7.5, as the reaction medium. Most epoxy coupling protocols require elevated temperatures (40–45°C) and much higher pH environments (pH 11) to open the ring and drive the reaction. This is especially true when coupling amine- or hydroxyl-containing ligands. Although the recommended procedure for Eupergit coupling does include a long reaction time (16–72 hr), the reaction is carried out only at room temperature. One possible explanation for the success of this procedure, even under mild conditions, is the presence of phosphate ions. It has been suggested by the manufacturer that phosphate ion catalyzes the reaction and allows it to proceed at physiological pH.

Handling

Eupergit C is supplied as a dry powder to prevent hydrolysis of its oxirane groups. During the manufacturing process, some residual acetone may be left in the support. Care should be taken to avoid electrostatic discharge and other sources of ignition. Handling the dry matrix in a fume hood is recommended. Eupergit can be added directly to any ligand-containing solution to initiate the coupling reaction. However, to aid in removal of trapped air bubbles and to allow better solute accessibility to the inner pores, the support should be thoroughly degassed prior to use by slurrying the beads in a suction filter flask and applying a mild vacuum. Then the beads may be washed free of residual acetone by using a Buchner funnel with a glass fiber filter pad. The only other precaution in handling the support is to avoid the use of magnetic stirring bars or any other method that could grind it against the walls of a reaction vessel.

1.3.3 Polystyrene And Its Derivatives

1.3.3.1 Poros

Structure and Properties

Poros supports, recently developed by PerSeptive Biosystems (Cambridge, Massachusetts), are said to represent a new development in chromatographic media. The polystyrene/divinylbenzene copolymer backbone of the matrix is unique in that it contains a network of large and small pores within each spherical bead (see Figure 1.12). The large pores, called "through pores" (600–800 nm), allow convective flow to occur directly through the bead itself. With conventional media, convective flow (caused by the mobile phase passing through the column) only occurs in the spaces between bead particles. Access to the majority of the active sites within the matrix structure is normally limited to molecular diffusion processes. In what PerSeptive has called "perfusion chromatography," the transport of solute into the Poros matrix can occur with much greater speed than in supports with only diffusive character. Once the flow reaches the inner parts of the bead structure, the solution can then rapidly diffuse into the smaller "diffusive pores" (50–100 nm). The result is the ability to increase the effective linear flow rate through a column without loss of capacity or reso-

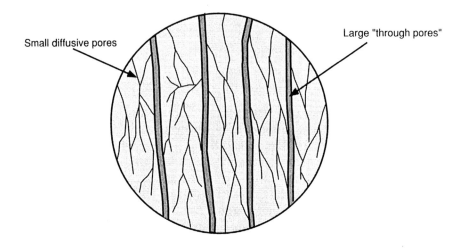

Hypothetical poros bead structure

FIGURE 1.12

The Poros matrix structure consists of large "through pores" that allow rapid flow into the interior of the bead and a network of "diffusive pores" that gives the support good capacity for affinity applications.

---

lution. Typically, an HPLC run can be reduced from 30–60 min to a few minutes or less.

Since the base matrix is made of a stable cross-linked polystyrene, the chemical and physical nature of the support is exceptionally robust. The media is resistant to extremes in pH (1–14) and is compatible with all common buffers and solvents used in HPLC environments. It can also withstand 0.5 N NaOH and 1.0 N HCl for cleaning and sterilizing purposes. The physical stability of the matrix is reflected by its maximum pressure limit of 3000 psi.

Currently, the Poros material is available in only two particle size ranges: 8–10 μm and 15–25 μm. Lack of a larger bead size makes the support somewhat restricted to HPLC-type separations. However, an HPLC system is probably the setup of choice for this support, since the perfusion properties of the matrix can only be exploited at high linear flows and high pressures. Process- or lab-scale equipment, often using peristaltic pumping systems, is not geared toward producing the high flow rates of several hundred or even several thousand centimeters per hour required by this matrix.

Immobilization Methods

The Poros matrix is constructed from a polymer-coated polystyrene/divinylbenzene bead. The polymer coating provides active sites for further modification and also blocks the harsh hydrophobic character of the styrene core.

Depending on the particular application, PerSeptive has used cross-linked polyethyleneimine as well as a proprietary polyhydroxylic polymer to coat the particles. The initial supports made using this technology consisted of only ion-exchange media for HPLC. Subsequent research has yielded good results in producing activated and ligand-derivatized versions of the matrix.

In particular, activation of the proprietary hydroxyl-containing polymer coating has yielded affinity supports based on the immobilization of protein A and iminodiacetic acid, as well as various specific immunoglobulins. Both reductive amination and tresyl-mediated coupling protocols have been used with success.

Handling

The Poros media are available only as prepacked columns for HPLC applications. Handling the current support material in bulk may pose problems because of its small particle size. Until larger sizes become available, the support will not be suitable for gravity flow or peristaltic pump operations.

1.3.3.2 Polystyrene Balls, Plates, and Devices

Structure and Properties

Polystyrene balls of dimension $1/8 - 5/8$ in. have been used extensively as solid-phase adsorbants in heterogeneous immunoassay systems. The balls are nonporous solid spheres that usually have etched surfaces to provide an increased surface area for affinity interactions. They are manufactured by injection molding of virgin, non-cross-linked polystyrene, and ground smooth to the desired dimensions. Their etched surfaces are further prepared through aqueous grinding with pumice in a tumbling operation. The balls are typically used individually in ELISA or RIA techniques, whereas multiples of them with immobilized enzymes or special reactive groups attached have been used to catalyze certain specific chemical modifications. A unique application in this respect is the Iodobeads product (Pierce Chemical, Rockford, Illinois) with its immobilized chloramine T derivative suitable for mediating the radioactive iodination of proteins.

The beads will settle quickly in most aqueous solutions (although in high salt they will remain suspended in solution). Their size allows them to be easily added to and separated from soluble reactants. This property provides a simple method to stop a timed reaction. Many diagnostic kits make use of this feature to stop enzyme–substrate reactions in immunoassay procedures. Although the size of the balls makes them unsuitable for most chromatographic separations, their continued use in immunoaffinity procedures lends significance to an understanding of the immobilization chemistry that can be used to attach ligands on their surface.

Another companion form of polystyrene that has found almost universal use in immunoassay systems is the 96-well microtiter plate. Unlike the surface of the polystyrene balls, the surface of the plates is extraordinarily smooth with excellent optical transparency in the visible spectrum. Ligands are often immobilized on these plates by simple hydrophobic adsorption and occasionally through covalent attachment. Most often these ligands are either antibodies or antigens useful in the development of a particular enzyme-linked assay. Regardless of immobilized ligand, the surface must remain uniform and unaffected in its optical clarity. Similar immobilizations have been done using multiwelled tissue culture plates. In this case, various cell adhesion factors are attached to the surfaces, as are ligands that target and bind certain cell-surface components.

Other forms of polystyrene that may be useful in certain situations include latex particles, rods, tubes, and specially formed apparatus. Assays involving an immobilized antibody or antigen attached to these devices have found use in some clinical determinations. The only form of polystyrene not considered in this book is the divinylbenzene cross-linked polystyrene often used in peptide or DNA synthesis, since the resultant immobilizations do not make use of affinity techniques.

Immobilization Methods

Unmodified polystyrene normally has no active sites or functional groups available for the direct immobilization of ligands. For this reason, the hydrophobic character of a polystyrene surface is often exploited to noncovalently immobilize ligands through passive adsorption. Usually high pH buffers are used (e.g., 0.1 M sodium carbonate, pH 9.25) to enhance the adsorption process. The polymer surface is subsequently blocked with a nonrelevant substance (e.g., bovine serum albumin) to prevent nonspecific interactions with any remaining hydrophobic sites. Using this approach, many proteins and antibodies can be efficiently coated on balls or microplates and used as affinity ligands just as if they were covalently coupled.

1.3 Synthetic Supports

For coupling small molecules and certain proteins, however, the technique of simple adsorption may not work. In many instances, hydrophilic substances do not adhere very well to the hydrophobic polystyrene surface. In other cases, the blocking reagent may compete more successfully for the hydrophobic surface than the desired coated substance, and ligand leaching may result. In most situations, covalent immobilization is much preferred over passive adsorption.

Fortunately, derivatives of the aromatic polymer surface may be prepared to provide aryl functionalities useful for covalent ligand attachment. Although many classical organic reactions may be used to modify an aromatic ring, care must be taken in choosing a method, since some solvent requirements may also dissolve the polymer. In some particularly unsuitable environments, a polystyrene ball will seem to melt or weaken at its surface and microplate wells may become cloudy and unusable. Many of the chemistries developed for preparing peptide synthesis resins from cross-linked polystyrene (mainly based on forming a chloromethyl derivative) have solvent requirements that are inappropriate for maintaining non-cross-linked polymer integrity.

For polystyrene balls, a number of methods are available for creating a covalent attachment between ligand and support. Most of these are based on electrophilic aromatic substitution reactions. For instance, brief treatment with chlorosulfonic acid will yield an amine-reactive surface acceptable for ligand immobilization. The primary site of sulfonation is the position *para* to the ethylene side chain of the polymer. The arylchlorosulfonyl groups thus formed will react with primary amines to give stable arylsulfonamides. Intermediary spacer molecules may be attached using this process or other active groups with differing reactivities may be made for coupling to non-amine-containing ligands. In this respect, hydrazide groups (which will couple to aldehydes) can be formed through direct reaction of the chlorosulfonyls with either hydrazine or adipic dihydrazide. Also, "Sanger Reagent" balls may be created by reacting an amine spacer with 1,5-difluoro-2,4-dinitrobenzene. These balls are highly reactive with amine nucleophiles.

Another method often used with polystyrene balls to generate functional groups on the surface is nitration followed by reduction to an aromatic amine. Brief treatment of the beads with HNO_3 and H_2SO_4 will yield a nitrated polymer derivative that can be subsequently reduced to the amine with either $SnCl_2$ or sodium hydrosulfite (pH 9.0). The aromatic amine can then be converted into an isocyanate, an isothiocyanate, or a diazonium salt for further reaction with ligands. Alternatively, the amino-polystyrene can be reacted with a cross-linking reagent that reacts with the arylamine on one end and with a functional group of the ligand on the other. The incorporation of a spacer molecule can also be accomplished by making use of the ring activating properties of the aromatic amine in the Mannich reaction (see Chapter 2). In this respect, an amine-containing spacer (or ligand for that matter) can be reacted with formaldehyde at elevated temperatures to provide substitution at the *ortho* position to the arylamine.

The derivatization of polystyrene microtiter plates is somewhat more difficult because optical clarity must be retained in the final product. Most of the just-mentioned chemistries that can be used on polystyrene balls or other devices cannot be used on microplates, because they will cloud or color the surface.

Methods to provide a covalent link between the plate surface and a ligand use one of two main approaches: (1) the intermediate noncovalent coating of a polymer that contains functional or reactive groups or (2) the direct covalent modification of the surface with a compound having such a functionality. The first method, passive coating of a functional polymer on the surface, again poses all the problems associated with noncovalent linkages discussed earlier. Although coatings of silicone and other polymers can be used to form a layer of functional groups on the wells, the polymer–ligand complex can still leach off the surface.

In one covalent modification technique developed by Nunc (Denmark), the polystyrene surface of the plate has been modified to provide secondary aliphatic amines at the end of a spacer molecule. The covalent link is accomplished through a graft copolymer process in which psoralen derivatives are chemically attached by a uv-catalyzed free radical reaction. The surface has been modified to the extent of 10^{14} functional molecules per square centimeter. The secondary amines are reactive with various types of cross-linking reagents containing *N*-hydroxy succinimide (NHS) esters. Thus, ligands may be immobilized through a two-step process using either homobifunctional or heterobifunctional cross-linkers. An example of this type of coupling, using the cross-linking reagent Sulfo-SMCC [sulfosuccinimidyl-4-(*N*-maleimidomethyl)cyclohexane-1-carboxylate] to immobilize an antibody molecule, is shown in Figure 1.13. The plates may also be coupled to a small molecule such as biotin (Fig. 1.14), which provides a surface that can immobilize biotinylated proteins when using avidin as a bridging molecule.

Micro Membranes (Newark, New Jersey) also has developed a surface modification technique that produces activated groups on the polystyrene plates. Although the chemical nature of their process is proprietary, it is presumably a graft copolymer derivative that leaves an amine-reactive portion available for coupling to proteins and other ligands. Unlike the Nunc plates just mentioned, the Micro Membrane plates can be incubated directly with ligand for immobilization and do not require an intermediate cross-linker treatment. Both technologies, however, result in a covalent linkage

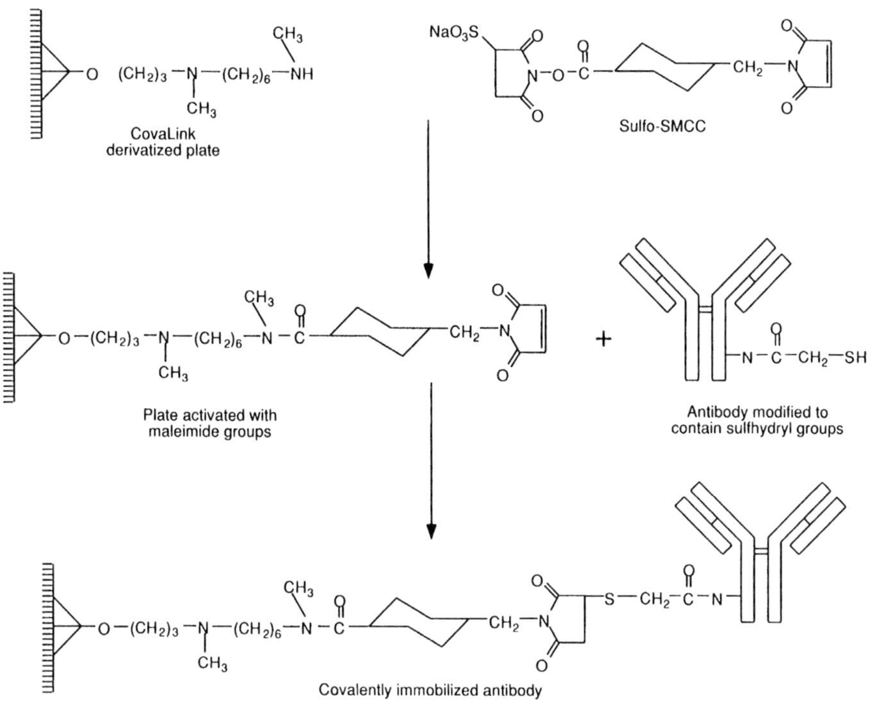

FIGURE 1.13
The cross-linking reagent Sulfo-SMCC [sulfosuccinimidyl-4-(*N*-maleimidomethyl)cyclohexane-1-carboxylate] can be used to derivatize the available secondary amine on CovaLink microplates (Nunc). The NHS ester end of the molecule reacts with the amine to form a tertiary amine linkage: this results in the maleimide end protruding outward to couple to sulfhydryl-containing ligands. An antibody molecule modified to contain free sulfhydryl groups [in this case, using the reagent SATA (*N*-succinimidyl *S*-acetylthioacetate), Pierce] can be immobilized to the plate surface by a thioether linkage.

between the surface and ligand and maintain the critical optical clarity necessary for ELISA.

1.3.4 Membranes

Structure and Properties

Membranes are similar in porous structure to synthetic beaded polymer supports, except they are constructed of large three-dimensional sheets with

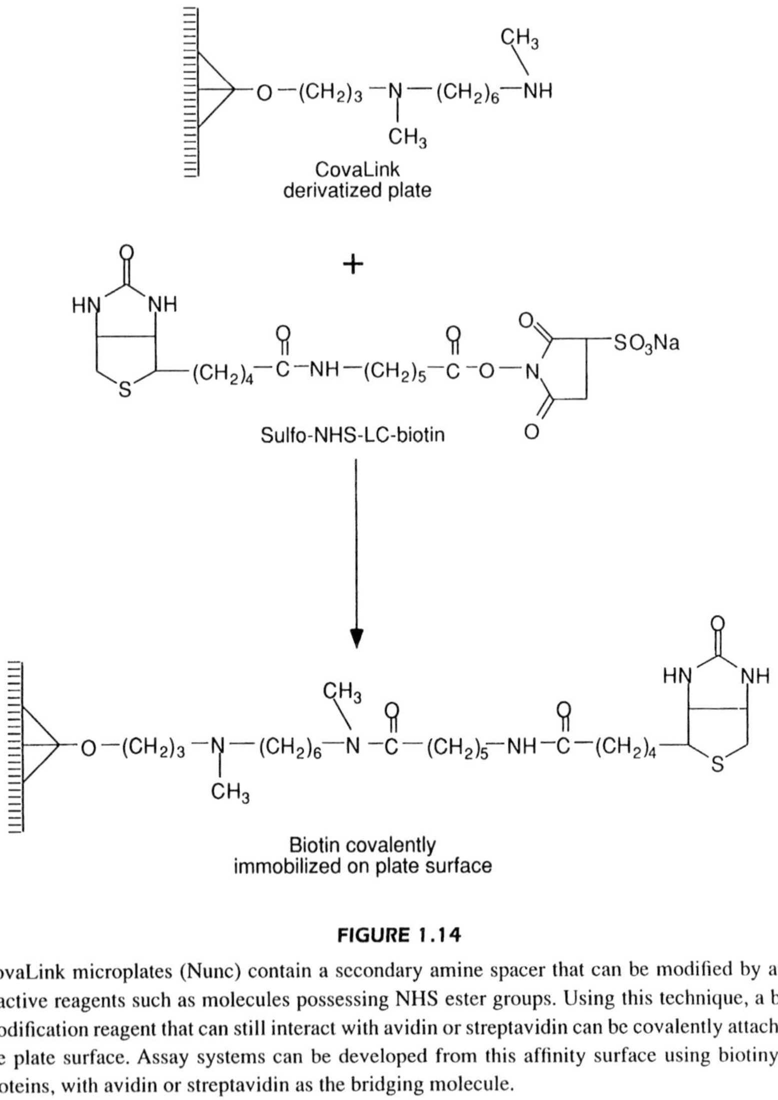

FIGURE 1.14

CovaLink microplates (Nunc) contain a secondary amine spacer that can be modified by amine reactive reagents such as molecules possessing NHS ester groups. Using this technique, a biotin modification reagent that can still interact with avidin or streptavidin can be covalently attached to the plate surface. Assay systems can be developed from this affinity surface using biotinylated proteins, with avidin or streptavidin as the bridging molecule.

---

a framework, on the microscopic scale, that resembles a sponge. The internal pore channels are connected, so a molecule entering one side of the membrane can pass through to the other side, provided it is small enough. The size of the pores can be controlled so that the membrane can restrict passage of any particle or molecule bigger than the largest pores. Therefore, filtration is by far the primary application of membrane technology.

1.3 Synthetic Supports

Microfiltration membrane structures consisting of pores between 0.1 and .01 μm are effective in filtering out relatively large particles and cells. In contrast, ultrafiltration membranes filter to a particle size of 0.001 μm, approximately the size of large protein molecules.

The chemical composition of membranes varies greatly. Sheets can be constructed from any of a number of primary polymers including cellulose, polyamide (nylon), polyacrylonitrile, polyvinylidene difluoride, polysulfone, polypropylene, polyester, polyethylene, and a host of composite resins consisting of combinations or derivatives of these polymers. Perhaps the most popular materials for use with proteins include the cellulose and nylon membranes, which can be derivatized relatively easily with various functionalities.

Polymeric membranes can be conjugated with virtually every type of affinity ligand that is immobilizable on beaded supports. Many activated and ligand-coupled membranes are already commercially available for use in affinity separations. Membrane affinity purifications are effected in a manner similar to beaded affinity chromatography separations, except cartridges are typically used instead of columns. A sample solution is passed through the membrane cartridge, using a pump or syringe, the unbound components are washed off, and finally any substance bound to the affinity ligand is specifically removed with a pulse of elution buffer. Unlike beads, membranes are strong enough that there is no concern about collapse of the support material, so flow rates can be as fast as the affinity interaction will allow. In addition, no tedious packing or repacking of the matrix is necessary, since it is permanently self-contained in the cartridge housing. Air bubbles, channeling, and column maintenance problems so common with beaded matrices are all of little concern when using membranes.

However, membrane separations often seem awkward to those people familiar with traditional column configurations. When using beaded supports, the column size can be adjusted simply by removing or adding gel to a column. With membranes, such flexibility is not possible. Although different membrane cartridge sizes are available, one is limited to what the manufacturers offer. Also, sample application can be a problem, especially when working with extremely small volumes.

The biggest problem with affinity separations on membrane-based supports is the minimal contact time a solution has with the membrane as it passes through its pores. A single sheet of membrane polymer can possess a large theoretical capacity per unit area for a particular protein, but the apparent capacity may fall far short of this maximum due to the limited time sample of interaction. In some membrane cartridge systems, multiple layers of membrane material are used to overcome these contact time limitations, causing a sample to pass through many layers of immobilized affinity ligand before it exits the cartridge. To further increase the apparent capacity levels, sample solutions can be recycled through a membrane until the maximal amount of binding is reached. Unfortunately, these problems can quickly eliminate any advantages a membrane-based system may have over a traditional column setup.

This contact time limitation is further exacerbated when eluting from a membrane affinity system. If the affinity interaction is strong and the contact time with the eluent is not sufficient to produce rapid elution of bound components, then the purified sample may be fully recovered only in unacceptably large volumes. Again, recycling a volume of elution buffer through the membrane usually overcomes this problem, but none of these additional steps are necessary with beaded chromatography supports.

On a positive note, much of these contact time limitations of binding and elution to affinity membranes can be overcome simply by using thicker layers of membrane material. Using cartridge "columns" of stacked membrane material, the deficiencies of membrane affinity separations might be greatly abated. For instance, the MAC affinity membrane system (Amicon) allows the use of 3,4,10, or even 20 membrane disks in one cartridge. These multilayer devices provide enough contact time and capacity to allow flow rates of up to 200 ml/min. Separation and elution times for some affinity systems, particularly the purification of IgG from ascites using immobilized protein A, can be reduced to only 7 min. These systems show particular promise when processing large sample volumes containing dilute protein concentrations. Pumping past a multilayered affinity membrane may be easier than using the traditional beaded chromatography support systems, since normal column maintenance concerns are eliminated.

One final caution about membrane affinity systems is necessary. Most manufacturers supply cartridges that are designed to fit on the end of a syringe. The syringe method of pushing a sample solution or buffer through the membrane may seem convenient and easy at first. However, this procedure requires constant attention to push the solutions through drop by drop, usually for an hour or so. Also, the variability in flow rates caused by this configuration can radically affect an affinity separation. When using traditional column setups with beaded supports, flow rate variations translate directly into apparent binding capacity fluctuations. In general, the faster the flow, the lower the contact time, which ultimately gives lower apparent capacities. Since syringe flow through a membrane is entirely manual and extremely variable, do not expect good reproducibility in separations. It is probably best to set up a peristaltic pump system to maintain flow accuracy and, thus, obtain optimal performance.

Membrane affinity supports also have considerable advantages over beaded matrices in blotting, detection, and immunoassay applications. For instance, activated membranes can be used to immobilize specific antibodies directed against some diagnostically useful antigen. Spotting a sample onto such a membrane will effect the binding of the antigen, and subsequent analysis by immunochemical techniques can be used to detect or quantify its presence. More advanced flow-through affinity immunoassays can be designed using membranes to speed up an assay system considerably, and even make possible laboratory testing of clinically important parameters (see Chapter 5, Section 4 for an in-depth discussion of this concept).

Immobilization Chemistry

Methods to covalently attach molecules to membrane materials are identical to those developed for beaded chromatography supports. A number of preactivated membranes in sheets and in cartridges are now commercially available. Cellulose membranes are commonly derivatized to provide formyl functional groups for the reductive-amination coupling of amine-containing ligands. Memtek activated membranes (now available from Amicon) contain aldehyde groups that can immobilize protein molecules in good yield. Acti-Disk membranes (FMC) can be obtained with an amine terminal spacer for coupling carboxylate ligands, in a glutaraldehyde-activated form for coupling primary amine-containing ligands, or as a hydrazide derivative for immobilizing aldehyde-containing ligands. Amicon, Millipore, and FMC offer membranes coupled with protein A for the purification of immunoglobulins. In addition, Gelman Sciences has a similar affinity membrane product called UltraBind that possesses formyl coupling functionalities. Pall Corporation offers a preactivated nylon membrane that will immobilize amine-containing ligands, but their coupling chemistry is kept confidential.

Typical immobilization yields for protein molecules (e.g., IgG) range from 100 $\mu g/cm^2$ to 300 $\mu g/cm^2$. These levels compare favorably with the density usually obtainable with passive adsorption of protein on hydrophobic membranes. However, covalent attachment has the advantage of being stable to detergent washing and repeated use without experiencing decreases in capacity due to leaching.

Custom activation of membrane material is also possible. When functional groups are present, as in the hydroxylic cellulose membranes, a variety of activation chemistries can be used (see Chapter 2). Membranes containing carboxylate groups can be used to immobilize amine-containing ligands through the carbodiimide reaction. Even on support materials that contain no readily available functionalities, photochemical cross-linking agents can be used to derivatize and activate the membrane. A photochemical phenyl azide group can be covalently linked to the membrane surface by light-catalyzed free radical addition. A cross-linking agent that containsa second reactive group on its other end can then be used to couple ligand molecules (Guire, 1978a,b, 1988, 1989, 1990; Guire and Chudzik, 1989).

2

ACTIVATION METHODS

2.1 GENERAL CONSIDERATIONS

Activation, as used in this manual, is defined as the process of chemically modifying a matrix so that the product of the process will react to form a covalent bond with a ligand of choice. The method selected for activation of an affinity matrix must be compatible with both the ligand and the matrix. The objective of activation is to attach a biospecific ligand efficiently and firmly to an insoluble matrix so that neither the ligand nor the matrix suffers any adverse effects and so that the activation process does not add any nonspecific characteristics to the separation system.

This chapter describes a number of viable activation chemistries. We have presented so many chemistries because there are so many different matrices of interest and so many ligands of interest and no one chemistry is best for all situations. Choosing the best available activation chemistry is not much different than initially choosing a matrix. The following criteria are offered to guide in selection.

When choosing an activation chemistry, we have made the fundamental assumption that it is almost always easier to adjust the matrix and activation to suit the ligand than to adjust the ligand to suit the matrix and the activation. Many successful examples in the literature discuss chemically modifying the ligand to better accommodate the activation chemistry. Our approach, however, favors examining the nonbiospecific functionalities available on a ligand and selecting a chemistry that is specific to and efficient with those nonrelevant groups. Notable exceptions to this approach are described in a subsequent chapter (see Chapter 3, Section 4, regarding oxidation of antibody ligands with periodate prior to coupling onto hydrazide supports). Fortunately, the selection of activation chemistries available allows a ligand-directed approach most of the time.

Criteria to Consider when Selecting an Activation Chemistry

1. *Will the activation chemistry result in a stable leak-resistant binding of the ligand to the matrix?*

Not all bonds have to be covalent to be effective. Certain antibodies, particularly polyclonals, will adhere very effectively to a hydrophobic surface for the duration of an immunoassay. If an affinity support is to be used repeatedly, the useful life of the device will be influenced by the rate of ligand leakage. If the target molecule is to be used as an injectable, special care must be taken to select a chemistry that results in a leak-free support under the conditions of use. Additionally, one must develop an extremely sensitive assay suitable for monitoring ligand leakage. Rate of ligand leakage from a particular matrix using a particular activation chemistry is usually best determined by attaching a high specific activity radiolabeled ligand and monitoring radioactivity eluted under running and storage conditions.

2. *Will the activation chemistry result in a support free of nonspecific effects?*

To examine only the nature of a successful coupling reaction is not sufficient. Most activation chemistries described in this chapter result in charge-free amide or thioether bonds. Equally important is the nature of the unsuccessful coupling reaction (i.e., the hydrolysis product of the activation chemistry). Activation methods that yield the base support on hydrolysis are preferable to those that hydrolyze to a charged function. For instance, N-hydroxy succinimide (NHS) ester-activated supports hydrolyze to a carboxylate function that can introduce ion-exchange effects (see Section 2.2.1.2). In contrast, carbonyl diimidazole (CDI) activation hydrolyzes to give the original hydroxylic matrix (Section 2.2.1.5).

3. *Will the activation chemistry alter the porosity or other properties of the matrix?*

Cyanogen bromide (CNBr), divinyl sulfone (DVS), and carbonyl diimidazole activation methods can (and usually do) result in extensive crosslinking of the matrix. This can be advantageous with some supports such as beaded agarose but the effect on beaded cellulose is that porosity is substantially reduced.

4. *Will the activation method result in rapid efficient coupling yields with the ligand?*

Coupling efficiency is primarily a function of the half-life of hydrolysis of the activated support in the coupling medium. The ideal chemistry has good reaction rates with the ligand and had a very long half-life of hydrolysis. This combination allows maximum loading of ligand, even if the ligand concentration is very low. This attribute is especially important when the ligand is scarce and expensive. Chemistries described in this chapter that have consistently high efficiencies are CNBr, reductive alkylation (amination), Mannich condensation, and azlactone.

5. *Is the activation chemistry "user friendly"?*
   1. Are the reagents readily available and inexpensive?
   2. Do I have a variety of options with respect to reaction conditions (i.e., aqueous, solvents, pH)?
   3. Will I have to contend with hazardous (toxic or flammable) reagents?
   4. Is the activation controllable (i.e., can I obtain any activation level I desire)?
   5. Can I scale the procedure up without difficulty?
   6. Is the activation chemistry reproducible?

Ultimately, the commercial availability of supports that are preactivated with one of the chemistries described in this chapter may provide the best guarantee of immobilization success while maintaining a good degree of reproducibility. Most manufacturers have the necessary quality control procedures to insure that each new lot of activated matrix contains approximately the same coupling potential as each previous lot. Ligand coupling is often no more difficult than mixing the activated support with the ligand solution in the recommended coupling buffer and reacting for the required amount of time. The majority of the activation chemistries described in this chapter are now commercially available from a number of manufacturers. Possible sources are presented in the introduction to each section.

2.2 PROCEDURES

2.2.1 Amine Reactive Chemistries

2.2.1.1 Cyanogen Bromide (CNBr)

Activation and Coupling Chemistry

One of the first methods introduced to activate solid supports was the cyanogen bromide activation procedure (Axen *et al.,* 1967). Since that time, CNBr has grown to be the most prevalent method for the activation of polysaccharide matrices and is still often the method of choice on both laboratory and industrial scales.

The mechanism of activation of polysaccharides by CNBr is shown in Figure 2.1. CNBr at high pH introduces cyanate esters and imidocarbonates into the matrix by reacting with the endogenous hydroxyl groups. Depending on the type of polysaccharide matrix used, the relative amounts of the two active groups may vary. In the case of agarose, cyanate esters are the major species formed. While cyclic imidocarbonates predominate on activated cross-linked dextrans and celluloses. The mechanism of activation by CNBr has been extensively studied by Kohn and Wilchek (1982).

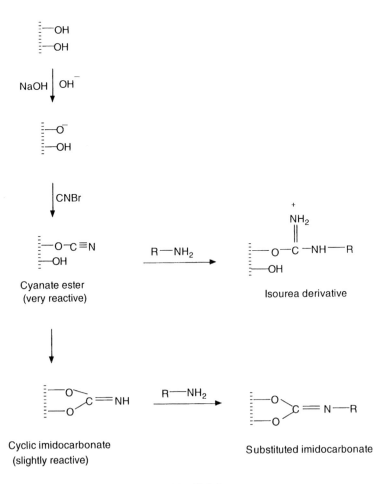

FIGURE 2.1

Mechanism of activation of polysaccharide matrix by CNBr and subsequent coupling of amine-containing ligands.

The CNBr activation method has many advantages. The method can be used almost universally to activate not only polysaccharide matrices, but also many other synthetic polymers containing hydroxyl groups. CNBr-activated supports can be used to couple to small ligands as well as to high molecular weight biopolymers containing primary amine groups. The procedure is relatively simple to carry out and is very reproducible. The method is especially mild for coupling sensitive biomolecules such as enzymes and antibodies. Hence, for many researchers, the CNBr method is the technique of choice for the preparation of affinity chromatographic supports.

The CNBr technique has some disadvantages. One of its main drawbacks is that it is often characterized by a small but constant leakage of coupled ligand. This leakage is caused primarily by the instability of the isourea bond formed between the activated support and an amine-containing ligand. Since isourea derivatives are positively charged at neutral pH, the adsorbent is also a weak anion exchanger. The resulting ion-exchange properties of the affinity matrix may cause nonspecific binding, especially when small ligands are immobilized. For these reasons, alternative amine coupling chemistries have grown in popularity in an attempt to overcome these deficiencies of the CNBr method.

The two activation methods described here work well for coupling amine-containing ligands. Both procedures yield an approximate activation level of 20–40 µmol active groups/ml gel.

ACTIVATION PROTOCOL

a. Traditional Procedure Using NaOH Titration (Cuatrecasas, 1970)

CAUTION: CNBr activation procedure should be carried out in a well ventilated hood since CNBr is highly toxic.

1. Wash Sepharose 4B (100 ml settled gel) with 1 L deionized water in a sintered glass funnel, suction dry to a wet cake, and transfer to a 500 ml beaker.
2. Suspend the gel in 100 ml deionized water and stir with a paddle. (**Caution:** Do not use magnetic stirring bars, since they can destroy the bead structure of Sepharose 4B.)
3. Add CNBr (20 gm) to the gel suspension. Maintain the pH of the reaction mixture at 11.0 by dropwise addition of 20% NaOH. Occasionally add ice to the reaction mixture to maintain the temperature around 25°C.
4. When all the CNBr dissolves (10–15 min), the rate of consumption of NaOH will decrease. At this time, pour the reaction mixture into a sin-

tered glass funnel containing ice. Quickly wash the gel with 1 L ice-cold water and 500 ml ice-cold coupling buffer (0.1 M sodium bicarbonate, pH 8.5). Due to the instability of the activated matrix, the gel should be used immediately to couple proteins or ligands.

b. The 2-min activation procedure at room temperature (March et al., 1974)

1. Wash 100 ml Sepharose 4B with water and then with 2 M sodium carbonate (activation buffer; no pH adjustment necessary). Suspend the gel in an equal volume of activation buffer and stir at room temperature in a fume hood.
2. In a fume hood, dissolve 10 gm CNBr in 5 ml acetonitrile.
3. Add the CNBr solution to the stirring gel and allow the activation reaction to continue for exactly 2 min at room temperature.
4. Quickly wash the activated support with 1–2 L ice-cold water; then with cold coupling buffer (0.1 M sodium bicarbonate, pH 8.5). Use the gel immediately in the ligand coupling protocol described next.

LIGAND COUPLING PROTOCOL

The following method is a generalized protocol for coupling ligands, spacers, or proteins to CNBr-activated supports. Specific examples can be found elsewhere in this book.

1. Suspend 100 ml CNBr-activated matrix in an equal volume of 0.1 M sodium carbonate buffer, pH 8.5, in which the ligand to be coupled has been dissolved. Other buffer salts can be used. Avoid amine-containing buffers such as Tris or other nucleophiles which will compete for the coupling sites. Coupling to CNBr-activated supports proceeds quite efficiently between pH 8.0 and 9.5. If the ligand to be immobilized is a protein, its concentration in the coupling buffer can be varied up to 20 mg/ml and still maintain at least 85% coupling yield. For small ligands, the quantity charged to the gel should be at least a three-fold molar excess over the concentration of activated groups on the matrix. For agarose, CNBr activation will generate 20–40 μmol active groups per ml gel.
2. Stir the gel at 4°C using a paddle stirrer for 24 hr.
3. Wash the coupled gel extensively with 0.1 M sodium carbonate buffer, pH 8.5, 1.0 M NaCl, and water to remove unreacted ligand.
4. Block the excess active groups on the gel by suspending it in 100 ml 1.0 M ethanolamine, pH 9.0, and stirring for 1.0 hr at room temperature.
5. Finally, wash the gel extensively with 1.0 M NaCl and water.

2.2.1.2 *N*-Hydroxy Succinimide Esters

Activation and Coupling Chemistry

The use of *N*-hydroxy succinimide esters as active groups to form amide bonds with primary amines no doubt arose from the peptide synthesis work of Anderson *et al.* (1964) and Sakakabara and Inukai (1965). Other "acidic" alcohols such as *p*-nitrophenol and trichlorophenol also have been described as capable of forming useful active esters (Sakakabara, 1965), but their utility is now largely historical as a result of the excellent water or solvent solubility of NHS.

Activation and use of matrices using NHS esters is similar to the use of carbonyl diimidazole described in Section 2.2.1.3. There are, however, some subtle differences (described in greater detail later) between the two methods that the potential user should consider.

There are two basic means for obtaining NHS-activated polyhydroxy supports: making them or buying them. BioRad (Richmond, California) offers two versions of NHS-activated cross-linked agarose supports, as shown in Figure 2.2. Affi-Gel 10 incorporates a noncharged spacer and is recommended for coupling neutral to basic proteins with isoelectric points from 6.5 to 11. Affi-Gel 15 bears a spacer arm with a positively charged nitrogen and is recommended for proteins with an isoelectric point lower than 6.5.

The structures of Affi-Gel 10 and 15 suggest that they are synthesized by a series of steps culminating with the "capping" of a selected primary

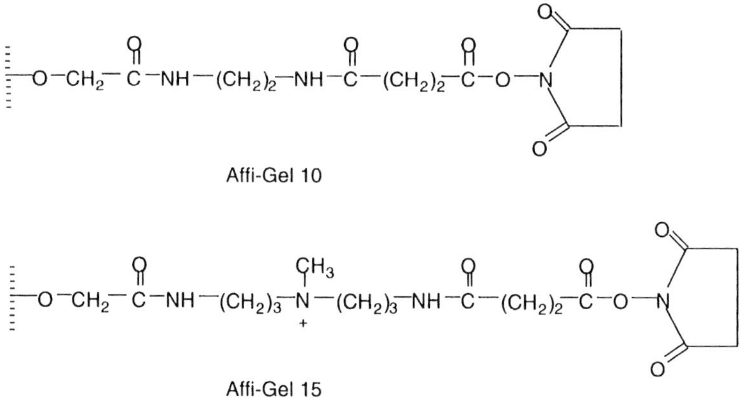

FIGURE 2.2

Commercially available *N*-hydroxysuccinimide active ester supports from BioRad.

amine spacer with the bifunctional reagent disuccinimidyl succinate. Affi-Gels are supplied preswollen as an isopropanol slurry.

For the occasional user who prefers to make an NHS-activated support, a fairly simple method for NHS activation of hydroxyl-containing matrices is offered by Wilchek and Miron 1985) (Fig. 2.3). The reagent, N,N'-disuccinimidyl carbonate (DSC) is a commercially available (Aldrich, Polysciences) and stable solid. Unlike the CDI activation method described in Section 2.2.1.3, there is little tendency for DSC to further cross-link beaded agarose supports.

ACTIVATION PROTOCOL (MODIFIED METHOD OF WILCHEK AND MIRON, 1985)

1. Sequentially exchange 100 ml Sepharose CL-6B into 100 ml dry acetone. Note: A *gradual* exchange of a water-swollen matrix such as cross-linked agarose is necessary to avoid damage or clumping of the particles. This is best accomplished by filtering the gel on a fritted glass funnel affixed to a water aspirator-evacuated filter flask. Wash the gel by slurrying

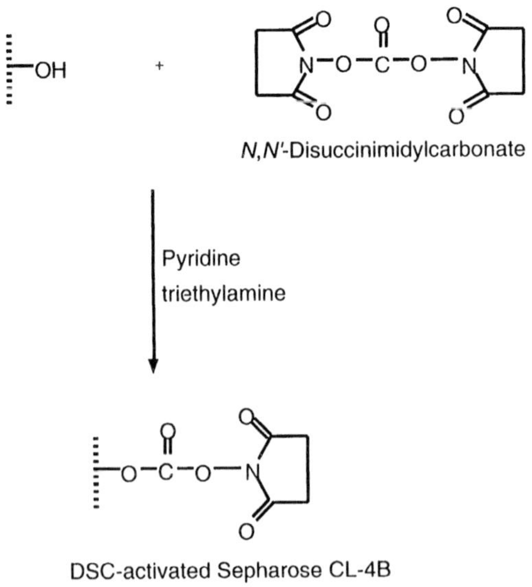

FIGURE 2.3

Preparation of N-hydroxysuccinimide-activated supports using N,N'-disuccinimidyl carbonate (DSC).

(with vacuum hose clamped) with 400–500 ml deionized water (to remove preservatives). A 1-in. wide spatula is a convenient tool with which to slurry the gel. Gently allow a vacuum to be created in the filter flask by attaching or unclamping the vacuum source; pulling the gel down to a firm, uncracked paste. Do *not* allow to dry or pack down! Repeat this process of slurrying and filtering with 400 ml quantities of 30/70 acetone/water (v/v), then 50/50, followed by 70/30, finally, filter three times with dry pure acetone. Periodically, the filter flask will need to be emptied and the water/solvent washes properly disposed of. We generally segregate our final acetone washes and use them to prepare the earlier water/acetone exchange formulations.
2. Resuspend the acetone damp gel in 100 ml dry acetone containing 8.0 grams (30 mmol) DSC. With paddle stirring, add 100 ml dry pyridine containing 7.5 ml (54 mmol) of anhydrous triethylamine (Pierce; Aldrich) dropwise over a period of 30–60 min. Continue stirring for an additional hour.
3. Filter and sequentially wash the activated gel on a fritted glass funnel with 3×200 ml dry acetone and finally with 3×200 ml dry isopropanol. For storage, prepare a slurry of known concentration in isopropanol (e.g., 50%) and transfer to a dry appropriately sized container equipped with a leakproof closure. Gel activated and stored in this manner at 4°C will remain fully active for several months.

NOTE: The wash protocol described here differs from that described by Wilchek (1985) in that it avoids the use of methanol. Our observation has been that methanol will slowly "trans-esterify" with NHS esters, yielding methyl esters. Ethanol and isopropanol do not share this tendency.

Cross-linked agarose activated by DSC in the proportions just described will typically contain 18–25 μmol/ml of active ester. This activity level is adequate for coupling most protein ligands and can be increased reproducibly by about 50%, or decreased, by adjusting the amount of DSC and triethylamine used during activation.

ACTIVATION OF GOLD FOIL ELECTRODES WITH NHS-ACTIVE ESTER

The structure of the bifunctional reagent dithiobis-succinimidyl propionate (DSP) is shown in Figure 2.4. DSP is commercially available (Pierce, Sigma). DSP has, in addition to active NHS groups, a disulfide linkage that will rapidly chemisorb to gold surfaces with a resulting stability exceeding that offered by covalent silane bonds with glass. NHS-activated gold

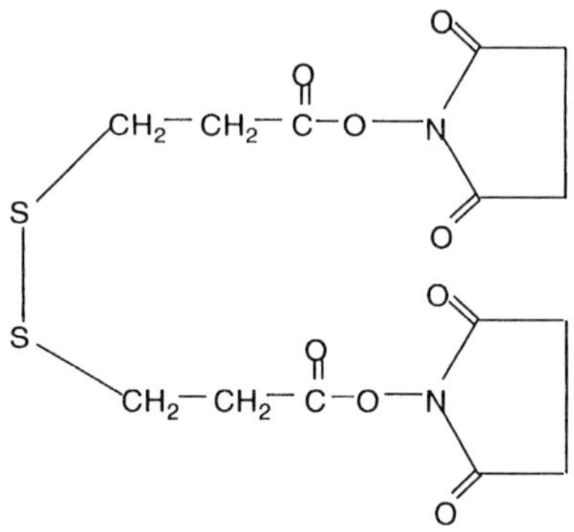

FIGURE 2.4
Chemical structure of dithiobis-succinimidyl propionate (DSP).

electrodes subsequently coupled with enzymes or redox-sensitive ligands no doubt will find widespread use in making certain electrochemical measurements.

Activation of gold foil surfaces according to the method of Katz (1990) is quick and efficient.

1. Incubate gold foil for 10 min at room temperature in 0.01 M isopropanolic DSP (400 mg/100 ml) solution.
2. After incubation, thoroughly rinse the foil with fresh isopropanol and ethanol.

Ligand Coupling Using NHS-Activated Matrices

In brief, a support activated as an NHS ester will react quite efficiently and, with the exception of thiols, nearly exclusively with primary amine-containing ligands to form a stable amide (peptide) bond (see Fig. 2.5). Efficiency of coupling compared with propensity of an activated matrix to hydrolyze should be a prime consideration when choosing an activation method. Activating a matrix to levels exceeding 100 μmol/ml is irrelevant if 90% of the activity will be lost to hydrolysis or be otherwise unused. NHS-activated gels prepared by the method described in this section or obtained commercially can routinely couple 10 mg protein per ml gel with efficiencies exceeding 80%.

An advantage offered by the DSC matrix of Wilchek and Miron (1985) over Affi-Gel or DSP–gold supports is that unused activity will ultimately

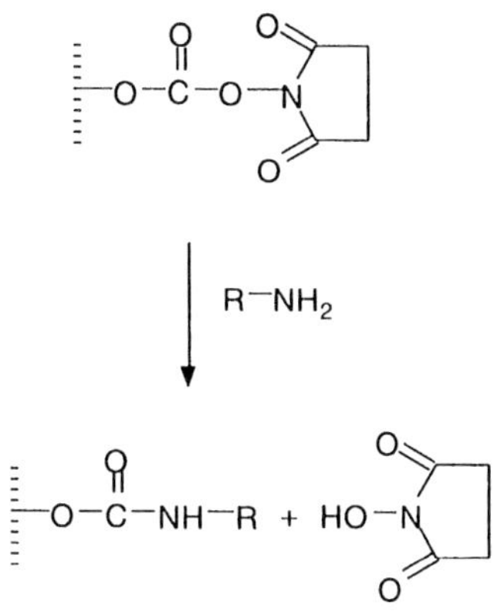

FIGURE 2.5
Coupling of amine-containing ligands to a DSC activated matrix.

revert via hydrolysis to the native support (Fig. 2.6). Spurious hydrolysis of Affi-Gel or DSP–gold will result in the formation of carboxylate groups that may contribute a weak ion-exchange character to the affinity matrix. In practice, the effect of a small level of carboxylate associated with the matrix can be obliterated by using buffers that have at least 0.05 M salt content.

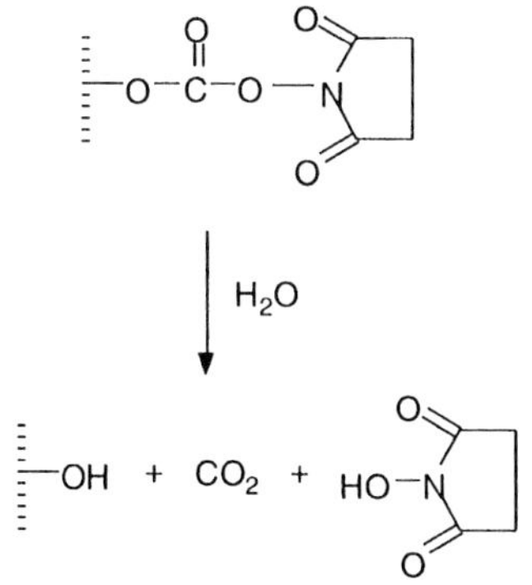

FIGURE 2.6
Hydrolysis of a DSC activated matrix.

LIGAND COUPLING PROTOCOL USING NHS-ACTIVATED MATRICES

Coupling of ligands to NHS-activated supports may be accomplished in either aqueous or nonaqueous solution. The choice of solvent system for any particular coupling reaction depends on considerations of ligand solubility/stability as well as economic and environmental considerations. Nearly all proteins are denatured by or insoluble in alcohols or acetone, making aqueous systems preferable for most large biomolecules. If gel storage solvents such as isopropanol are not completely removed and replaced with buffer, proteins (especially antibodies) will precipitate rather than couple on the gel surface.

Attachment of primary amine-containing ligands to NHS-activated supports requires an uncharged primary amine function. Thus, for efficient coupling, aqueous solvents should be buffered at pH values of 7.5–9.0. Unfortunately, as the pH and the rate of coupling reaction increase, so does the rate of hydrolysis of the activated matrix.

Perhaps the most critical step in the production of an affinity support is the exchange of the activated matrix from a solvent to an aqueous system. The moment water is introduced to the gel, hydrolysis begins. Thus, this exchange should be performed quickly. This is not meant to imply that panic is required to fashion an affinity support. On the contrary, NHS-activated supports are very easy to work with because of half-lives of hydrolysis at pH 8 in excess of 20 min at 4°C. The practitioner should be aware of any hydrolysis effects, especially as the procedure is scaled up. The failure to exactly reproduce an affinity support when increasing scale invariably can be traced to time differences experienced in solvent/buffer exchange.

Naturally, with nonaqueous solvents such as acetone [dioxane or tetrahydrofuran (THF) are also suitable, but more expensive, hydrolysis of active sites with water is not an issue. Compared with aqueous reactions, coupling efficiencies are much higher in nonaqueous coupling systems since competing hydrolysis reactions are eliminated.

Often mixed aqueous/solvent systems are appropriate, since many small primary amine-containing organic ligands are obtained commercially as the more stable water-soluble hydrochloride salt. Frequently, adjusting the pH of amine hydrochloride solutions to the desirable 8–9 range will cause precipitation of the free base. One can add sufficient solvent to the mixture to achieve a homogeneous solution without adverse effects.

a. Protocol for Nonaqueous Coupling

1. In a well-ventilated hood, sequentially exchange the gel into acetone and prepare a 50% slurry of the activated gel in acetone. An appropriately

sized beaker or Erlenmeyer flask with an overhead stirring paddle is an adequate reaction vessel. Dissolve, while stirring at room temperature, the amine-containing ligand to a level equivalent to a 10-fold excess of gel activity (~250 μmol ligand/ml DSC gel).

2. Add to the gel/ligand slurry anhydrous triethylamine to a level that is equimolar to the amount of ligand charged to the slurry. (Anhydrous triethylamine has a molecular weight of 101 and a density of about 0.72. 100 μl triethylamine is therefore equal to about 720 μmol). Stir 1 hr at room temperature.

3. Filter the ligand-loaded gel on a fritted glass funnel and slurry/filter wash the gel three times with 2X gel volume quantities of fresh acetone. During the final slurry, gradually add degassed deionized water to gently reswell the matrix. Continue washing with degassed deionized water 3–4 times with 4X gel volume quantities, taking care not to pull air into the matrix pores. Prior to using the gel to pack an affinity column, it is preferable to slurry equilibrate the support on the fritted funnel with the starting buffer. Although this pre-equilibration step does not obviate the need for column equilibration, any shrinkage or swelling of the support in the running buffer should occur before column packing. If the gel, or portions of the gel, is to be stored for extended periods, prepare a known slurry concentration (33% is easy to work with) in degassed deionized water containing 0.02% sodium azide (in order to inhibit bacterial or fungal growth) prior to storage at 4°C in a well-sealed container.

b. Protocol for Aqueous Coupling

The procedure given here is offered with reference to protein ligands, but it can be successfully employed with any water-soluble primary amine.

1. Dissolve the ligand to be coupled in a volume of aqueous coupling buffer equal to the volume of active gel being processed. Cool to 4°C. A 10-fold molar excess (over the level of active groups on the gel) for small ligands works well. For proteins, the higher the concentration, the higher the loading achieved on the gel. Optimal loadings are achieved with 10–20 mg/ml protein, but very satisfactory supports can be constructed using only 1–2 mg/ml. Economic considerations usually influence the choice of protein concentration. Suitable buffers for coupling to NHS-activated supports include 0.1 M MOPS (pH 7.0), 0.1–0.2 M phosphate (pH 7.5), 0.1–0.2 M NaHCO$_3$ (pH 8.0; prepare fresh), 0.1 M sodium borate (pH 8.5). Do not use Tris or glycine buffers since they contain primary amine functions!

2. To a stirred volume of isopropanol or acetone slurry of NHS-activated support, add two volumes of ice-cold degassed deionized water. Slurry/

filter wash the gel three times with 2X gel volumes of ice-cold degassed deionized water. The objective here is to exchange water for solvent under conditions that minimize hydrolysis of the active gel. Direct exchange with cold coupling buffer is not recommended since (1) buffer salts may precipitate when mixed with solvent and (2) hydrolysis is slower in cold water than it is in cold buffer, pH 8–9.
3. Transfer the cold water-equilibrated gel cake to the buffered ligand-containing solution. (A wash bottle containing a little cold buffer is useful to effect complete transfer.) Stir using an overhead paddle stirrer at 4°C for 4–16 hr.
4. Wash the coupled gel with fresh cold coupling buffer using the slurry/filter technique until unbound ligand is no longer detected in the wash effluent. Store the support at 4°C as a slurry (33% or 50%) in 0.02% sodium azide in a well-sealed container.

2.2.1.3 Carbonyl Diimidazole

Activation and Coupling Chemistry

N, N'-Carbonyl diimidazole (CDI) is a highly reactive carbonylating reagent that was first shown to be an excellent amide bond forming agent in peptide synthesis (Paul and Anderson, 1962) and later used to activate both carboxyl groups and hydroxyls in the immobilization of amine-containing ligands (Bartling *et al.*, 1973; Hearn, 1987).

The activation of a carboxylate group with CDI proceeds to give an intermediate amide with imidazole as the active leaving group. In the presence of a primary amine-containing compound, the imidazole is then displaced to form a stable amide bond between the matrix and ligand (Fig. 2.7). Lysozyme has been immobilized on carboxylated polystyrene using this method (Bartling *et al.*, 1973).

With hydroxyl-containing matrices, CDI will react to form an intermediate imidazolyl carbamate that can react with N-nucleophiles to give an N-alkyl carbamate linkage. Proteins normally couple through their N-terminal (α-amine) and lysine side-chain (ϵ-amine) functionalities. The final matrix-ligand bond is an uncharged urethane-like derivative with excellent chemical stability (Fig. 2.8). Active imidazolyl carbamates may also react with hydroxyl groups to form active carbonates. Some sites of primary activation in an agarose matrix probably form cross-linked carbonates because of the proximity of hydroxyls within the gel. These active carbonates can also react with primary amines to give the same product as the original imidazolyl carbamate. CDI-activated matrices are commercially available through Pierce Chemical (Rockford, Illinois).

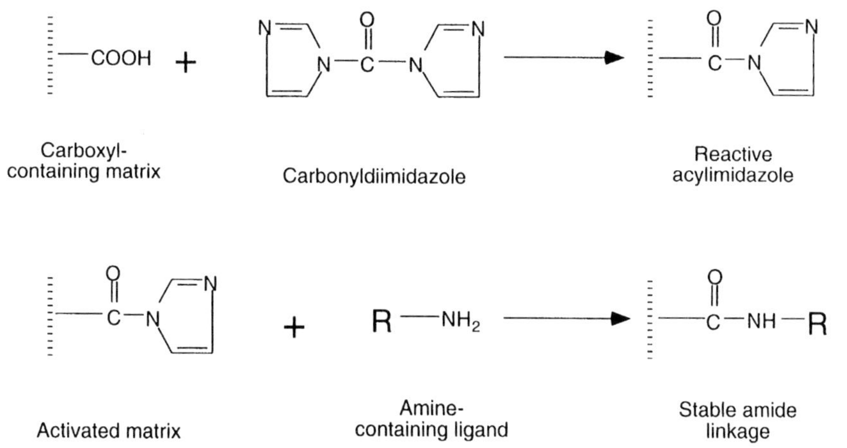

FIGURE 2.7
Carboxyl-containing matrices can be activated with carbonyl diimidazole (CDI) to give highly reactive acylimidazoles. Amine-containing ligands will rapidly couple to these reactive groups to yield stable amide linkages.

The lack of a charge in the ligand linkage and the subsequent resistance of the chemical bond to ligand leakage are both major factors in advocating the CDI method over CNBr activation. An affinity support free of extraneous charges as a result of the coupling chemistry is important for the maintenance of true biospecificity and for minimization of nonspecific adsorption effects. Low ligand leakage is important to minimize affinity support degradation and to prevent contamination of the final product with ligand.

A hydroxyl-containing matrix that is CDI-activated is stable in nonaqueous solutions for years. The support also will have an excellent half-life of hydrolysis, even in the ligand coupling environment. Unlike some activation chemistries that degrade rapidly and have half-lives on the order of minutes (e.g., CNBr and NHS ester), imidazole carbamates have half-lives measured in hours. An agarose support activated with CDI will take up to 30 hr at pH 8.5–9.0 for complete loss of activity. After the release of CO_2 and imidazole, the hydrolyzed product reverts to the original hydroxyl of the matrix (Fig. 2.9), leaving no residual groups to potentially create sites for nonspecific binding.

Optimal coupling conditions of pH, buffer composition, time, and temperature are generally a function of the ligand to be immobilized. The coupling reaction proceeds at greatest efficiency when the ligand is reacted at

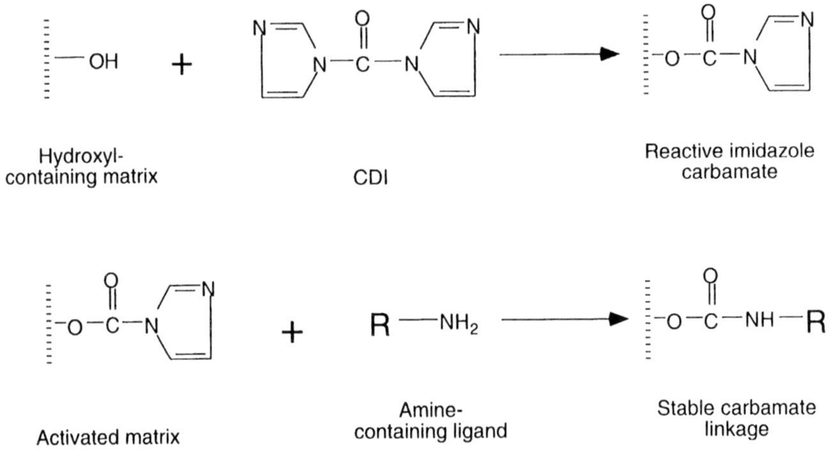

2. Activation Methods

```
    —OH    +    [imidazole]–N–C(=O)–N–[imidazole]    ⟶    —O–C(=O)–N–[imidazole]

Hydroxyl-              CDI                                 Reactive imidazole
containing matrix                                          carbamate


    —O–C(=O)–N–[imidazole]    +    R—NH₂    ⟶    —O–C(=O)–NH—R

Activated matrix              Amine-              Stable carbamate
                              containing ligand   linkage
```

FIGURE 2.8
Carboxyl-containing matrices can be activated with carbonyl diimidazole (CDI) to give highly reactive acylimidazoles. Amine-containing ligands will rapidly couple to these reactive groups to yield stable amide linkages.

---

a pH 1 higher than its pI or pK_a. Immobilization can be achieved directly in an organic solvent environment if the ligand to be coupled demonstrates poor solubility in aqueous systems. The advantage of an organic coupling reaction is that there is no competing hydrolysis of the active groups, so substitution yields can be very high.

A few precautions should be noted when performing a CDI activation and coupling experiment. First, CDI itself is extremely unstable in aqueous environments, much more than the active imidazolyl carbamate that is formed after matrix activation. Therefore, the activation step must be done in a solvent that is free of water. If unacceptable amounts of water are present, CDI will be immediately broken down to CO_2 and imidazole. The evolution of bubbles on addition of CDI to a solvent–gel slurry is a sign of a high water content. Sometimes this water may come from incomplete washing of the support into the organic solvent prior to activation. If the gel is originally supplied as an aqueous slurry, extensive washing with solvent is often necessary to completely remove the last traces of water.

A second precaution to be taken is to carry out the activation step in a fume hood away from sources of ignition. Most CDI activation protocols use flammable or toxic solvents and care should be taken in handling and disposing of them.

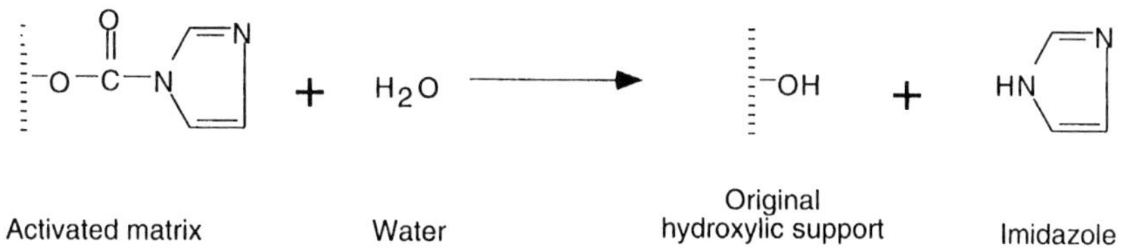

$$\text{Activated matrix} \;\; \text{-O-C(=O)-N(imidazolyl)} + H_2O \longrightarrow \text{-OH (Original hydroxylic support)} + \text{HN(imidazole) (Imidazole)}$$

FIGURE 2.9
Hydroxyl-containing matrices can be activated with CDI to give reactive imidazolyl carbamates. Amine-containing ligands will couple to these active groups to yield stable carbamate linkages.

Finally, if a ligand is to be immobilized that is sensitive to the presence of solvent, complete removal by washing the activated gel with several volumes of ice-cold water is recommended before adding the support to the ligand solution.

ACTIVATION PROTOCOL

The following protocol is for the preparation of 1 L of affinity matrix. If smaller or larger quantities of gel are desired, proportionally adjust the amount of each component or wash step mentioned. CDI may be obtained from Pierce Chemical (Rockford, Illinois) or Aldrich (Milwaukee, Wisconsin). The substance should be highly purified and white to buff in color. A good qualitative check for CDI activity is to add a small amount to several milliliters of water. If the solution fizzes vigorously, the CDI is active enough to use for gel activation. Do not, however, use CDI that is yellow, since some of the colored contaminants may adhere to the gel during activation. The solvents used should be at least ACS grade and low in water content ($\leq 0.1\%$). In addition, the matrix used for activation should be stable in the organic activation medium. If agarose is used, it should be cross-linked variety.

1. Measure out 1 L of support material and wash it with several volumes of water to remove any preservatives or storage solutions. For gels supplied dry and unswollen, such as the Sephadex series, direct addition to a solvent that will promote gel swelling, for example, dimethyl sulfoxide (DMSO), is the easiest approach.
2. In a fume hood, wash the support into acetone or a similar compatible solvent [e.g., dioxane, DMSO, dimethylformamide (DMF)] by sequential equilibration with increasing amounts of solvent. For example, first

wash with several gel volumes of 30% acetone/70% water, then with 70% acetone/30% water, and finally with 100% acetone. Wash the gel with at least five additional gel volumes of acetone to remove any residual water that may be trapped in the pores. Do not allow the gel to dry between acetone washes.

3. Suspend the gel as a 50% slurry in acetone (i.e., 1 L gel plus 1 L solvent) and stir with a paddle stirrer in a fume hood. Add 100 gm CDI and activate by stirring 1 hr at room temperature.
4. Transfer the gel to a filter funnel and wash with several volumes of solvent to remove the imidazole generated during activation. The CDI-activated gel can then be stored in dry acetone under nitrogen for at least 1 year while maintaining activity or it can be immediately coupled with ligand. With cross-linked agarose as the matrix, this amount of CDI addition should provide at least 100 μmol active groups for coupling ligands/ml gel.

LIGAND COUPLING PROTOCOL

The following method is a generalized protocol for coupling ligands, spacers, or proteins to a CDI-activated support. Specific examples can be found elsewhere in this book.

1. For ligands that are soluble in the organic solvent used in the activation step, coupling may be done entirely in a nonaqueous environment for the highest resulting density of immobilized groups. To perform an aqueous coupling, one of two procedures may be followed. (1) For ligands that can tolerate some solvent remaining in the gel, remove most of the solvent by gentle suction under vacuum. As the gel is freed of solvent, break the gel into small finely divided pieces, but be careful not to dry the support fully. Suction should be maintained only to the point at which solvent no longer drips from the support. (2) For ligands that would precipitate or be harmed by the presence of some residual solvent (i.e., some proteins), remove most of the solvent as just described, and then perform several quick washes with ice-cold water.
2. Dissolve the ligand to be coupled in a suitable buffer or solvent. The concentration of the ligand should take into account the highest possible substitution yield based on the activity of the gel. A 10-fold molar excess for small ligands works well. For proteins, the higher the concentration, the better the coupling yield. Protein solutions of 10–20 mg/ml in the coupling buffer work well. Prepare enough ligand solution to equal the volume of gel to be prepared. Suitable buffers should be formulated at pH 8.5–11. Some examples include 0.1 M sodium borate, pH 8.5, or

0.5 M sodium carbonate, pH 9–10. Buffers of pH 9–10 work best but, for proteins that are unstable at that high a pH, a buffer of pH 8.5 will give good results. Avoid buffers or other components that contain amines that will compete with the ligand for coupling (for example, Tris or glycine).

3. Mix the ligand solution with the activated gel and stir for at least 24 hr at 4°C. At pH less than 9.0, allow the reaction to continue for 30 hr to remove any remaining active sites by either coupling or hydrolysis.
4. For solvent coupling reactions, wash the gel with solvent to remove any unreacted ligand. Then sequentially wash the support back to an aqueous environment. For buffer coupling protocols, wash the support with coupling buffer, 1 M NaCl, and water. Store the affinity support as an aqueous slurry containing a preservative, preferably at 4°C.

2.2.1.4 Reductive Amination

Activation and Coupling Chemistry

Aldehydes and ketones can react with primary and secondary amino groups to form reversible Schiff bases. This interaction is enhanced at slightly alkaline pH, but is readily broken down by equilibrium dynamics. These Schiff bases can be reduced and stabilized as covalent linkages by using a reducing agent such as sodium borohydride or sodium cyanoborohydride (Borch *et al.*, 1971). Once reduced, these adducts form alkylamine bonds which are highly stable for use in immobilization techniques (Fig. 2.10).

Generation of aldehyde groups on a matrix is the first step in using this reaction for ligand immobilization. With simple polysaccharide supports such as agarose-based gels, aldehyde groups may be created by mild oxidation of adjacent diols with sodium *meta*-periodate ($NaIO_4$). The periodate breaks the carbon–carbon bonds between the adjacent hydroxyl residues in sugar monomers, creating two aliphatic aldehydes (Fig. 2.11). The activation method results in a high density of formyl groups on an agarose matrix that can then be used in reductive amination coupling procedures.

Other matrices that lack periodate oxidizable diols nonetheless can be modified with an intermediate compound that provides the necessary chemical structure. We have found, for instance, that a hydroxylic support such as Trisacryl GF-2000LS can be modified with the epoxide compound glycidol. This modification technique has been used previously to derivatize agarose (Hoyer and Shainoff, 1980; Shainoff, 1980; Shainoff and Dardik, 1981). After coupling, glycidol provides two adjacent hydroxyls that are easily oxidized with sodium periodate to form an aldehyde (Fig. 2.12). This

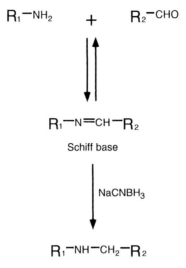

FIGURE 2.10
The interaction between a formyl group and a primary amine forms a reversible Schiff base. Reduction with sodium cyanoborohydride specifically reduces these Schiff bases to secondary amines without reducing the aldehyde.

---

route provides a higher density of formyl groups on the support surface than the direct oxidation of a polysaccharide matrix.

Alternatively, glycidol may be used to modify an amine-containing support to provide similar functionality (Fig. 2.13A). In this approach, a diamine spacer is first lengthened through the epoxide coupling of glycidol. So modified, the support will contain a spacer that terminates in vicinal diols. Periodate oxidation again provides a formyl function at the end of this spacer. Figure 2.13B shows the ligand coupling reaction using a periodate-oxidized glycidol spacer.

Both sodium borohydride and sodium cyanoborohydride have been used to reduce the Schiff bases formed between a ligand and an aldehyde-containing matrix. However, sodium cyanoborohydride is the preferred reagent. Cyanoborohydride is a milder reductant that will rapidly reduce the Schiff bases but, unlike sodium borohydride, will not reduce aldehyde groups (Borch *et al.*, 1971). With borohydride, as the Schiff bases are being reduced to couple ligand, potential formyl coupling sites are also being destroyed. Thus, cyanoborohydride actually helps drive the coupling reaction by leaving the aldehydes alone and removing the Schiff base complexes as

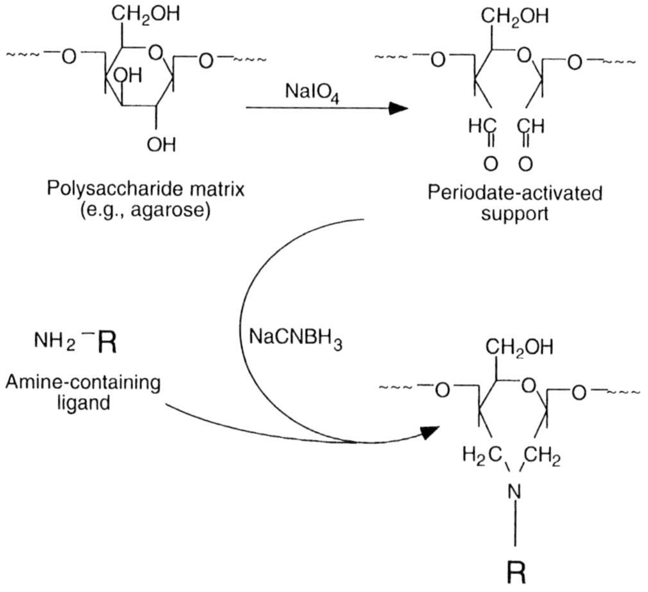

2.2 Procedures

FIGURE 2.11
Oxidation of polysaccharide matrices with sodium *meta*-periodate cleaves the carbon–carbon bond between adjacent hydroxylic groups and produces two formyl functionalities. Reductive amination with an amine-containing ligand results in secondary or tertiary amine linkages.

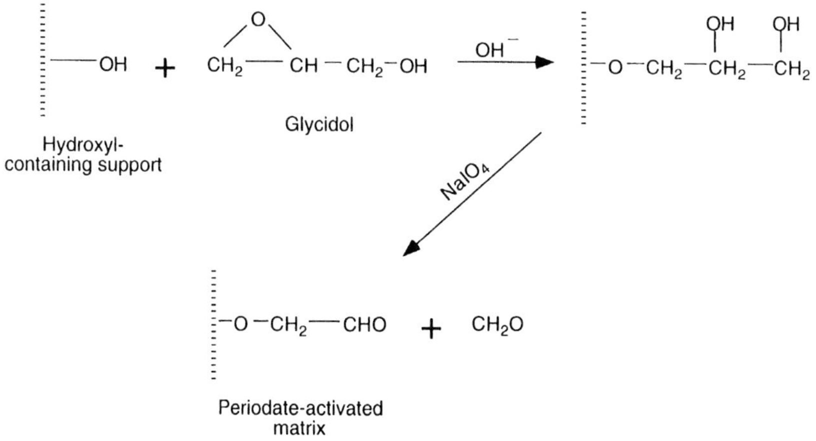

FIGURE 2.12
The glycidol modification of a matrix material introduces adjacent hydroxyls that can be oxidized by sodium meta-periodate to form aldehyde groups suitable for reductive amination coupling.

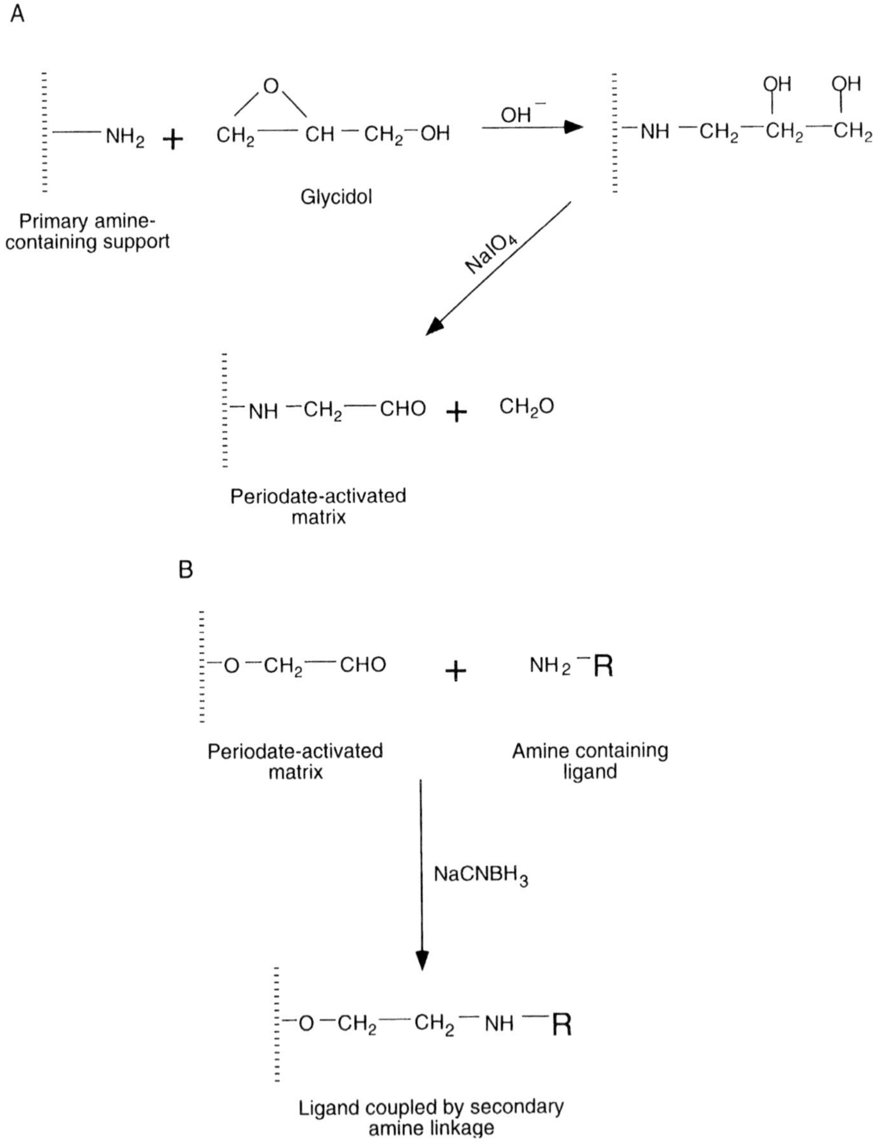

FIGURE 2.13

(A) The modification of an amine-containing matrix by glycidol can be used to create a periodate-oxidizable support for subsequent reductive-amination coupling of ligands. Capping an amine-containing matrix with glycidol results in a support with an indigenous periodate-oxidizable spacer that provides adequate room for affinity interactions once a ligand is coupled. (B) Reductive-amination coupling of ligands to a glycidol-modified matrix that had been previously oxidized with periodate to produce the necessary formyl coupling groups.

TABLE 2.1
Coupling Efficiency of Various Proteins Using Reductive Amination on Periodate-Oxidized Agarose

Protein	pI	Molecular weight	Coupling efficiency %
Ovalbumin	4.7	45,000	52
α-1 Antitrypsin	4.0	—	66
Fetuin	3.3	—	74
IgG (bovine)	—	150,000	79
Cytochrome C	9.0–9.4	11,700	79
Myoglobin	6.8–7.8	16,900	84
α-Lactalbumin	5.2	35,000	84
Transferrin	5.9	85,000	88
Hemoglobin	6.8	64,500	91
Ribonuclease A	9.5	13,700	92
Lysozyme	11.0	14,000	96

soon as they are formed. The stability of the two reagents is also markedly different. Borohydride will decompose at acid pH, whereas sodium cyanoborohydride works effectively in a range of pH 4–10.

A number of studies have been done in our laboratories to determine the coupling efficiency of proteins immobilized on periodate-oxidized cross-linked agarose (Domen et al., 1990). Tables 2.1 through 2.4 illustrate the excellent yields that can be expected for most proteins, regardless of the pH of the coupling reaction or the molecular weight or the pI of the protein.

TABLE 2.2
Immobilization Efficiency with Respect to pH for Reductive Amination Coupling of Human IgG to 2 ml Periodate-Oxidized Agarose

pH	Coupling efficiency of 9.58 mg human IgG (%)
3	50.8
4	91.8
5	92.7
6	89.1
7	87.3
8	85.3
9	94.9
10	98.4

TABLE 2.3
Coupling Efficiency of BSA on Periodate Oxidized Agarose via Reductive Amination

Time mixing (hr)	Temperature mixing (°C)	Time standing (hr)	Temperature standing (°C)	Coupling efficiency (%) 2 mg	Coupling efficiency (%) 20 mg
1	RT	—	—	64	62
1	RT	1	RT	77	71
2	RT	—	—	79	73
2	RT	2	RT	84	79
2	RT	4	RT	85	82
2	RT	16	RT	88	87
2	RT	2	4	79	74
2	RT	4	4	81	76
2	RT	16	4	81	82

RT, Room temperature

TABLE 2.4
Leakage of ^{125}I Labeled IgG from Immobilized IgG Affinity Supports Prepared by Various Coupling Methods

Support	Total radioactivity in 1 ml gel (cpm)	Total counts (cpm) leaked in 28 days (cpm)	Leakage per day (%)
CNBr–agarose	6.20×10^4	57800	0.03
CDI–agarose	0.47×10^4	4948	0.04
Tresyl–agarose	0.43×10^4	26306	0.22
NHS-activated	0.62×10^4	74846	0.43
Periodate/Reductive amination	1.00×10^4	6948	0.02

Table 2.4 shows the results of a radiolabel study in which several popular activation and coupling methods are compared with respect to ligand stability. Coupling through reductive amination is among the most stable of immobilization chemistries.

One disadvantage is that sodium cyanoborohydride is highly toxic and should be handled with extreme care. All operations should be done in a fume hood to prevent inhalation of cyanide gases. All solutions and washes resulting from the use of this compound should be handled as toxic waste and disposed of properly.

ACTIVATION PROTOCOL

a. Direct Periodate Oxidation of a Support

The following protocol is designed to prepare 1 L activated matrix. Proportionally adjust the quantities of reagents and washes to make other amounts.

1. Measure out 1 L of support material containing vicinal hydroxyl groups (for example, cross-linked agarose). Wash the gel thoroughly with water to remove preservatives and storage buffers.
2. Dissolve 42.8 gm sodium *meta*-periodate in 1 L of water (0.2 M NaIO$_4$). Add the wet gel cake to this solution and mix with a paddle stirrer.
3. Allow the reaction to continue for 90 min at room temperature.
4. Wash the formyl gel with at least 5 L water to stop the oxidation. The reaction may also be stopped by adding an equal molar quantity of glycerol to the solution prior to washing. Avoid allowing the reaction to continue for extended periods, since degradation of the support may occur with continued oxidation. With some matrices, such as cellulose

and non-cross-linked agarose, the polymeric structure may totally dissolve if the reaction is prolonged. Trial and error may be necessary with some sensitive polysaccharide matrices to discover the best combination of reaction time and periodate concentration to produce an active and viable support.

The aldehydes created by this procedure are stable enough to allow the gel to be stored for long periods without a decrease in coupling potential. However, to prevent bacterial growth, the gel should be stored at 4°C with a preservative (e.g., 0.02% NaN_3).

b. Use of Glycidol to Create a Periodate-Oxidizable Matrix

This protocol is useful for producing formyl coupling groups on supports that contain no endogenous vicinal diols. This chemistry also can be used with an amine-containing spacer to make an aldehyde support with a linker arm protruding from the matrix.

1. Measure out 1 L support material (Trisacryl GF-2000 or Toyopearl HW-65f both work well in this protocol) and thoroughly wash it with water to remove storage solutions. Finally, wash the gel with 1 L 1 N NaOH.
2. Suspend the gel in an equal volume of 1 N NaOH: add 100 ml glycidol (Aldrich) and 1 gm $NaBH_4$ with stirring.
3. React the gel overnight at room temperature while mixing with a paddle stirrer.
4. In the morning, wash the gel extensively with water, 1 M NaCl, and again with water. The glycidol-modified gel is now ready for periodate oxidation as just described.

LIGAND COUPLING PROTOCOL

The following method is a generalized protocol for coupling ligands, spacers, or proteins to a periodate-activated support. Specific examples can be found elsewhere in this book.

1. Suspend 100 ml periodate-oxidized matrix in an equal volume of 0.1 M sodium phosphate buffer, pH 7.0, in which the ligand to be coupled has been dissolved. Alternative buffer salts and pH values may be used, depending on the ligand solubility and stability requirements. Reductive amination proceeds quite efficiently between pH 4 and pH 10. The only precaution is to avoid amine-containing buffers such as Tris or other nucleophiles that will compete for the coupling sites. If the ligand to be immobilized is a protein, its concentration in the coupling buffer can be

varied up to about 20 mg/ml gel and still maintain at least 85% coupling yield. For small ligands, the quantity charged to the gel should be at least a three-fold molar excess over the concentration of aldehydes on the matrix. For cross-linked agarose, periodate oxidation will give 20–40 µmol formyl groups per ml gel.

2. Stir the gel in a fume hood using a paddle stirrer.
3. Add 0.6 gm $NaCNBH_3$ (toxic) with stirring and react overnight at room temperature.
4. The coupled gel should be washed extensively with water, 1 M NaCl, and again with water to remove unreacted ligand and sodium cyanoborohydride.

c. Use of Glutaraldehyde to Activate Supports Containing Amine or Amide Groups

Glutaraldehyde can be used to immobilize amine-containing ligands through several different reaction schemes (Weston and Avrameas, 1971; Ternynck and Avrameas, 1972). In one approach, the aldehyde groups of glutaraldehyde are used in a reductive amination procedure with cyanoborohydride similar to the procedure just described. In a second method, the double bonds of a glutaraldehyde polymer can be made to react with either amide or amine residues in a ligand or matrix. A third option is to react glutaraldehyde with an immobilized hydrazide-containing matrix. One aldehyde end will form a hydrazone linkage with the matrix, while the other end is left free and available for coupling to an amine-containing ligand. All these procedures yield secondary amine linkages with primary amine-containing ligands (see Fig. 2.14).

The procedure for activating amide groups on a matrix with glutaraldehyde and coupling to amine-containing ligands is given here.

1. Wash 100 ml amide-containing gel (such as polyacrylamide or Ultrogel AcA, available from IBF) with water and then with 0.5 M potassium phosphate, pH 7.6 (coupling buffer). Remove excess buffer until the gel is a wet cake.
2. Add 100 ml 25% aqueous glutaraldehyde to the gel, mix, and readjust the pH to 7.6 if necessary.
3. Stir the reaction for 18 hr at 37°C (a round-bottom flask with a heating mantle is convenient for maintaining the temperature).
4. Wash the gel thoroughly with water and coupling buffer.
5. To couple an amine-containing ligand, suspend the gel in 100 ml coupling buffer containing 5–10 mg/ml ligand (e.g., a protein). React overnight at 4°C.

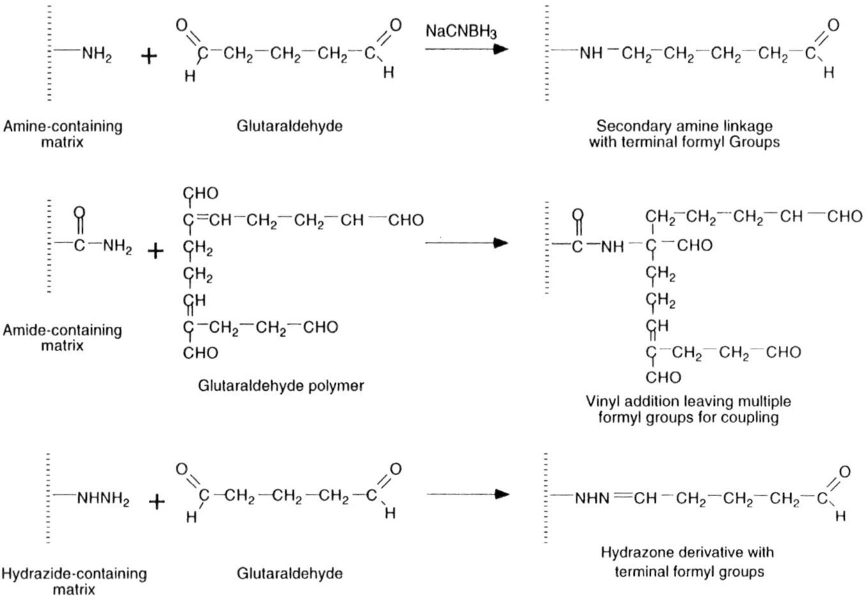

FIGURE 2.14

Three routes to a glutaraldehyde activated matrix. In the first reaction, a primary amine-containing support is reductively alkylated with an excess of glutaraldehyde to produce terminal formyl groups. The use of sodium cyanoborohydride reduces the intermediate Schiff bases without affecting the free aldehyde end. The second reaction makes use of a glutaraldehyde polymer containing double bonds to modify a matrix containing amide bonds. In this case, free-radical addition leaves the formyl groups free for subsequent coupling to ligand. This reaction also works with amine-containing supports. The last reaction illustrates the glutaraldehyde modification of a hydrazide-containing matrix. One formyl group reacts to create a hydrazone linkage, whereas the other is left free to couple with a ligand.

6. Wash the support thoroughly with coupling buffer and water. Eliminate any remaining aldehydes and other double bonds with sodium borohydride (500 mg) dissolved in 100 ml 0.1 M sodium borate, pH 9.0. React with the gel for 30 min at room temperature.
7. Finally wash the gel with water, 1 M NaCl, and again with water.

The procedure for using glutaraldehyde to activate an amine- or hydrazide-containing support and subsequently coupling to amine-containing ligands using reductive amination follows:

1. Wash 100 ml amine- or hydrazide-containing matrix (such as a support containing a spacer arm terminating in a primary amine or a hydrazide

functionality) with water and then with 0.1 M sodium phosphate, 0.15 M NaCl, pH 7.0 (coupling buffer). Drain to a moist cake.
2. Add 100 ml 12.5% glutaraldehyde (w/v) in coupling buffer to the gel. If a hydrazide-containing matrix is used, mix the reaction for 2 hr at room temperature and proceed to Step 3. If an amine-containing matrix is being modified, add 0.6 gm sodium cyanoborohydride (toxic; use a fume hood) to the gel and mix for at least 4 hr.
3. Thoroughly wash the activated support with coupling buffer.
4. Add the washed gel to 100 ml solution containing an amine ligand dissolved in coupling buffer (for a protein, use a concentration of 1–20 mg/ml; for a small ligand, 1–5 mg/ml is often enough).
5. Add 0.6 gm sodium cyanoborohydride (use fume hood) and react for at least 4 hr at room temperature.
6. Wash the gel with coupling buffer, water, 1 M NaCl, and water.

2.2.1.5 FMP Activation

Activation and Coupling Chemistry

Ngo (1986) has described a promising new reagent called FMP (2-fluoro-1-methylpyridinium toluene-4-sulfonate) to activate hydroxylic matrices. FMP, like so many of the activating reagents described in this section, converts any primary hydroxyl groups within a matrix into activated forms containing a good leaving group.

FMP-activated gels can be used successfully to couple ligands containing either sulfydryl or amine groups in slightly alkaline (pH 8–9) aqueous solutions or in organic solvents. The resulting linkages formed from reacting an FMP-activated gel and amino or thiol containing ligands (see Fig. 2.15) are stable as well as nonionic. In addition, because the leaving group on ligand coupling, 1-methyl-2-pyridone, has a molar extinction coefficient of 5900 at 297 nm, the coupling reaction can be followed conveniently by simple spectrophotometry.

ACTIVATION PROTOCOL

The activation is carried out in a fume hood because of the volatile organic solvents and amines used in the procedure.

1. Wash Sepharose CL-4B (100 ml settled gel) successively with 1 L each of water, 25:75, 50:50, 75:25 acetone:water mixtures, acetone, dry acetone, and dry acetonitrile.
2. Place the gel in a dried beaker containing 100 ml dry acetonitrile and 2 ml dry triethylamine (Pierce, Rockford, Illinois).

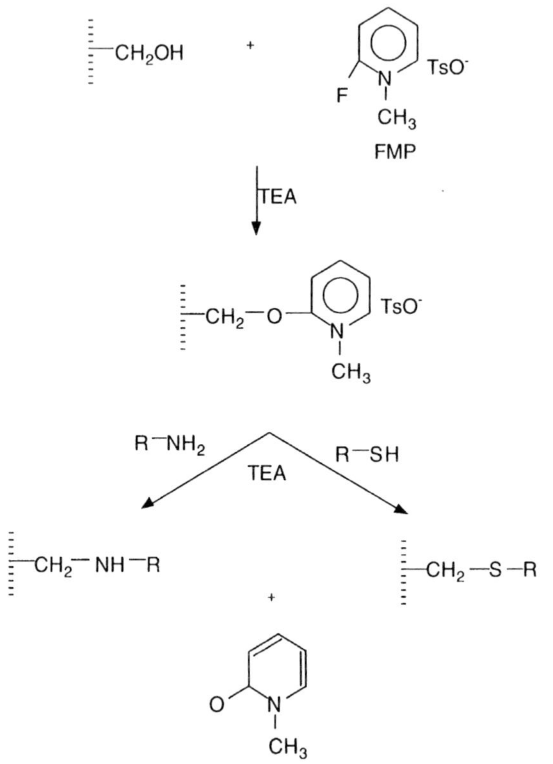

FIGURE 2.15

Mechanism of activation of hydroxylic matrices by 2-fluoro-1-methylpyridinium toluene-4-sulfonate (FMP) and subsequent coupling of amino- and thiol-containing ligands.

---

3. Dissolve FMP (6 gm) in 300 ml dry acetonitrile containing 3 ml dry triethylamine.
4. Vigorously stir the gel suspension and add the FMP solution to the stirred gel suspension in 10 ml portions.
5. Wait 15 min after the last addition of FMP solution; then transfer the activated gel in a sintered glass filter funnel and wash successively with 1 L each of acetonitrile, acetone, and then 75:25, 50:50, 25:75 acetone:2 mM HCl mixtures, and finally 2 mM HCl. FMP-activated gel can be stored in 2 mM HCl at 4°C.

LIGAND COUPLING PROTOCOL

The following method is a generalized protocol for coupling ligands, spacers, or proteins to FMP-activated supports.

1. Suspend 100 ml FMP-activated support in an equal volume of 0.2 M sodium carbonate buffer, pH 8.5, in which the ligand to be coupled has been dissolved. Other buffer salts such as phosphate can be used. Avoid amine-containing buffers such as Tris or other nucleophiles that will compete for the coupling sites. Coupling to FMP-activated supports proceeds quite efficiently between pH 7.0 and pH 9.0. If the ligand to be immobilized is a protein, its concentration in the coupling buffer can be varied up to 20 mg/ml and still maintain at least 85% coupling efficiency. For small ligands, the quantity charged to the gel should be at least a three-fold molar excess over the concentration of activated groups on the matrix. For agarose, FMP activation will generate 20–40 μmol of active groups per ml gel.
2. Stir the gel at 4°C using a paddle stirrer for 24 hr.
3. Wash the coupled gel extensively with 0.2 M sodium carbonate buffer, pH 8.5, 1.0 M NaCl, and water to remove unreacted ligand.
4. Block the excess active groups on the gel by suspending it in 100 ml 0.1 M Tris-HCl, pH 8.0, and stirring for 2 hr at room temperature.
5. Finally, wash the gel extensively with 1.0 M NaCl and water.

2.2.1.6 EDC-Mediated Amide Bond Formation

Activation and Coupling Chemistry

Carbodiimides can be used to facilitate the formation of amide bonds between a carboxylate group and an amine (Hoare and Koshland, 1966, 1967). The water soluble carbodiimide EDC [1-ethyl-3-(3-dimethylaminopropyl) carbodiimide] has been widely used in conjugation reactions, peptide synthesis, and as an immobilization reagent in the preparation of affinity gels (Lowe and Dean, 1974; Stewart and Young, 1984). EDC is one of the so-called "zero-length" cross-linkers, since it mediates the formation of amide linkages without leaving a spacer molecule (Grabarek and Gergely, 1990). N-substituted carbodiimides can react with a carboxylic group to form a highly reactive and short lived O-acylisourea derivative. With proteins, this complex is formed from C-terminal and Glu or Asp side-chain carboxyl groups. This activated species can react with a primary amine to form a peptide bond, with a sulfhydryl group to generate a thiol ester linkage, or with a molecule of water to hydrolyze to the original carboxyl group (Fig. 2.16). The rate of hydrolysis is very rapid, with a rate constant at pH 4.7 of about 2–3 s^{-1}. Thus, for all practical purposes, the activated species is not stable enough to be isolated, but must be reacted with ligand as soon as it is formed.

The immobilization of ligands using EDC can be done is two ways. An amine-containing support can be used to couple a carboxyl-containing

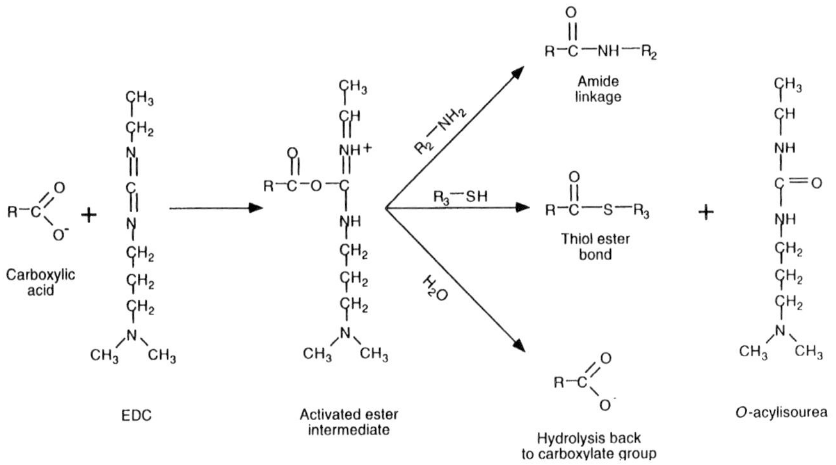

2. Activation Methods

FIGURE 2.16

The water soluble carbodiimide EDC [1-ethyl-3-(3-dimethylaminopropyl) carbodiimide] reacts with carboxylate groups to form an active ester intermediate. This short-lived ester can react with primary amines to form amide bonds, with sulfhydryl groups to form thioesters, or with water to hydrolyze back to the carboxylate. The isourea by-product is soluble in aqueous solutions and can be removed simply by washing the matrix with water or buffer.

ligand or a matrix with a carboxyl group available can be used to couple an amine-containing ligand. In most situations, a spacer molecule containing the appropriate terminal group is coupled first, then the EDC reaction is done to immobilize the ligand of choice (Fig. 2.17). Most literature references to EDC-mediated conjugation or immobilization schemes give an optimal pH for this reaction of 4.75. We have found that the EDC reaction proceeds quite well at any pH between 4.5 and 7.5. However, avoid the use of buffers that contain free amines, sulfhydryls, or carboxyl groups, since these will compete with ligand coupling.

Although EDC-mediated coupling reactions are quite efficient under most conditions, Staros et al. (1986) have developed a two-step approach to enhance the yield of amide bond formation. In this procedure, the addition of N-hydroxysulfosuccinimide (sulfo-NHS) to the carbodiimide reaction (sulfo-NHS and EDC are available from Pierce, Rockford, Illinois) results in the formation of an intermediate sulfo-NHS ester, which in turn reacts with the amine functionality (Fig. 2.18). The advantage of this approach lies in the increased stability of the active ester over the O-acylisourea intermediate formed with just EDC and a carboxyl group. The

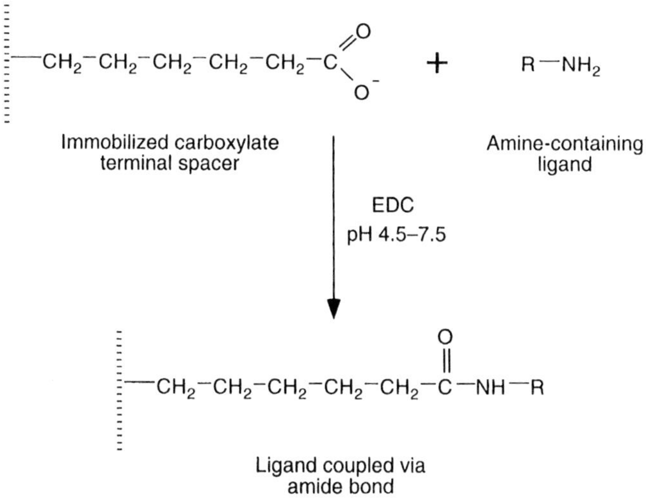

2.2 Procedures

$$-CH_2-CH_2-CH_2-CH_2-CH_2-C\begin{smallmatrix}\diagup O \\ \diagdown O^-\end{smallmatrix} \quad + \quad R-NH_2$$

Immobilized carboxylate terminal spacer   Amine-containing ligand

EDC
pH 4.5–7.5

$$-CH_2-CH_2-CH_2-CH_2-CH_2-\overset{\overset{O}{\|}}{C}-NH-R$$

Ligand coupled via amide bond

FIGURE 2.17

Spacers terminating in carboxylate groups can be coupled to primary amine-containing ligands using the water soluble carbodiimide EDC. The formation of an amide linkage proceeds rapidly and with good yields. The opposite reaction of a carboxylate ligand with an amine-containing matrix is also possible.

---

increased half-life of the sulfo-NHS esters translates into a substantially increased coupling yield or a greater degree of amide bond formation. This two-step protocol is done at slightly basic pH to encourage active ester formation and to provide an optimum coupling environment for the ester to subsequently react with amine. The only precaution we have for this procedure is to note the tendency for ligands to precipitate if they contain both carboxylate and amine functions. For instance, most proteins will severely precipitate in the presence of EDC and sulfo-NHS, presumably because of extraordinarily efficient intermolecular cross-linking. Therefore, if the ligand to be coupled contains both amines and carboxyl groups, avoid using the sulfo-NHS enhancement technique.

ACTIVATION AND LIGAND COUPLING PROTOCOL

It is assumed that, prior to this procedure, the matrix of choice has been modified (by coupling a spacer molecule) to contain either an amine or a carboxyl group appropriate for ligand immobilization. See Section 3.1.1, Chapter 3, for further details on how to form the intermediate gel.

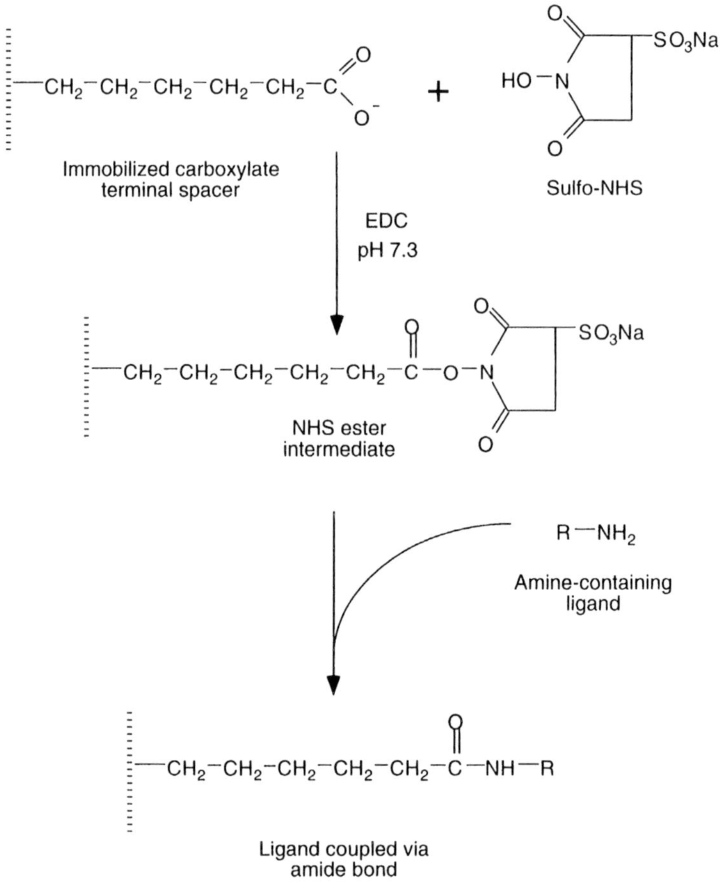

FIGURE 2.18
To increase the efficiency of amide bond formation the use of the water soluble carbodiimide EDC, along with sulfo-N-hydroxysuccinimide (sulfo-NHS) creates an active sulfo-NHS ester intermediate. This intermediate is less prone to hydrolysis than the EDC ester precursor, thus giving greater yields in the reaction with amine-containing ligands.

1. Wash 1 L carboxyl- or primary amine-containing support with water and then with coupling buffer. If the one-step EDC protocol is used, this buffer may be 0.1 M MES, pH 4.75, 0.1 M sodium phosphate, pH 7.3, or any other buffer with a pH somewhere between these two values. If sulfo-NHS is to be added to enhance the coupling yield, use the phosphate buffer.

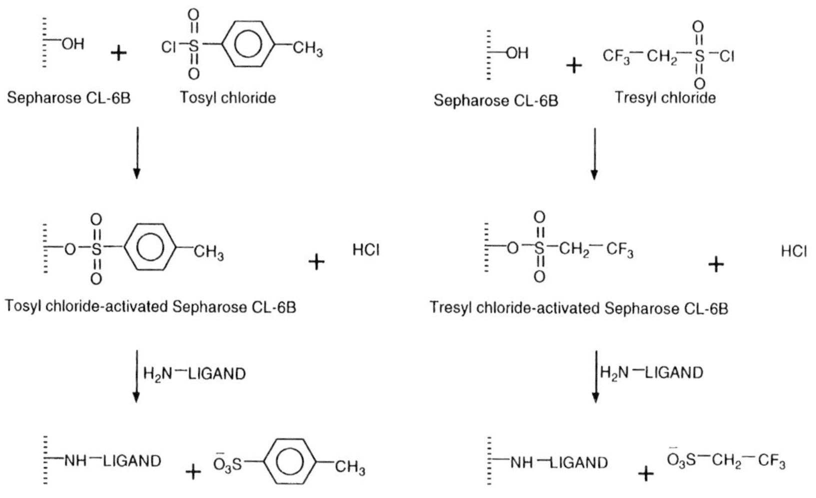

2.2 Procedures

FIGURE 2.19
Mechanism of activation of hydroxylic matrices by tosyl chloride and tresyl chloride and subsequent coupling of amine-containing ligands.

2. Suspend the gel in an equal volume of coupling buffer containing the dissolved ligand to be immobilized, and stir with a paddle stirrer.
3. For the one-step protocol, add 30 gm EDC and continue stirring for at least an additional 3 hr at room temperature. For the two-step reaction, add 30 gm EDC and 5 gm sulfo-NHS. React as before.
4. Wash the support with coupling buffer, water, 1 M NaCl, and water (2–3 L each). Store the affinity support as an aqueous slurry containing a suitable preservative (0.02% sodium azide).

2.2.1.7 Organic Sulfonyl Chlorides: Tosyl Chloride and Tresyl Chloride

Activation and Coupling Chemistry

Nilson and Mosbach (1980, 1981, 1987) introduced organic sulfonyl chlorides as activating reagents to activate agarose and other hydroxylic matrices. Organic sulfonyl chlorides, such as *p*-toluenesulfonyl chloride (tosyl chloride) and 2,2,2-trifluoroethanesulfonyl chloride (tresyl chloride) convert hydroxyl groups of the matrices into active sulfonates (Fig 2.19). Sulfonates are good leaving groups that, after reaction with nucleophiles, form stable linkages between the nucleophile and the initial hydroxyl group-carrying carbon.

Tosylated and tresylated gels can be used to couple ligands containing sulfhydryl and/or amino groups, although sulfhydryl groups show the highest reactivity. The ligand coupling step can be performed in both aqueous and organic solvents (DMF). Tresylated gels are very efficient for immobilization of ligands at neutral pH and at 4°C, whereas tosylated gels require coupling at pH 9.0–10.5.

ACTIVATION PROTOCOL USING TOSYL CHLORIDE

CAUTION: Tosyl chloride activation procedure should be carried out in a well-ventilated hood. Use only dry acetone during activation to prevent hydrolysis of tosyl chloride.

1. Wash Sepharose 4B (100 ml settled gel) with 1 L water in a sintered glass filter funnel. Then successively wash the gel with 1 L each of 30:70, 60:40, and 80:20 acetone:water mixtures, wash twice with acetone, and finally three times with dry acetone.
2. Transfer the gel to a dried beaker containing tosyl chloride (6 gm) dissolved in 100 ml dry acetone. While stirring, add 10 ml dry pyridine to neutralize the liberated HCl.
3. Continue stirring for 1 hr at room temperature.
4. Wash the activated gel twice with 1 L each of acetone, 30:70, 50:50, 70:30 of 1 mM HCl:acetone, and finally with 1 mM HCl. Tosylated Sepharose 4B can be stored for several weeks at 4°C in 100 ml 1 mM HCl without losing coupling efficiency.

LIGAND COUPLING PROTOCOL

The following method is a generalized protocol for coupling of ligands, spacers, or proteins to tosyl chloride-activated supports.

1. Suspend 100 ml tosyl chloride-activated matrix in an equal volume of 0.25 M sodium carbonate buffer, pH 9.5, in which the ligand to be coupled has been dissolved. Other buffer salts can be used. Avoid amine-containing buffers such as Tris or other nucleophiles that will compete for the coupling sites. Coupling to tosyl chloride-activated supports proceeds quite efficiently between pH 8.5 and pH 9.5. If the ligand to be immobilized is a protein, its concentration in the coupling buffer can be varied up to 5 mg/ml and still maintain at least 85% coupling yield. For small ligands, the quantity charged to the gel should be at least a threefold molar excess over the concentration of activated groups on the ma-

trix. For agarose, tosyl chloride activation will generate 20–40 μmol of active groups per ml gel.
2. Stir the gel at 4°C using a paddle stirrer for 24 hr.
3. Wash the coupled gel extensively with 0.25 M sodium carbonate buffer, pH 9.5, 1.0 M NaCl, and water to remove unreacted ligand.
4. Block the excess active groups of the gel by suspending the gel in 100 ml 0.1 M Tris-HCl, pH 8.0, and stirring for 2 hr at room temperature.
5. Finally, wash the gel extensively with 1.0 M NaCl and water.

ACTIVATION PROTOCOL USING TRESYL CHLORIDE

CAUTION: Tresyl chloride activation should be carried out in a well-ventilated hood. Use only dry acetone during activation to prevent hydrolysis of tresyl chloride.

1. Wash Sepharose 4B (100 ml settled gel) with 1 L water in a sintered glass filter funnel. Then successively wash the gel with 1 L each of 30:70, 60:40, and 80:20 acetone:water mixtures, wash twice with acetone, and finally three times with dry acetone.
2. Place the gel in a dried beaker containing 100 ml dry acetone and 10 ml dry pyridine (pyridine is to neutralize the liberated HCl during activation). While stirring, add dropwise 1 ml tresyl chloride over a period of 1 min.
3. Continue the stirring for 10 min at room temperature.
4. Wash the activated gel twice with 1 L each of acetone, 30:70, 50:50, 70:30 of 1 mM HCl:acetone, and finally with 1 mM HCl. Tresylated Sepharose 4B can be stored for several weeks at 4°C in 100 ml 1 mM HCl without losing coupling efficiency.

LIGAND COUPLING PROTOCOL

The following method is a generalized protocol for coupling ligands, spacers, or proteins to tresyl chloride-activated supports.

1. Suspend 100 ml a tresyl chloride-activated matrix in an equal volume of 0.2 M sodium phosphate buffer, pH 7.5, in which the ligand to be coupled has been dissolved. Other buffer salts can be used. Avoid amine-containing buffers such as Tris or other nucleophiles that will compete for the coupling sites. Coupling to tresyl chloride-activated supports proceeds quite efficiently between pH 7.5 and pH 8.0. If the ligand to be immobilized is a protein, its concentration in the coupling buffer can be

varied up to 20 mg/ml and still maintain at least 85% coupling yield. For small ligands, the quantity charged to the gel should be at least a threefold molar excess over the concentration of activated groups on the matrix. For agarose, tresyl chloride activation will generate 20–40 μmol active groups per ml gel.

2. Stir the gel at 4°C using a paddle stirrer for 24 hr.
3. Wash the coupled gel extensively with 0.2 M sodium phosphate buffer, pH 7.5, 1.0 M NaCl, and water to remove unreacted ligand.
4. Block the excess groups on the gel by suspending the gel in 100 ml 1.0 M ethanolamine, pH 9.0, and stirring for 1.0 hr at room temperature.
5. Finally, wash the gel extensively with 1.0 M NaCl and water.

2.2.1.8 Divinylsulfone

Activation and Coupling Chemistry

Divinylsulfone, a bifunctional cross-linking reagent, can be used to activate agarose and other hydroxylic matrices (Porath, 1974). DVS introduces reactive vinyl groups into the matrix that will couple to amines, alcohols, sulfhydryls, and phenols (Fig. 2.20). DVS-activated gels are more reactive than epoxy-activated gels, therefore, the coupling proceeds rapidly and completely. DVS activated gels are very useful for immobilizing sugars through their hydroxyl groups. However, immobilized ligands prepared by the DVS method are unstable at alkaline pH. The amine-linked gels are not stable above pH 8.0 and the hydroxyl-linked gels are unstable above pH 9 or 10. The primary instability of this procedure, however, is caused by the lability of the ether bond formed between DVS and the matrix at the time of activation.

ACTIVATION PROTOCOL

CAUTION: DVS activation should be carried out in a well-ventilated hood since DVS is highly toxic.

1. Wash Sepharose 4B (100 ml settled gel) with 1 L water in a sintered glass funnel, suction dry to a wet cake, and transfer to a 500 ml beaker.
2. Suspend the gel in 100 ml 0.5 M sodium carbonate, and stir the suspension by paddle stirring.
3. Slowly add, dropwise, DVS (10 ml) with constant stirring over a 15 min period. After addition is completed, stir the gel suspension for 1 hr at room temperature.

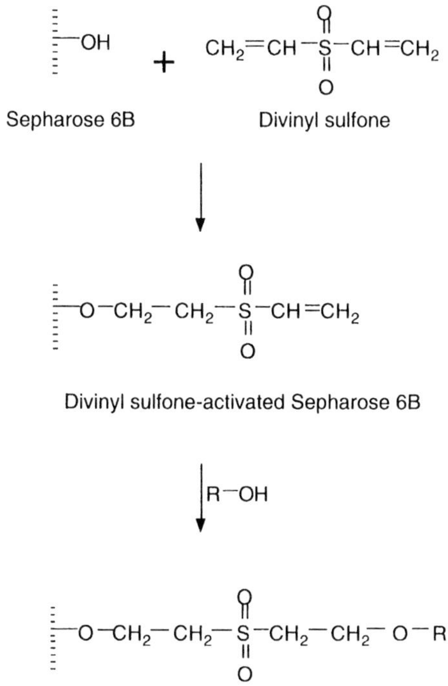

FIGURE 2.20

Mechanism of activation of polysaccharide matrices by divinyl sulfone (DVS) and subsequent coupling to hydroxyl-containing ligands.

---

4. Extensively wash the activated gel with water until the filtrate is no longer acidic. At this stage, the gel can be used to couple to ligands or stored for future use. For storage, extensively wash the activated gel with acetone and keep as a suspension in acetone at 4°C.

LIGAND COUPLING PROTOCOL

The following method is a generalized protocol for coupling of sugars and thiol-containing ligands to a DVS-activated support. Specific examples can be found elsewhere in this book.

1. Suspend 100 ml DVS-activated matrix in an equal volume of a 20% solution of sugar or thiol compound dissolved in 0.5 M sodium carbonate. Coupling to DVS-activated supports proceeds quite efficiently between

pH 10 and 11. For agarose, DVS activation will generate 40–50 µmol active groups per ml gel.
2. Stir the gel at room temperature using a paddle stirrer for 24 hr.
3. Wash the coupled gel extensively with 0.5 M sodium carbonate and water.
4. Block the excess active groups on the gel by suspending the gel in 100 ml 0.5 M sodium bicarbonate containing 5 ml 2-mercaptoethanol and stir for 2 hr at room temperature.
5. Finally, wash the gel extensively with 1.0 M NaCl and water.

2.2.1.9 Azlactone

Activation and Coupling Chemistry

Azlactones, or oxazolones, are cyclic anhydrides of N-acylamino acids and have been used extensively in organic synthesis (for review see Rao and Filler, 1986). The formation of a five-membered azlactone of particularly useful functionality for immobilization purposes can be accomplished through the reaction of a carboxylate group with a-methyl alanine using a two-step process (Fig. 2.21). One method of forming azlactone beads makes use of this process in the polymerization of monomers to first yield a carboxyl group on the matrix. In the second step, the azlactone ring is formed in anhydrous conditions through the use of a cyclization catalyst (Fig. 2.22A, B). Suitable cyclization agents that will drive this reaction include acetic anhydride, alkyl chloroformates, and carbodiimides. The process of forming these active groups and of making beaded polymeric supports containing them has been thoroughly described in patents assigned to 3M Corporation (U.S. patent no. 4,871,824 and 4,737,560). These support materials are now available under the tradename Emphase.

In the manufacturing process, the highly cross-linked azlactone beads are produced by a copolymerization of the monomers vinyldimethyl azlactone and N, N'-methylene-bis(acrylamide) (Fig. 2.23), yielding a preformed azlactone in one step. By controlling the azlactone/bis monomer ratio, supports with 1 meq/gm (100 µmol/ml) to 3 meq/gm (300 µmol/ml) azlactone functionality have been produced (Coleman *et al.,* 1990). The preactivated azlactone gel manufactured by 3M (Minneapolis, Minnesota) has an activity of about 1 meq/gm. No difference in protein immobilization yields is observed between the low and highly activated forms of the support, presumably because of steric considerations.

The azlactone group is highly reactive with nucleophiles such as amines, thiols, and, to a lesser extent, alcohols. The ring opening reaction will couple ligands from pH 4–9 to give stable amide linkages with an endogenous

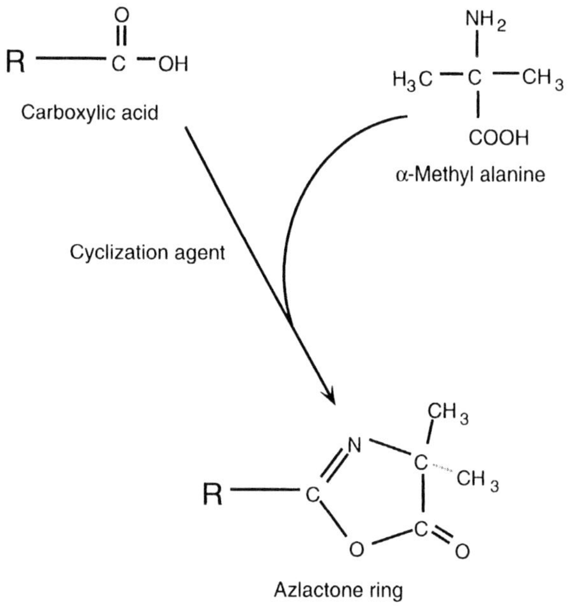

FIGURE 2.21
The reaction of a carboxylate group with a-methylalanine in the presence of a suitable cyclization agent produces an azlactone ring. In nonaqueous solutions, carbodiimides can be used to form these active oxazolones.

---

spacer molecule (Fig. 2.24). Although the azlactone group is labile in an aqueous environment, the hydrolysis rate is very slow compared with the rate of ligand coupling. Proteins and other small ligands are coupled in extraordinarily high yields in less than 1 hr. For instance, the maximum immobilization density reportedly achieved for protein A is over 30 mg/ml gel, whereas human IgG has been immobilized to a density of over 20 mg/ml.

Since the azlactone functionalities are labile in aqueous environments, the gel in its active form is supplied dry. Dry matrices often trap substantial quantities of air and may be difficult to completely rehydrate. Hydration of azlactone beads can be best accomplished by suspending in the coupling buffer and applying a mild vacuum to eliminate entrapped air.

The gel in its active form is slightly hydrophobic because of the azlactone ring structures. This surface environment causes the repulsion of some hydrophilic proteins, especially in the vicinity of their charged amino or

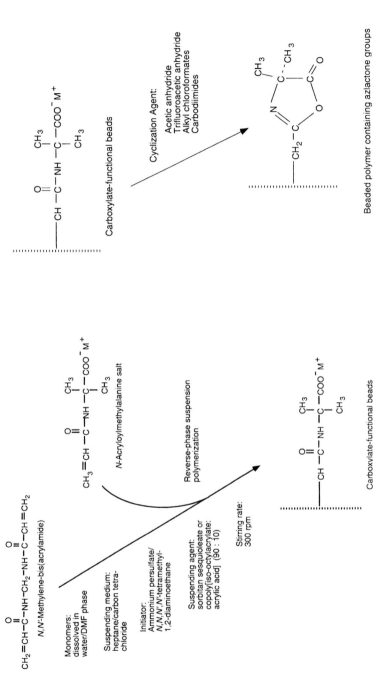

FIGURE 2.22

(A) One route to the preparation of a beaded polymer containing azlactone groups is through the reaction of the water-soluble crosslinking agent N,N'-methylene-bis(acrylamide) with N-acryloylmethylalanine salt. The details of the reverse-phase suspension polymerization are from the associated patents issued to 3M Corporation. The carboxylate-beaded polymer resulting from this reaction can then be used to create the final azlactone support. (B) The carboxylate intermediate matrix formed from the polymerization of N,N'-methylene-bis(acrylamide) with N-acryloylmethylalanine salt can be cyclized to the azlactone functionality with the appropriate catalyst.

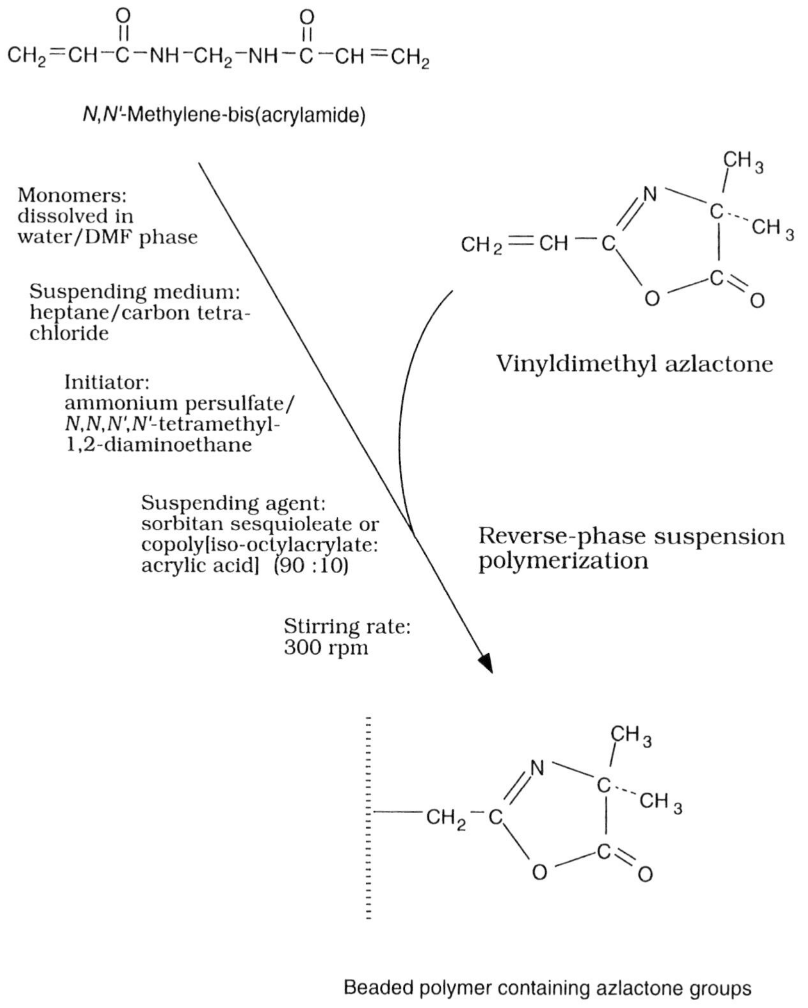

$$CH_2{=}CH{-}\overset{\overset{O}{\|}}{C}{-}NH{-}CH_2{-}NH{-}\overset{\overset{O}{\|}}{C}{-}CH{=}CH_2$$

N,N'-Methylene-bis(acrylamide)

Monomers:
dissolved in
water/DMF phase

Suspending medium:
heptane/carbon tetra-
chloride

Initiator:
ammonium persulfate/
N,N,N',N'-tetramethyl-
1,2-diaminoethane

Suspending agent:
sorbitan sesquioleate or
copoly[iso-octylacrylate:
acrylic acid] (90:10)

Stirring rate:
300 rpm

Vinyldimethyl azlactone

Reverse-phase suspension polymerization

Beaded polymer containing azlactone groups

FIGURE 2.23

The most direct route to a beaded azlactone matrix is to use the monomer vinyldimethyl azlactone with the crosslinking agent N,N'-methylene-bis(acrylamide). In a patented improvement issued to 3M Corporation, the preformed azlactone monomer obviates the need to form it from the acryloyl-methylalanine precursor as in Fig. 2.22B.

---

carboxyl groups. A consequence of this effect is occasional difficulty in getting some proteins to approach close enough to couple to the azlactone ring. This effect can be overcome, however, by the addition of a lyotropic agent (chosen from the Hofmeister series) to the coupling buffer. Sodium sulfate in the concentration range of 0.8 M to 1.5 M has been found to be effective.

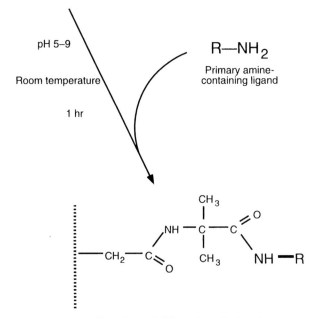

FIGURE 2.24

The azlactone group rapidly reacts with an amine-containing ligand at physiological pH. The ring opening results in an indigenous spacer coupled to the ligand via a stable amide linkage.

Primary amine-containing proteins and other ligands will couple to the gel in 1 hr at room temperature. Once the azlactone ring has opened and conjugated with the ligand, the gel surface environment is completely hydrophilic and displays low nonspecific binding character. Excess active sites should be blocked with a suitable amine-containing molecule, such as ethanolamine, after coupling the ligand. Under optimal conditions, protein coupling yields should be greater than 90%.

LIGAND COUPLING PROTOCOL

1. Dissolve the ligand to be coupled in a suitable buffer with a pH of 4 and 9. The buffer should be free of primary amines or sulfhydryls, because they will compete for the azlactone coupling sites. For proteins, a good suggested buffer is 25 mM sodium phosphate, 1.5 M sodium sulfate, pH 7.4 (for antibodies, reduce the sodium sulfate level to 0.8 M to avoid precipitation). For small ligands, replace the sodium sulfate with 0.15 M NaCl. If the ligand is stable in a 0.1% Triton X-100 solution, then including the detergent in the coupling buffer will aid in the hydration of the dried beads when they are added. The amount of protein added to the coupling buffer can be highly varied but, for maximal immobilization densities, a concentration of 10–30 mg/ml will work best. Small ligand concentration should be at least twice the theoretical azlactone activity level (i.e., 2×100 µmol/ml swollen gel).
2. If detergent is included in the coupling buffer, add the dried azlactone beads directly to the solution with stirring at a concentration of 100 mg per 2 ml ligand solution. If no detergent is added to the coupling buffer, first hydrate the dried beads in 25 mM sodium phosphate, 0.1% Triton X-100, pH 7.4, at a concentration of 100 mg beads/2 ml buffer. Add the beads, stirring to the solution. After several minutes of hydration, remove the excess buffer by filtration and add the hydrated support to the ligand solution.
3. Allow the coupling reaction to continue for at least 1 hr at room temperature. Stir by using a paddle stirrer or mix by end-over-end rocking. Longer reaction times are not detrimental, and 4°C reactions, when required, work well.
4. Wash the gel with water to remove the reactants; then add a volume of 1 M ethanolamine, pH 9.0, equal to the volume of swollen gel. Stir or mix for an additional 30 min to block unreacted active sites.
5. Wash extensively with PBS, pH 7.4, 1.0 M NaCl, and water. Store the affinity support at 4°C in water or buffer containing a preservative.

2.2.1.10 Cyanuric Chloride (Trichloro-s-Triazine)

Trichloro-s-triazine (TsT) is a symmetrical heterocyclic compound containing three reactive acyl-like chlorines (Fig 2.25). This reagent and its derivatives are extensively used in the dye industry to form strong covalent bonds between chromophores and fabrics.

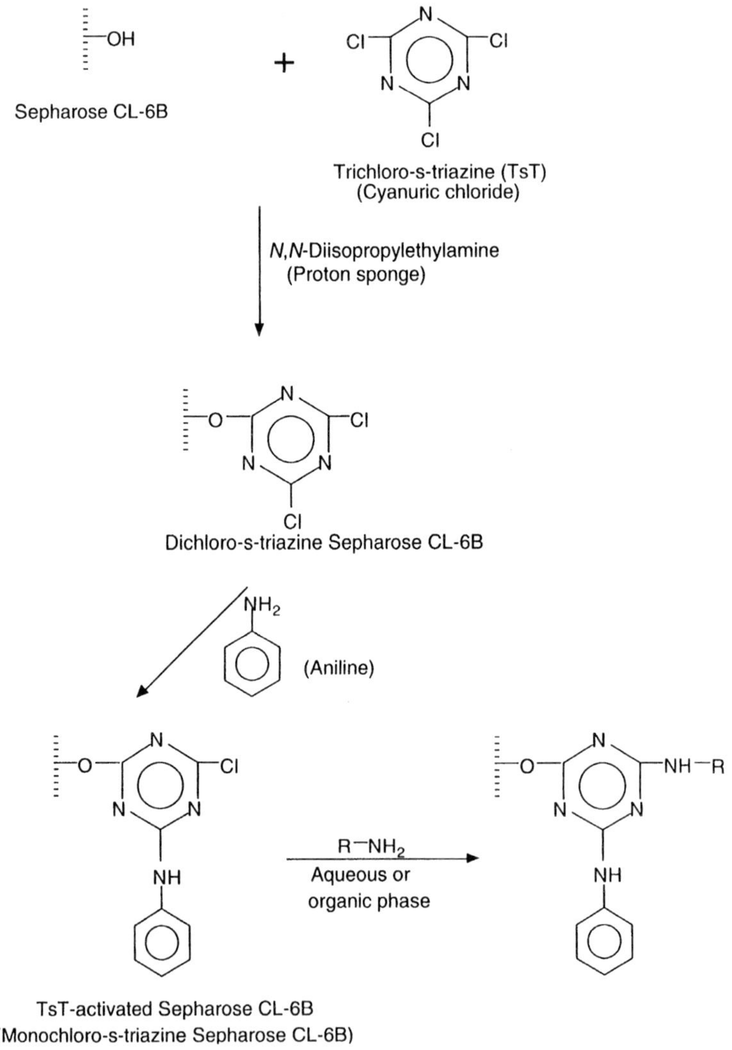

FIGURE 2.25

Mechanism of activation of polysaccharide matrices by cyanuric chloride and subsequent coupling to amine-containing ligands.

ACTIVATION PROTOCOL

The following activation protocol is adapted from Finlay et al. (1978). The activation is carried out in a fume hood because of the volatile organic solvents and amines used in the procedure.

1. Slowly wash Sepharose CL-6B (100 ml settled gel) on a sintered glass funnel with 500 ml water/acetone (70/30) followed by 500 ml water/acetone (30/70), and finally with 1 L acetone. This step is necessary to transfer the gel from aqueous to organic phase without damage to the gel structure.
2. Suspend the gel in 100 ml acetone and transfer to a 500-ml three-necked round-bottom flask equipped with a water-jacketed condenser, a thermometer, and a glass-sealed stirrer.
3. Heat the reaction flask to 50°C using a heating mantle. Gently stir the gel suspension while adding 20 ml 2.0 M N,N-diisopropylethylamine (Aldrich, Milwaukee, Wisconsin) in acetone.
4. After 30 min, add 20 ml 1.0 M trichloro-s-triazine (recrystallized; for recrystallization, see Chapter 3, Section 3.6.7.1) in acetone.
5. Gently stir the reaction at 50°C for 1 hr.
6. Slowly wash the activated support on a sintered glass funnel with 1 L acetone to remove unreacted cyanuric chloride and N,N-diisopropylethylamine.
7. Stir the activated dichloro-s-triazine-Sepharose CL-6B with 200 ml 2.0 M aniline in acetone at room temperature for 30 min. Under these conditions, aniline displaces only one of the two remaining chlorines on the triazine ring, leaving the final chlorine of the monochloro-s-triazine-Sepharose CL-6B available for replacement by a ligand containing a suitable nucleophilic group.
8. Slowly wash the gel, TsT-activated Sepharose CL-6B (now monochloro-s-triazine-Sepharose CL-6B), with 1 L acetone to remove excess aniline. The gel is ready for coupling.
9. The activated gel can be stored in acetone at 4°C.

LIGAND COUPLING PROTOCOL

TsT-activated Sepharose CL-6B can be used directly for coupling in acetone to react with acetone-soluble nucleophiles. For coupling of water-soluble nucleophiles, the activated gel is transferred to aqueous phase by washing with ice-cold acetone:water solutions of increasing water content before mixing with coupling buffer; then the gel used immediately. TsT-activated Sepharose couples efficiently to small ligands and proteins.

a. Coupling Proteins to TsT-Activated Sepharose CL-6B

1. Transfer the TsT-activated Sepharose CL-6B (10 ml settled gel) to aqueous phase as follows. Slowly wash the gel in a sintered glass funnel with 50 ml each of ice-cold acetone:water mixture (70:30), acetone:water mixture (30:70), water, and coupling buffer (0.1 M sodium borate buffer, pH 8.5, containing 0.15 M NaCl).
2. Suction dry the gel to a moist cake, and add it to a protein solution (25–50 mg protein dissolved in 10 ml coupling buffer).
3. Gently rock the reaction mixture for 24 hr at room temperature or at 4°C. The efficiency of coupling is 50% less at 4°C than at room temperature.
4. Wash the gel on a sintered glass funnel with 100 ml coupling buffer, suction dry to a moist cake, and add 10 ml 1.0 M Tris-HCl buffer, pH 9.0. Mix the gel suspension for 1 hr at room temperature. This step is to block the remaining triazine chlorides not consumed during coupling.
5. Finally, wash the gel with 100 ml each coupling buffer and water.
6. The immobilized protein gel can be stored in 50% glycerol containing 0.05% sodium azide at 4°C or at –20°C.

b. Coupling Small Ligands to TsT-Activated Sepharose CL-6B

The protocol just given for coupling protein can be used for coupling small ligands to TsT-activated Sepharose CL-6B. For ligands insoluble in aqueous buffers, the ligands can be dissolved in acetone and coupled to TsT-activated Sepharose CL-6B in acetone.

2.2.2 Sulfhydryl Reactive Chemistries

2.2.2.1 Iodoacetyl and Bromoacetyl Activation Methods

Activation and Coupling Chemistry

Bromoacetyl- and iodoacetyl-containing matrices are excellent activated supports for the immobilization of ligands containing sulfhydryl groups. The matrices will provide extremely stable thioether bonds between the ligands and matrices (Fig. 2.26). The coupling reaction with sulfhydryl ligands is very fast (15 min) and can be carried out at pH 8.0–8.5 (see Tables 2.5 and 2.6).

ACTIVATION PROTOCOL

The following procedure makes use of an intermediate spacer gel that contains a terminal amine group. In this case, diaminodipropylamine is used as

2.2 Procedures

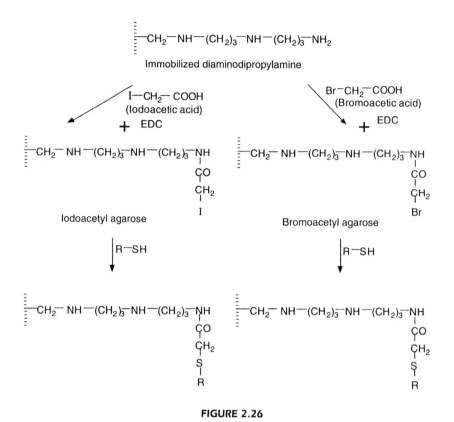

FIGURE 2.26

Reaction sequence for the preparation of iodoacetyl and bromoacetyl agarose.

the spacer, but other spacers may also be suitable. (See Section 3.1.1, Chapter 3, for additional choices.)

1. Wash diaminodipropylamine–/ /–Sepharose 4B (100 ml settled gel) with 1 L water in a sintered glass funnel, suction dry to a wet cake, and transfer to a 500-ml beaker.
2. Suspend the gel in 50 ml water, and stir the suspension using a paddle.
3. Dissolve bromoacetic acid or iodoacetic acid (12 gm) in 50 ml water, adjust the pH to 4.5 using 50% NaOH, and add the solution to the gel suspension.
4. While stirring, add EDC (10 gm) to the reaction mixture, and maintain the pH at 4.5 for 1 hr by adding first 50% HCl and then 50% NaOH.
5. Continue the reaction for a total of 2 hr.
6. Filter the reaction mixture and wash extensively with water. Bromoacetyl- and iodoacetyl-agarose can be stored in 10 mM EDTA con-

TABLE 2.5
Coupling of –SH/S–S-Containing Proteins to Iodoacetyl Activated Agarose

Protein	Molecular weight	–SH groups/ molecule	S–S groups/ molecule	Coupling efficiency (%)
Ceruloplasm	150,000	1, 3	—	75
Aldolase	147,000	7, 28	—	87
BSA	66,000	0.7	17	25
HSA	66,000	0.7	17	72
Ovalbumin	45,000	3, 4	1	40
β-Lactoglobin	36,000	2	2	63
Trypsin	24,000	0	6	13
Thioredoxin	11,700	2	1	20
[Arg[8]] Vasopressin	1,084	0	1	90

taining 0.02% sodium azide, pH 7.2, and at 4°C protected from light. Alkyl halide-containing compounds are extremely light sensitive.

2.2.2.2 Maleimide

Activation and Coupling Chemistry

Maleimide chemistry has been used extensively as a tool in cross-linking studies to facilitate the conjugation of sulfhydryl groups in proteins or peptides (Imagawa *et al.*, 1982; Yoshitake *et al.*, 1982; Brown *et al.*, 1988; Bhatia *et al.*, 1989). Usually, maleimide groups are incorporated into one end of a heterobifunctional cross-linking reagent and used in a two-step reaction to conjugate one functionality of a protein to the sulfhydryls of another. One of the most popular cross-linkers in this respect is SMCC (succinimidyl-4-(N-maleimidomethyl) cyclohexane-1-carboxylate; Pierce, Rockford, Illinois), which has an amine reactive NHS ester on one end and a maleimide group on the other. The excellent stability and slow hydrolysis

TABLE 2.6
Immobilization of IgG and F(ab')$_2$ to Iodoacetyl-Activated Agarose

Human IgG	Coupling efficiency (%)
Reduced whole molecule	90
Reduced F(ab')$_2$	88

rate of the SMCC maleimide end allows for primary modification of one protein through its amine groups (via the NHS ester), a quick purification step, and a secondary coupling to the sulfhydryl group of another protein (Fig. 2.27).

Amine-bearing matrices also can be modified with SMCC to generate a terminal maleimide group for coupling to sulfhydryl-containing ligands. A support is first modified with a spacer molecule containing a free primary amine (see Chapter 3). This modified matrix is then reacted with SMCC (or preferably its water-soluble analog sulfo-SMCC) to generate the available maleimides. After a quick wash step, small ligands or proteins can be added and immobilized through their sulfhydryl groups (Fig. 2.28). The coupling

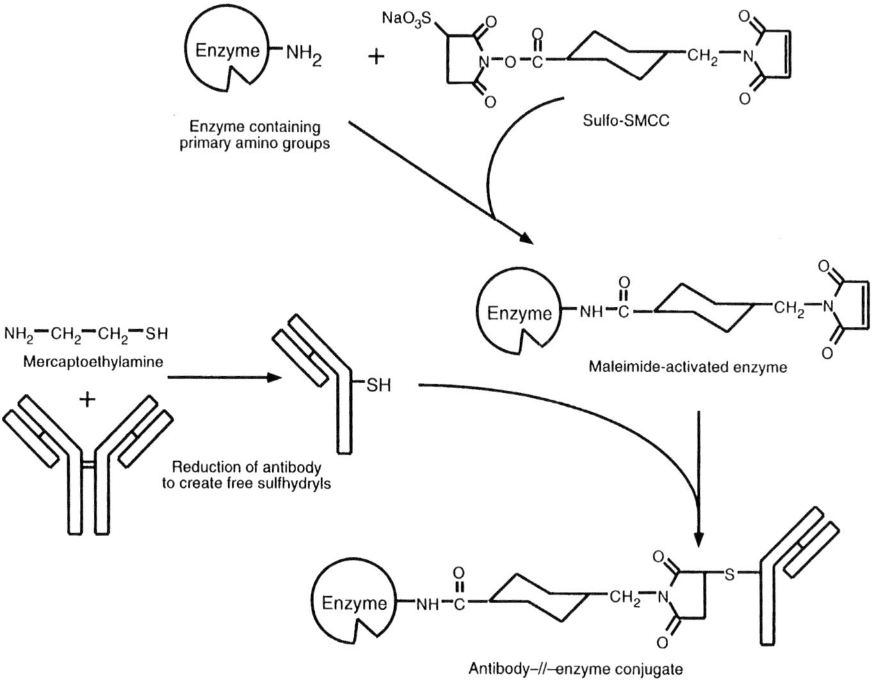

FIGURE 2.27

Sulfo-SMCC long has been used to conjugate protein molecules in a two-step reaction process. The stability of the maleimide group allows one protein, for example, an enzyme, to be modified first through the NHS ester end to the amine residues of the protein. At this point, the modified enzyme contains covalently attached cross-linker molecules with terminal maleimide groups. After a brief purification step, a second protein, for example, a reduced antibody molecule as shown here, can be added and conjugated through its available sulfhydryl groups.

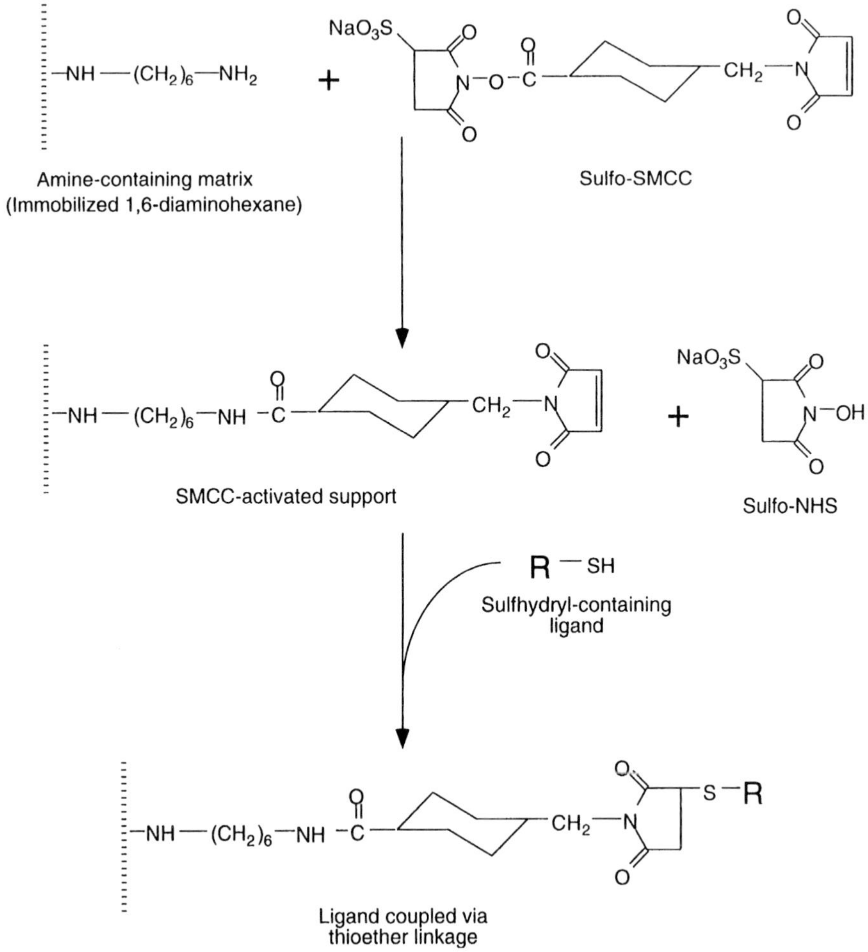

FIGURE 2.28

The reaction of Sulfo-SMCC with an amine-containing matrix results in an intermediate activated support containing terminal maleimide groups. Subsequent coupling with a sulfhydryl-containing ligand produces covalent attachment via a thioether linkage.

---

reaction is rapid and results in a stable linkage between ligand and matrix. Isolation of the maleimide active species is possible if the solid support material can be immediately frozen and freeze-dried without damaging the matrix. SMCC maleimides are stable for at least 1 year in a dry state.

ACTIVATION PROTOCOL

It is assumed that prior to this procedure the matrix of choice has been modified (by coupling a spacer molecule) to contain a terminal primary amine appropriate for the NHS ester conjugation reaction. See Section 3.1.1, Chapter 3, for further details on how to form the intermediate gel. The quantity of gel used in this example is small because of the expense of the cross-linking reagent.

1. Wash 10 ml primary amine-containing support with water and then ice-cold activation buffer. The buffer of choice to promote efficient NHS ester coupling is 0.1 M sodium phosphate, 0.15 M NaCl, pH 7.2. Other buffer salts may be used as long as they contain no amines that will compete with the conjugation reaction. Suspend the gel in an equal volume of this cold buffer and mix by rocking at 4°C. The low temperature will reduce the hydrolysis rate of the active groups on the cross-linking reagent.
2. To the suspended gel, add 218 mg sulfo-SMCC (approximately 50 μmol/ml gel) and mix to dissolve.
3. Continue mixing the reaction for 1 hr at 4°C.
4. Quickly wash the support with cold coupling buffer, 1 M NaCl, and again with buffer to remove uncoupled cross-linker and the reaction by-products. The matrix now will contain a terminal maleimide group and should be used immediately for coupling a sulfhydryl-containing ligand.

LIGAND COUPLING PROTOCOL

1. Suspend the 10 ml maleimide-activated gel just prepared in an equal volume of cold coupling buffer (0.1 M sodium phosphate, 0.15 M NaCl, pH 7.2) containing the ligand to be immobilized. If the ligand is a protein with available –SH groups, dissolving it at a concentration of 10–20 mg/ml will give excellent coupling yields. For small molecules, add at least 50–100 μmol of ligand per ml buffer.
2. Mix the reaction by rocking, and allow it to continue at 4°C for at least 2 hr.
3. Wash the affinity support with coupling buffer, 1 M NaCl, and water. Store the gel as an aqueous slurry containing a preservative (0.02% sodium azide) at 4°C.

2.2.2.3 Pyridyl Disulfide

Pyridyl disulfide-activated supports are noteworthy for their versatility, selectivity, and ease of preparation. The following protocols for activation

104   2. Activation Methods

and ligand coupling are modifications or hybrids of previously described procedures (Cuatrecasas, 1970; Brocklehurst *et al.*, 1974; Porath and Axen, 1976).

ACTIVATION PROTOCOL

a. Preliminary Steps

1. Prepare a sulfhydryl-containing matrix as described in Section 8 of Chapter 3.
2. Synthesize 2,2'-dipyridyl disulfide. The reagent, 2,2'-dipyridyl disulfide required for this procedure can be synthesized very readily at a considerable cost savings over purchasing material from the usual suppliers. In addition to cost savings, we find the quality of material synthesized by the following method is routinely superior to commercial because yellow impurities that adhere strongly to the final activated gel are avoided. If commercially obtained 2,2'-dipyridyl disulfide is yellow in color, we suggest applying the recrystallization technique offered in Step 4 prior to use as a gel-activating reagent.

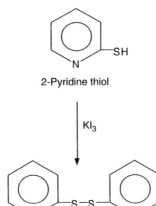

FIGURE 2.29

The reaction of Sulfo-SMCC with an amine-containing matrix results in an intermediate activated support containing terminal maleimide groups. Subsequent coupling with a sulfhydryl-containing ligand produces covalent attachment via a thioether linkage.

a. In an appropriately sized beaker equipped with a means for vigorous stirring (a magnetic stir bar and plate is satisfactory), prepare approximately a 10% slurry of a weighed amount of 2-mercaptopyridine in water. Add sufficient 4 N NaOH to achieve a pH of 8 (wide range pH paper is a suitable indicator).
b. While maintaining the pH at 7–9 (by periodic dropwise addition of 4 N NaOH), add in small increments or a steady slow stream a slight molar excess of iodine as a KI_3 solution. The product, 2,2'-dipyridyl disulfide, will form initially as a water-insoluble oil that will spontaneously crystallize to a yellow mass. Filter the crude product and wash with water.
c. Dissolve the crude product in a minimum amount of room temperature acetone to yield a slightly yellow solution. Add water dropwise to the acetone solution until a faint permanent haze of oil appears.
d. Add activated carbon (Darco G-25) to the acetone solution (0.5 gm carbon per gm theoretical yield) and stir briefly. Filter the Darco-treated solution (water white) and precipitate the purified product with three volumes of water. Filtered water-wet needles of 2,2'-dipyridyl disulfide may be used directly for the matrix activation procedure without further drying. Drying at room temperature in a vacuum desiccator yields pure white needles melting at 59–60°C in nearly quantitative yield.

b. Gel Activation (Porath and Axen, 1976)

1. Exchange the thiol-containing matrix into acetone:water (1:1) and prepare a 50% slurry of the matrix in this solvent.
2. For each ml thiol support containing 10–50 μmol –SH, add 100 mg 2,2'-dipyridyl disulfide as a solution in acetone:water (1:1). Stir at room temperature for 30 min.
3. Wash the activated gel with acetone:water (1:1) to remove excess 2,2'-dipyridyl disulfide and the yellow by-product 2-mercaptopyridine. Finally, wash the gel with 1 mM EDTA. Gel prepared and washed in this manner is stable and fully active for several months.

c. Coupling of Thiol-Containing Ligands (Brocklehurst et al., 1974)

Pyridyl disulfide-activated matrices are especially useful for reversibly immobilizing thiol-containing proteins (Fig. 2.31). Thiol-containing target protein may be separated from nonthiol impurities by temporary immobilization. Ligand attachment is best accomplished by adding the buffered

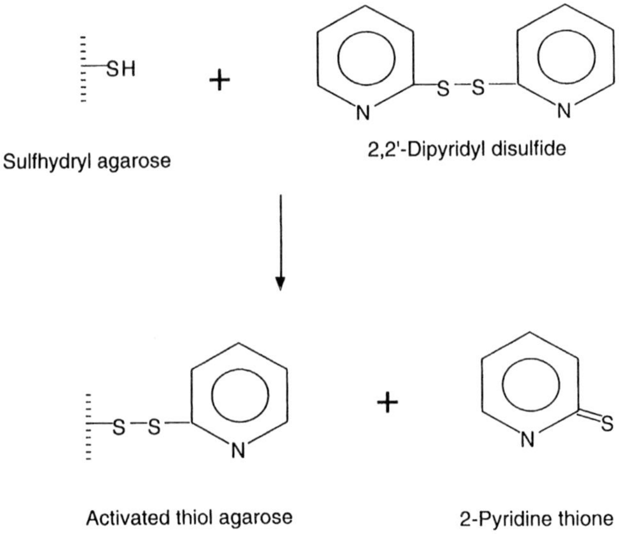

FIGURE 2.30
Synthesis of 2,2′-dipyridyl disulfide.

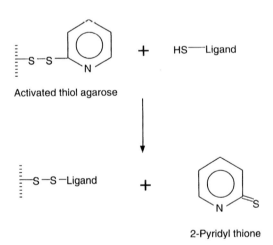

FIGURE 2.31
Activation of a thiol matrix by 2,2′-dipyridyl disulfide.

ligand solution to a column of the activated support. Rapid attachment of ligand can be affected at either pH 4.0 or pH 8.0.

1. For a column containing 100 ml activated thiol matrix, equilibrate the gel with either 0.1 M Tris-HCl, pH 8, or 0.1 M sodium acetate, pH 4, both containing 0.3 M NaCl, 1 mM EDTA.
2. Add the thiol-containing ligand to the equilibrated column as a solution in the chosen equilibration buffer. For protein ligands, the support as prepared in the activation section will have a binding capacity of 2–4 mg per ml for the protein papain.
3. Wash the loaded column with 0.1 M Tris, pH 8.0, containing 0.3 M NaCl, 1 mM EDTA until the absorbance of the washes is less than 0.02 at 280 nm. Protein loaded and washed free of nonthiol impurities may be released from the support by eluting with the wash buffer containing 50 mM L-cysteine.

2.2.2.4 Divinylsulfone

Divinylsulfone activation chemistry can be used to immobilize sulfhydryl-containing ligands with good efficiency. The activation and coupling procedures are identical to those outlined in the discussion of DVS activation in Section 2.2.1.8.

2.2.2.5 Epoxy or Bisoxirane Activation

The epoxy (or bisoxirane) activation chemistry can be used to immobilize sulfhydryl-containing ligands with good efficiency. The activation and coupling procedures are identical to those outlined in the discussion of epoxy activation in Section 2.2.4.1.

2.2.2.6 TNB-Thiol

Activation and Coupling Chemistry

TNB-thiol agarose is a versatile affinity matrix in which the ligand of interest can be linked to the matrix through a cleavable connector arm containing a disulfide bond (Jayabaskaran *et al.*, 1987). The matrix can be prepared from an amine-containing matrix by successive reaction with 2-iminothiolane, 5,5´-dithio-bis-(2-nitrobenzoic acid), and a sulfhydryl ligand of choice (Fig. 2.32).

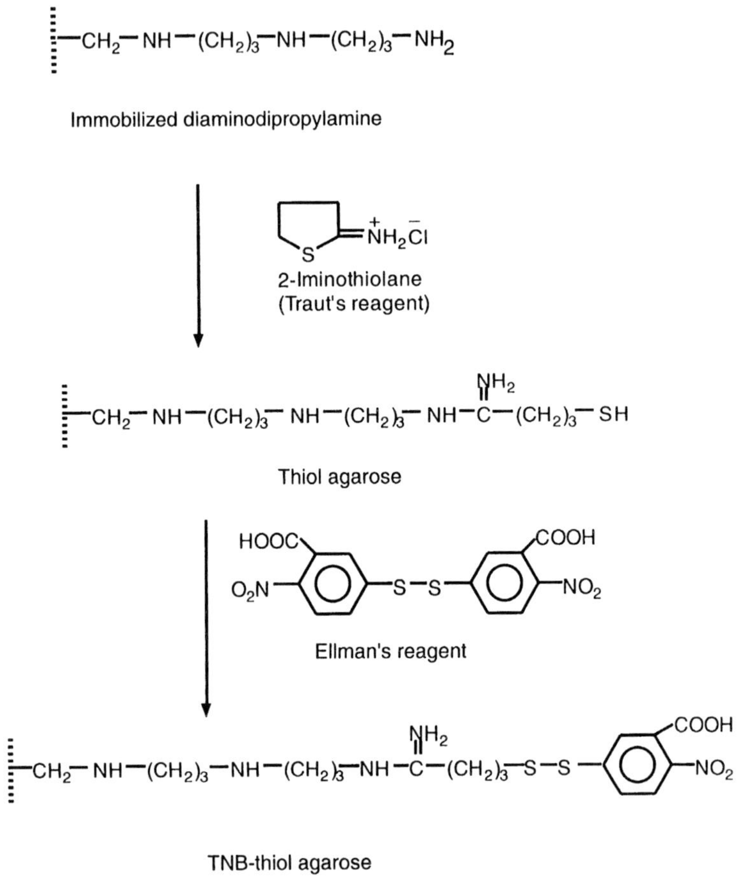

FIGURE 2.32
Reaction sequence for the preparation of 5-thio-2-nitro-benzoic acid (TNB)-thiol agarose.

ACTIVATION PROTOCOL

The following procedure makes use of an intermediate spacer gel that contains a terminal amine group. In this case, diaminodipropylamine is used as the spacer, but other spacers may also be suitable. See Section 3.1.1, Chapter 3, for additional choices.

1. Successively wash diaminodipropylamine–/ /–Sepharose 4B (100 ml settled gel) with 1 L each of water and 0.1 M sodium borate buffer, pH 8.5, in a sintered glass filter funnel, suction dry to a wet cake, and transfer to a 500-ml beaker.

2. Suspend the gel in 100 ml ice-cold 20 mM 2-iminothiolane in 0.1 M sodium borate buffer, pH 9.5.
3. Stir the reaction mixture with a paddle for 1 hr at 4°C.
4. Filter and wash the gel with 1 L 0.1 M sodium phosphate, pH 6.8, until the filtrate is devoid of thiol as indicated by reaction with Ellman's reagent [5,5'-dithio-bis-(2-nitrobenzoic acid; see Section 4.1.9, Chapter 4].
5. Immediately convert the resulting thiol agarose to TNB-thiol agarose as follows.
   a. Mix the thiol agarose with 100 ml 10 mM Ellman's reagent in 0.1 M sodium phosphate, pH 6.8, and stir for 5 min at room temperature.
   b. Wash the gel successively with 1 L each of 0.1 M sodium phosphate, pH 6.8, water, and 0.5 M acetic acid.

TNB-thiol agarose can be used to immobilize proteins containing sulfhydryl groups and other low molecular weight thiols.

LIGAND COUPLING PROTOCOL

The following method describes the coupling of F(ab')$_2$ to TNB-thiol agarose [preparation of F(ab')$_2$-disulfide-agarose].

1. Pack 0.5 ml TNB-thiol agarose in a disposable polypropylene column as described in Chapter 4.
2. Equilibrate the column by washing with 5.0 ml 0.1 M sodium phosphate buffer, pH 6.8, containing 1 mM EDTA (Pi/EDTA buffer).
3. Pass through the TNB-thiol agarose column a solution (0.3 ml) containing 1.0 mg F(ab')$_2$ fragment (freshly prepared; see Section 3.4.2.5, Chapter 3) in Pi/EDTA buffer.
4. After 30 min at room temperature, wash the column with 5.0 ml Pi/EDTA buffer. Measure the absorbance of the column effluent at 440 nm to determine the release of TNB groups.
5. F(ab')$_2$-disulfide-agarose can be used directly in immunoaffinity chromatography to isolate the corresponding antigen.

COUPLING LOW-MOLECULAR-WEIGHT THIOLS

The following is a general protocol for coupling low-molecular-weight thiols to the TNB-thiol support.

1. Pack 5.0 ml TNB-thiol agarose in a disposable polypropylene column as described in Chapter 4.
2. Equilibrate the column by washing with 25 ml 0.1 M sodium phosphate buffer, pH 6.8, containing 1 mM EDTA (Pi/EDTA buffer).

3. Pass through the TNB-thiol agarose column, at room temperature, 15 ml 2 mM solution of low molecular weight thiol compound dissolved in Pi/EDTA buffer.
4. Wash the column with 25 ml Pi/EDTA buffer. Measure the absorbance of the column effluent at 440 nm to determine the release of TNB groups.
5. The immobilized disulfide-linked low-molecular-weight-thiol-containing gel can be used directly for affinity chromatography.

2.2.3 Carbonyl Reactive Chemistries

2.2.3.1 Hydrazide

Activation and Coupling Chemistry

Hydrazide immobilization chemistry permits the coupling of aldehyde- or ketone-containing ligands through the formation of stable hydrazone linkages. Glycoproteins in particular may be immobilized using this procedure if they are previously oxidized with sodium periodate to generate formyl groups on their carbohydrate chains. The method is a powerful way to immobilize proteins and leave critical active sites free. (For a recent review, see O'Shannessy and Wilchek, 1990.)

Two methods may be used to make a hydrazide-containing support. In the first procedure, an aldehyde-containing gel is generated by oxidation with sodium periodate (see previous text for protocol) (Hoffman and O'Shannessy, 1988). This intermediate is then reacted with adipic dihydrazide to produce the active hydrazide-containing support (Fig. 2.33). In the second method, a spacer arm is initially coupled to a matrix, leaving a terminal carboxyl group. Adipic dihydrazide is then coupled to this gel using the EDC conjugation chemistry described previously (Fig. 2.34) (Bayer *et al.,* 1987). The spacer provides greater steric accommodation for some proteins.

The first procedure often results in a gel that is highly cross-linked and damaged when the adipic dihydrazide is coupled. Presumably, the reactivity of the two hydrazide ends for the formyl groups of the matrix is so great that extensive cross-linking occurs more often than do "one-on" hits. However, when using the EDC coupling method to attach the bis-hydrazide spacer to carboxylic groups, this phenomenon is not observed. For this reason, only the EDC procedure is described here.

The hydrazide gel may be used to immobilize a variety of glycoproteins. The coupling efficiency of a particular protein depends on the amount and accessibility of its carbohydrate side chains. A study of the coupling yields of eight glycoproteins revealed that most can be expected to react at over

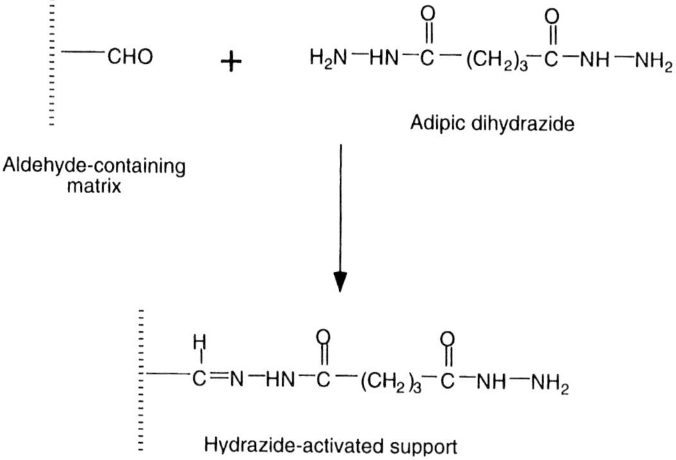

2.2 Procedures

⋮—CHO + $H_2N-HN-\overset{\overset{O}{\|}}{C}-(CH_2)_3-\overset{\overset{O}{\|}}{C}-NH-NH_2$

Aldehyde-containing matrix   Adipic dihydrazide

↓

⋮—$\overset{H}{\underset{|}{C}}=N-HN-\overset{\overset{O}{\|}}{C}-(CH_2)_3-\overset{\overset{O}{\|}}{C}-NH-NH_2$

Hydrazide-activated support

FIGURE 2.33

A matrix containing aldehyde groups can be reacted with an excess of adipic dihydrazide to form an activated support capable of immobilizing formyl-containing ligands. This process of producing a hydrazide support often results in severe cross-linking due to the bifunctional nature of adipic dihydrazide.

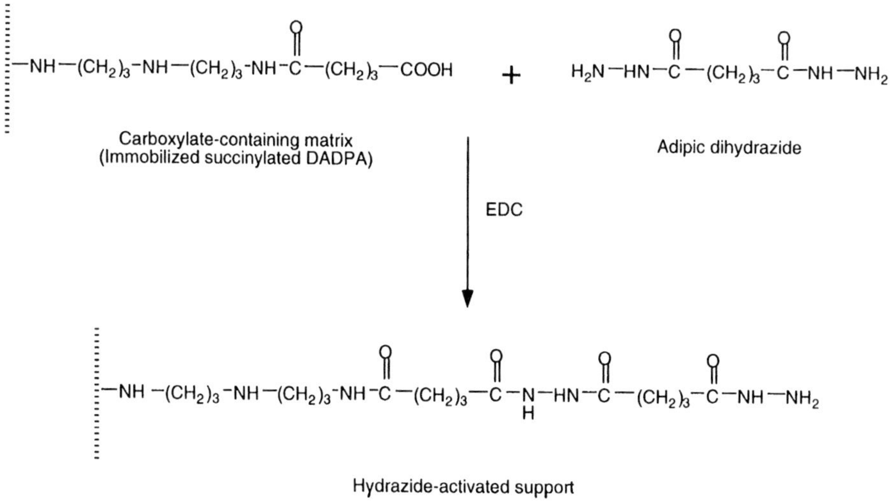

⋮—$NH-(CH_2)_3-NH-(CH_2)_3-NH-\overset{\overset{O}{\|}}{C}-(CH_2)_3-COOH$ + $H_2N-HN-\overset{\overset{O}{\|}}{C}-(CH_2)_3-\overset{\overset{O}{\|}}{C}-NH-NH_2$

Carboxylate-containing matrix (Immobilized succinylated DADPA)   Adipic dihydrazide

EDC ↓

⋮—$NH-(CH_2)_3-NH-(CH_2)_3-NH-\overset{\overset{O}{\|}}{C}-(CH_2)_3-\overset{\overset{O}{\|}}{C}-\underset{H}{N}-HN-\overset{\overset{O}{\|}}{C}-(CH_2)_3-\overset{\overset{O}{\|}}{C}-NH-NH_2$

Hydrazide-activated support

FIGURE 2.34

Adipic dihydrazide can be reacted with a carboxylate spacer (in this case succinylated DADPA) with the use of the water soluble carbodiimide EDC. One hydrazide forms an amidine linkage with the matrix, while the terminal hydrazide can be used to couple to aldehyde-containing ligands.

TABLE 2.7
Coupling Efficiency of Various Glycoproteins to Hydrazide Activated Agarose

Protein	Molecular weight	Coupling efficiency (%)
Collagen (type VI)	163,000	63
Human IgG	150,000	74
Avidin	66,000	95
Chorionic gonadotropin	59,000	91
Fetuin	48,700	87
Ovalbumin	45,000	86
α1-Acid glycoprotein	44,100	90
Pepsin	34,700	47

80% efficiency of immobilization (Table 2.7). Since the hydrazone bonds are stable without the use of reductants, sensitive proteins are more likely to retain activity after immobilization. A comparison of the immobilization of antibodies to a hydrazide gel with the use of either reductive amination or iodoacetyl chemistry revealed the highest retention of antigen binding activity using the hydrazide method (Domen et al., 1990).

ACTIVATION PROTOCOL

It is assumed that, prior to this procedure, the matrix of choice has been modified (by coupling a spacer molecule) to contain a terminal carboxyl group. Several options are available for the chemical structure and length of the spacer molecule. For instance, one can prepare the intermediate carboxyl-containing matrix by periodate oxidation of agarose, coupling of diaminodipropylamine (DADPA) by reductive amination, or succinylation using succinic anhydride (see Fig. 2.34) After attachment of adipic dihydrazide, the result is a gel with a 23-atom spacer arm with an active group on the end, ready to couple any carbonyl-containing ligands. See Section 3.1.1 for further details on how to form this intermediate gel.

1. Wash the succinylated DADPA gel (100 ml) with three bed volumes of 0.1 M MES buffer, pH 4.75.
2. Suction the gel free of excess buffer, and add it to a stirring solution of 0.5 M adipic dihydrazide in 100 ml 0.1 M MES buffer (in a fume hood).
 Caution: Adipic dihydrazide is toxic.
3. Add 3 gm EDC and react with stirring for 3 hr.

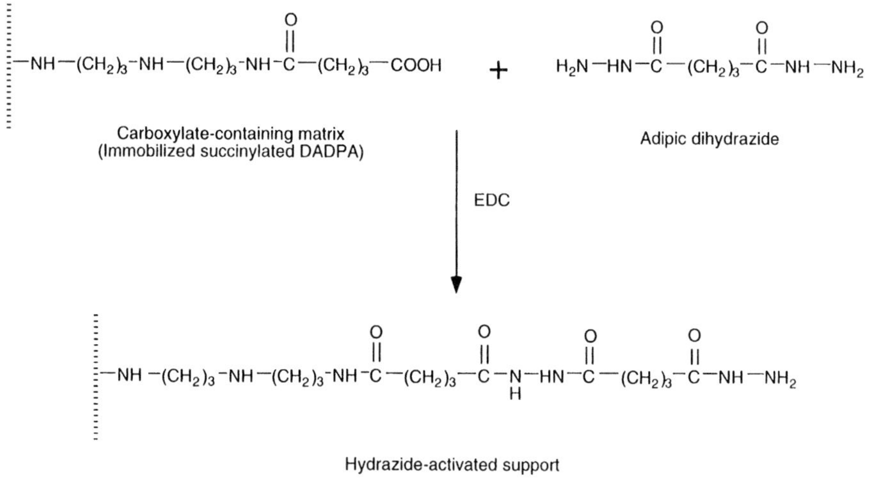

2.2 Procedures

$-NH-(CH_2)_3-NH-(CH_2)_3-NH-\overset{O}{\underset{\|}{C}}-(CH_2)_3-COOH$ + $H_2N-HN-\overset{O}{\underset{\|}{C}}-(CH_2)_3-\overset{O}{\underset{\|}{C}}-NH-NH_2$

Carboxylate-containing matrix
(Immobilized succinylated DADPA)

Adipic dihydrazide

↓ EDC

$-NH-(CH_2)_3-NH-(CH_2)_3-NH-\overset{O}{\underset{\|}{C}}-(CH_2)_3-\overset{O}{\underset{\|}{C}}-\underset{H}{N}-HN-\overset{O}{\underset{\|}{C}}-(CH_2)_3-\overset{O}{\underset{\|}{C}}-NH-NH_2$

Hydrazide-activated support

FIGURE 2.34

Adipic dihydrazide can be reacted with a carboxylate spacer (in this case succinylated DADPA) with the use of the water soluble carbodiimide EDC. One hydrazide forms an amidine linkage with the matrix, while the terminal hydrazide can be used to couple to aldehyde-containing ligands.

---

4. Wash the gel thoroughly with 2–3 gel volumes of water, 2–3 gel volumes of 1 M NaCl, and, finally, 3–4 gel volumes of water.
5. Check for successful coupling of adipic dihydrazide using the TNBS qualitative test for the presence of hydrazides (see Section 4.1.3, Chapter 4).

LIGAND COUPLING PROTOCOL

A hydrazide-activated support may be used to couple small ligands that have reactive formyl groups or to immobilize proteins that have oxidizable carbohydrate residues (Fig. 2.35). .

a. Oxidation of Glycoproteins

1. Dissolve or dilute 1–10 mg glycoprotein in 1 ml 0.1 M sodium phosphate buffer, pH 7.0 (coupling buffer).
2. Add the protein solution to an amber vial containing 5 mg sodium *meta*-periodate. Swirl gently to dissolve the oxidizing agent.

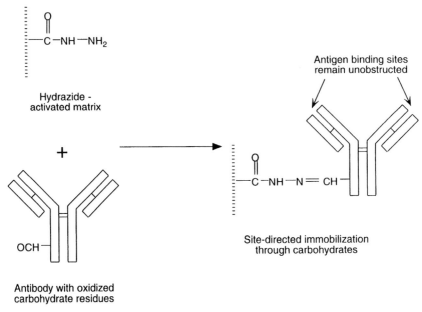

FIGURE 2.35
A hydrazide-activated matrix can be used in a powerful site-directed immobilization technique with glycoproteins. In this case, an antibody molecule is first oxidized with sodium *meta*-periodate to yield formyl coupling groups on its carbohydrate side chains. Reaction of the oxidized antibody with a hydrazide support causes immobilization through hydrazone linkages at known locations away from the antigen binding sites.

3. Incubate the sample for 30 min at room temperature. The reaction is light sensitive and should be performed in the dark.
4. Stop the reaction by desalting and buffer exchange. Equilibrate a 5-ml Sephadex G-25® column with coupling buffer. Apply the oxidized sample to the column and allow it to enter the gel bed. Apply a 0.5-ml rinse of coupling buffer and allow it also to enter the gel bed. Finally apply 2 ml coupling buffer and collect the eluent. This fraction contains the oxidized protein. Alternatively, monitor the protein peak by absorbance at 280 nm.

b. Immobilization Reaction

1. Mix the 2-ml fraction containing the oxidized glycoprotein with 2 ml hydrazide gel that has been washed previously with coupling buffer.
2. React with stirring for a minimum of 6 hr.

3. Wash the gel thoroughly with coupling buffer. Store at 4°C in a solution containing a preservative (0.02% sodium azide).

2.2.3.2 Reductive Amination

Reductive amination (alkylation) is a selective and versatile technique for the direct covalent attachment of aliphatic aldehydes, ketones, or reducing carbohydrates to amino groups of proteins as well as to amine-containing matrices (Means and Feeney, 1968; Gray, 1974; Baues and Gray, 1977; Kamicker *et al.*, 1977; Schwartz and Gray, 1977; Lee and Lee, 1979; Matsumoto *et al.*, 1981). The process is accomplished by treating proteins or amine-containing matrices with aliphatic aldehydes, ketones or reducing carbohydrates and a reducing agent, usually sodium cyanoborohydride. Cyanoborohydride anion selectively reduces the imminium salt (Schiff base) formed between an aldehyde or ketone and an amine (Borch *et al.*, 1971) (Fig. 2.36).

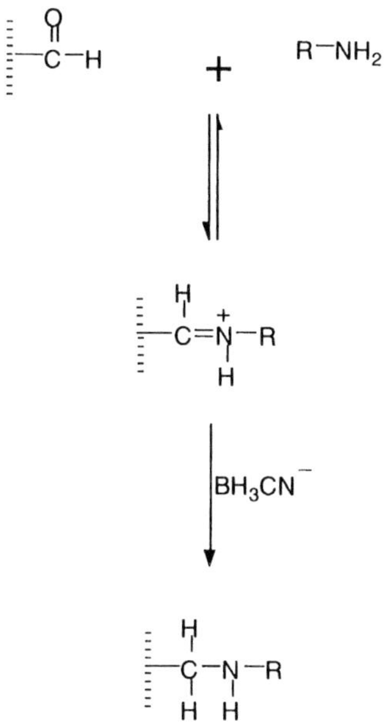

FIGURE 2.36

Reaction mechanism for the direct covalent attachment of reducing carbohydrates to proteins and amine-containing matrices by reductive amination using cyanoborohydride in aqueous solution at pH 7.0.

A variety of reducing disaccharides (see Table 2.8) has been coupled to bovine serum albumin (Gray, 1974; Kamicker *et al.*, 1977; Schwartz and Gray, 1977; Lee and Lee, 1979) as well as to amine-containing matrices (Gray, 1974; Baues and Gray, 1977; Matsumoto *et al.*, 1981). Reductive amination proceeds readily in aqueous solutions over a broad pH range to give high degrees of substitution.

The coupling reaction of reducing disaccharides to amine-containing matrices by reductive amination with sodium cyanoborohydride is very slow and requires at least several days. The slow reaction may be attributed to the minute concentration of the aldehyde form of the sugar in solution. This may be the only drawback of this reductive amination procedure. However, this disadvantage may be acceptable, considering the fact that chemically stable affinity adsorbents with particularly high binding capacities for lectins are prepared with a mild and very simple procedure. Affinity adsorbents prepared by coupling disaccharides to amine-containing matrices have been used successfully for the affinity chromatographic purification of lectins (Baues and Gray, 1977; Matsumoto *et al.*, 1981) and glycosidases (Uy and Wold, 1977; Dey *et al.*, 1982; Dey, 1984). Since the ring form of the reducing sugar is destroyed (Fig. 2.37), disaccharides must be used to obtain a monosaccharide determinant in the affinity gel.

LIGAND COUPLING PROTOCOL

The following general procedure may be used for the coupling of reducing disaccharides to aminoethyl polyacrylamide or to amino-agarose (Gray, 1974; Baues and Gray, 1977; Matsumoto *et al.*, 1981).

CAUTION: Carry out the reaction in a well-ventilated hood since $NaCNBH_3$ is highly toxic.

1. Suspend 20 ml aminoethyl polyacrylamide (for preparation, see Section 3,1,1,4, Chapter 3) or amino-agarose (for preparation, see Section 3.1.1, Chapter 3) in 20 ml 0.2 M K_2HPO_4, pH 9.0, and gently mix with 500 mg disaccharide and 250 mg $NaCNBH_3$ using a paddle stirrer for 1 hr at room temperature. We routinely use a 125-ml Erlenmeyer flask as the reaction vessel.
2. Cover the reaction vessel using Parafilm and incubate the reaction mixture at room temperature for at least 4 weeks.
3. Filter the gel using Whatman No. 1 filter paper on a Buchner funnel and wash with 200 ml 0.2 M K_2HPO_4, pH 9.0, and 500 ml 0.2 M sodium acetate.

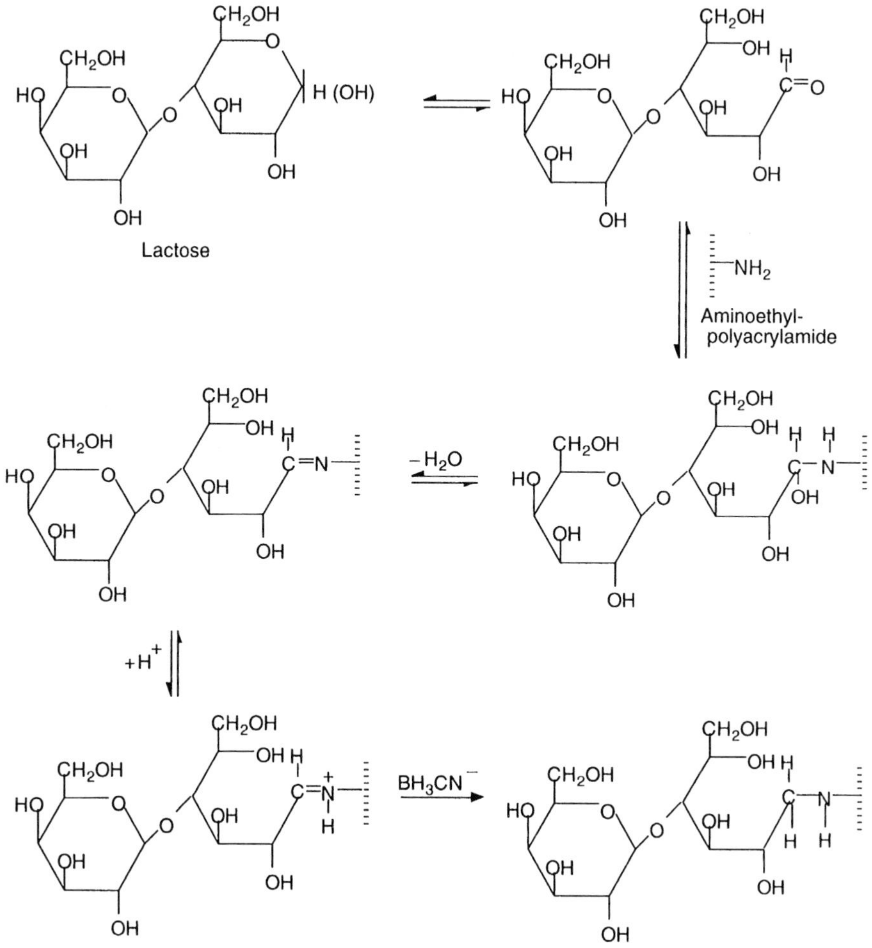

FIGURE 2.37
Reaction intermediates in the coupling of lactose to amine-containing matrices.

4. Block the excess amino groups on the gel by *N*-acetylation as follows. Suspend the gel in 20 ml 0.2 *M* sodium acetate and cool to 4°C.
5. Add 4 ml acetic anhydride to the gel suspension and mix at 4°C for 30 min.
6. After 30 min, add another 4 ml acetic anhydride to the reaction mixture and mix at room temperature for 30 min.
7. Filter the reaction mixture using Whatman No. 1 filter paper on a Buchner funnel and wash sequentially with 200 ml each of water, 0.1 N

NaOH, water, and 0.15 M NaCl. Store the gel in 0.15 M NaCl containing 0.02% sodium azide at 4°C. A high degree of substitution (30–40 µmol carbohydrates coupled per ml gel) can be achieved by this method.

2.2.4 Hydroxyl Reactive Chemistries

2.2.4.1 Epoxy (Bisoxirane) Activation Method

Activation and Coupling Chemistry

Bisoxiranes can be used to introduce oxirane (epoxy) groups in many hydroxylic polymers. 1,4-Butanediol diglycidyl ether is the reagent most commonly used to introduce these active groups (Sundberg and Porath, 1974). At alkaline pH, this compound readily reacts with hydroxylic polymers to yield derivatives containing a long-chain hydrophilic spacer molecule with a reactive oxirane on the end (Fig. 2.38). The bond between the bisoxirane and the matrix becomes a stable ether bond, while the other end provides the ligand coupling potential. This terminal epoxy then can react with ligands containing hydroxyl, amine, or thiol groups.

Bisoxirane-activated (epoxy-activated) gels are extensively used to couple mono- and oligosaccharides and ligands containing secondary amines (e.g., iminodiacetic acid). At neutral pH, sulfhydryl groups couple to epoxy-activated gels more readily than do amino groups. The epoxy activation method provides an extremely stable linkage between the ligand and the matrix. By using a long-chain bisoxirane, one can obtain a long hydrophilic spacer arm that separates the ligand from the matrix, a potentially useful characteristic in affinity chromatography (see Section 3.1.1).

ACTIVATION PROTOCOL

CAUTION: Bisoxirane activation should be carried out in a well-ventilated hood since 1,4-butanediol diglycidyl ether is toxic.

1. Wash Sepharose 4B (100 ml settled gel) with 1 L water in a sintered glass funnel, suction dry to a wet cake, and transfer to a 500-ml beaker.
2. Suspend the gel in 75 ml 0.6 N NaOH containing 150 mg sodium borohydride and stir with a paddle.
3. Slowly add 75 ml 1,4-butanediol diglycidyl ether with constant stirring.
4. Stir the reaction mixture at room temperature for 10 hr (or overnight).
5. Extensively wash the activated gel with water to remove excess reagent. This washing should proceed until there is no longer evidence of an oily film on the surface of the gel representing the remaining epoxy com-

pound. To aid in the complete removal of bisoxirane groups, we have found that washing the gel with acetone quickens the washing. The gel may be washed back into water for ligand coupling.

LIGAND COUPLING PROTOCOL

The following method is a generalized protocol for coupling ligands to epoxy-activated supports. Specific examples can be found elsewhere in this book.

NOTE: Since high pH must be used to couple hydroxyl (pH 11–12) and amino ligands (pH > 9.0), epoxy-activated supports are not suitable for some base-sensitive ligands (e.g., many proteins) or unstable support materials (e.g., glass or silica). However, epoxy-activated supports can be used to couple thiol-containing proteins at lower pH (7.5–8.5) since thiol groups are better nucleophiles than amine and hydroxyl groups.

1. Suspend 100 ml epoxy-activated matrix in an equal volume of a buffer with a pH of 9–13 (carbonate, borate, or phosphate buffers can be used; for coupling carbohydrates, use a pH of 11–13) in which 0.5–1 mmol/ml small ligand and 5–10 mg/ml protein is dissolved. If necessary, up to 50% organic solvent (dioxane, dimethylformamide) can be used to dissolve the ligand.
2. Stir the gel at room temperature (for proteins) or at 45°C (for small ligands) for 24–48 hr.
3. Wash the gel extensively with 1.0 M NaCl and water to remove unreacted ligand.
4. Block the excess active groups on the gel by suspending the gel in 100 ml 1.0 M ethanolamine, pH 9.0, and stirring at room temperature for 6 hr.
5. Finally, wash the gel extensively with 1.0 M NaCl and water.

2.2.4.2 Divinylsulfone

Divinylsulfone (DVS) activation can be used to couple hydroxyl-containing ligands with good efficiency. See the discussion of DVS activation in Section 2.2.1.8.

2.2.4.3 Cyanuric Chloride

Cyanuric chloride activation can be used to couple hydroxyl-containing ligands with good efficiency. See the discussion of the cyanuric chloride activation method in Section 2.2.1.10

2.2.5 Active Hydrogen Reactive Chemistries

2.2.5.1 Diazonium

Activation and Coupling Chemistry

Diazonium chemistries long have been used in organic synthesis for conjugations involving active aromatic hydrogens or other compounds susceptible to electrophilic attack. Diazonium derivatives of chromatography supports can react with ligands containing available phenolic or imidazole groups, rapidly creating reversible diazo linkages (Inman and Dintzis, 1969; Cuatrecasas, 1970). The activated support is usually prepared from spacer molecules that terminate in a *p*-aminobenzylalkyl group. Two paths to such an immobilized diazonium are shown in Figures 2.39 and 2.40. In both methods, an immobilized *p*-nitrophenyl derivative is first reduced by sodium dithionite to the aminophenyl and then diazotized using an acidic solution of sodium nitrite. As the diazotization reaction proceeds, the gel will become colored, beginning with an orange/brown shade, sometimes becoming a deep brown or even black.

Once created, the diazonium derivative usually is used immediately without further washing because of the instability of the active group in aqueous environments. Ligands may be added directly to the activation reaction if the pH is first adjusted to the appropriate alkaline conditions. Coupling reactions are optimally performed at pH 8 for histidyl residues and pH 9–10 for tyrosyl or phenol compounds. Figure 2.41 shows the chemistry of the immobilization reaction. Ligands immobilized by this method often give highly colored complexes on the matrix, typically dark brown or even black. Diazo-coupled ligands also can be released from the matrix by reduction with 0.1 M sodium dithionite in 0.2 M sodium borate, pH 9 (Fig. 2.42). As the bonds are broken, the gel will return to its original color.

From a practical perspective, diazonium chemistry for the immobilization of ligands can be replete with problems. The rate of reaction of the diazonium species is so rapid that often a significant portion of the active groups will cross-link with the matrix before addition of ligand. The potential for cross-linking is due to the reactivity of the active hydrogens on the precursor molecules that contain terminal *p*-aminophenyl groups (Fig. 2.43). Even without addition of ligand, a diazonium-activated matrix will turn dark brown to black within an hour (or more quickly if exposed to bright light) and lose significant activity by formation of intermolecular cross-links. For this reason, the reproducibility of the immobilization is often poor with this method. We also have experienced that some peptide molecules couple poorly despite the fact that they contain tyrosyl residues theoretically able to interact with the diazonium groups. We have used

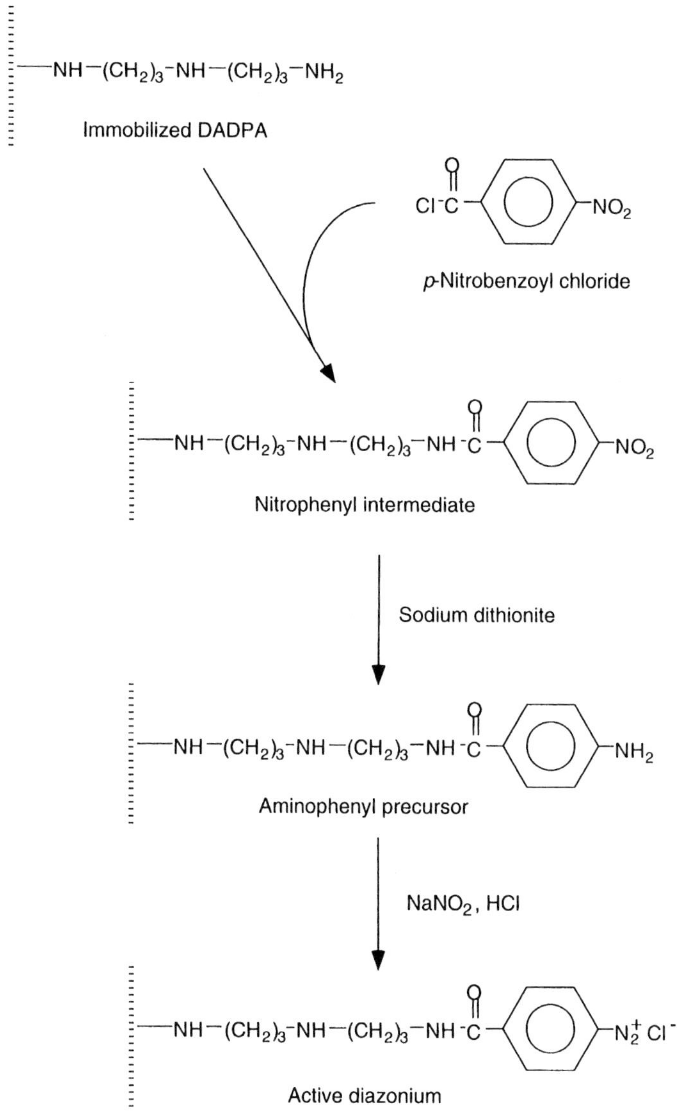

2.2 Procedures

─NH─(CH$_2$)$_3$─NH─(CH$_2$)$_3$─NH$_2$

Immobilized DADPA

Cl─C(=O)─C$_6$H$_4$─NO$_2$

p-Nitrobenzoyl chloride

─NH─(CH$_2$)$_3$─NH─(CH$_2$)$_3$─NH─C(=O)─C$_6$H$_4$─NO$_2$

Nitrophenyl intermediate

Sodium dithionite

─NH─(CH$_2$)$_3$─NH─(CH$_2$)$_3$─NH─C(=O)─C$_6$H$_4$─NH$_2$

Aminophenyl precursor

NaNO$_2$, HCl

─NH─(CH$_2$)$_3$─NH─(CH$_2$)$_3$─NH─C(=O)─C$_6$H$_4$─N$_2^+$ Cl$^-$

Active diazonium

FIGURE 2.39

The synthesis of a diazonium-activated matrix can be accomplished through a multistep process, starting with the reaction of *p*-nitrobenzoyl chloride with an amine-containing support. The nitro group of the resulting derivative is reduced to the aromatic amine by reduction with sodium dithionite. The immobilized aminophenyl compound is converted to the diazonium derivative by the addition of cold acidic sodium nitrite.

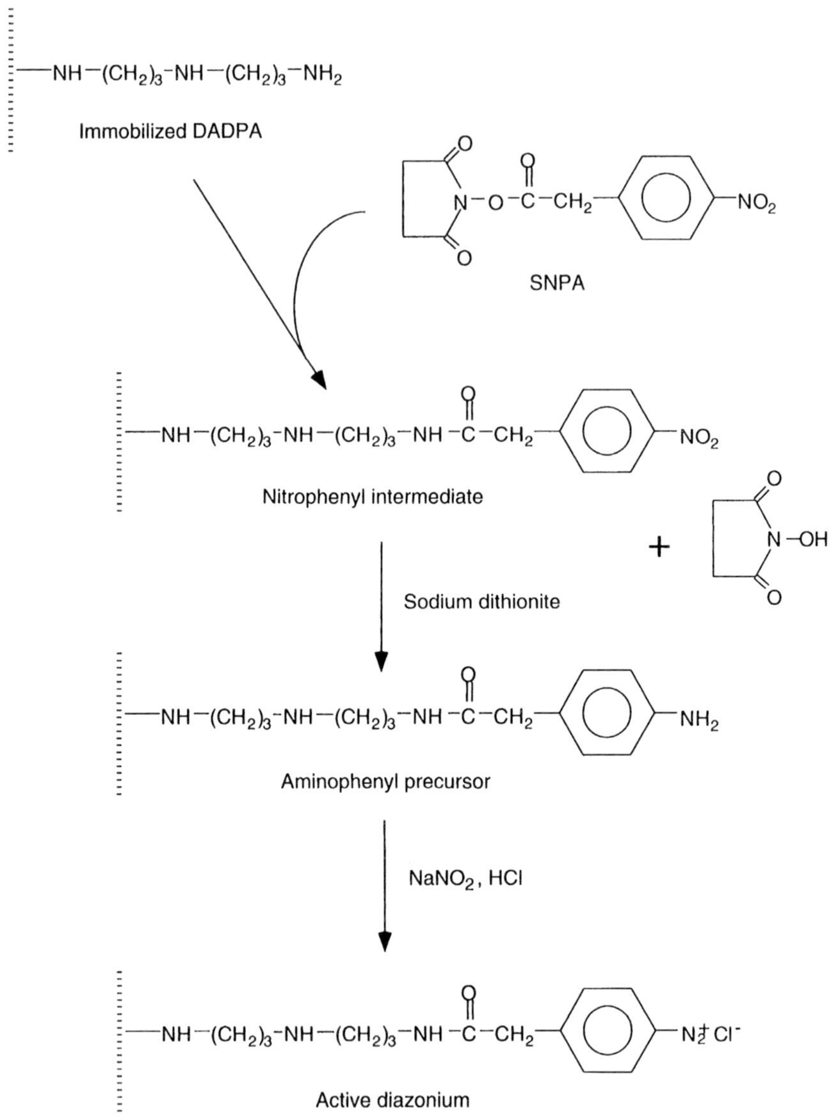

FIGURE 2.40
Another route to a diazonium-activated matrix begins with the reaction of the crosslinking agent SNPA (succinimidyl nitrophenylacetate) with an amine-containing spacer. Reduction of the nitrophenyl group with dithionite results in the aminophenyl precursor. The reactive diazonium can then be formed by sodium nitrite in dilute HCl.

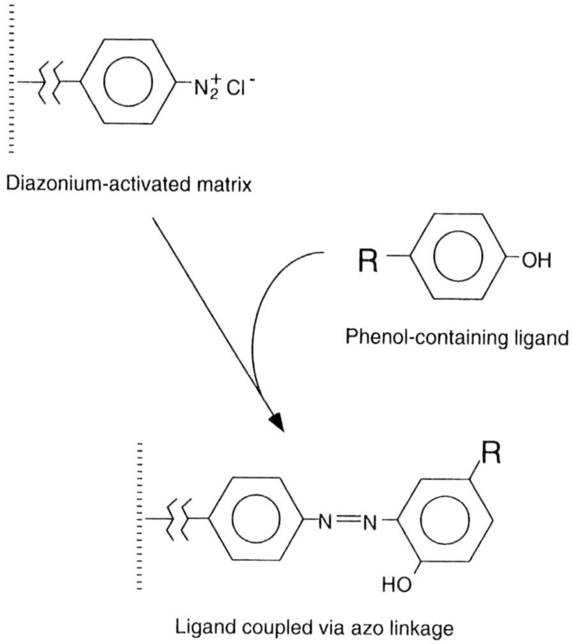

FIGURE 2.41

The highly reactive and short-lived diazonium group will react with active hydrogens, such as the *ortho* and *para* position hydrogens on a phenolic ligand, to yield a colored azo compound.

---

diazonium chemistry only as a last resort. When no other functionalities are present on a ligand that would allow an alternative immobilization scheme, we have used the Mannich condensation method described next to couple active hydrogens, before resorting to diazonium chemistry.

ACTIVATION AND LIGAND COUPLING PROTOCOL

The following protocol describes the synthesis of a diazonium derivative from a series of intermediate gels, the first one containing a spacer molecule that terminates in a primary amine group. We have found that immobilized diaminodipropylamine (DADPA) on agarose works well as the starting support for this method. This amine-containing matrix is reacted with the cross-linking reagent SNPA (*N*-succinimidyl-*p*-nitrophenylacetate; Pierce, Rockford, Illinois) to introduce terminal nitrophenyl groups. The

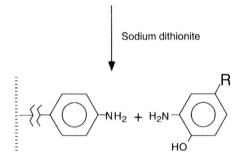

FIGURE 2.42

Ligands coupled to diazonium supports can be released by azo bond cleavage using sodium dithionite. As the bonds are broken, the support will return to its original color.

---

aromatic nitro groups are then reduced to amines to create a support that can be activated to form the desired diazonium functionalities.

a. Formation of the Intermediate Support

1. Wash 100 ml DADPA–*l*–Sepharose CL-6B into 100% DMF by sequential washing using increasing concentrations of DMF in water. (Use a fume hood throughout this procedure to protect from DMF vapors.) The gel is finally suspended as a 50% slurry in DMF and stirred with a paddle.
2. Dissolve 3 gm *N*-succinimidyl-*p*-nitrophenylacetate (SNPA) in a minimum quantity of DMF and add it to the stirred gel. The solution of SNPA in DMF will be a bright pink to red color.
3. React for 3 hr at room temperature.
4. Wash the gel extensively with DMF, 40% DMF, water, and finally 0.1 M sodium borate, pH 9.0.

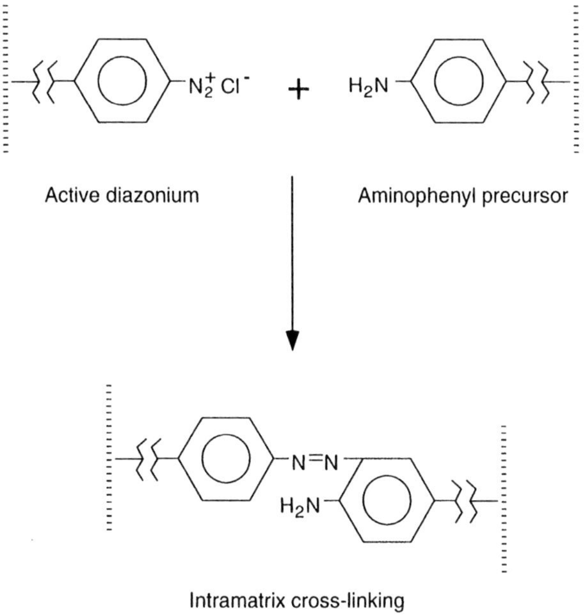

FIGURE 2.43
The process in preparing a diazonium-activated support is inherently prone to side reactions. The aminophenyl precursor is able to react with the diazonium groups as they are formed. The matrix usually becomes highly colored from extensive cross-linking even without added ligand.

---

5. Suspend the gel as a 50% slurry in the borate buffer, and add 12 gm sodium dithionite with stirring to reduce the aromatic nitro groups to amines.
6. React for 1 hr at room temperature.
7. Wash the support thoroughly with borate buffer and water.

b. Formation of Diazonium Groups and Coupling of Ligand

The following protocol makes use of small polystyrene minicolumns to wash an aliquot of gel and as a reaction chamber for the coupling reaction. After the affinity ligand has been immobilized, the columns also serve as convenient reservoirs for subsequent chromatographic purifications. Section 4.2.1.1, Chapter 4, describes the manipulations involved in use of these columns.

1. Pack 2 ml aminophenyl spacer gel in a 5-ml polystyrene disposable column containing a bottom porous polyethylene disk, but no top disk.
2. Wash the gel with 5 ml ice-cold 0.3 N HCl, place the bottom cap on the column, and suspend the gel in 2 ml 0.3 N HCl.
3. Add 0.5 ml sodium nitrite ($NaNO_2$) solution (50 mg/ml in cold water) to the gel suspension. Place the top cap on the column.
4. React for 15 min at 4°C while the column is rocked to mix the gel.
5. Quickly wash the gel with 5 ml 0.3 N HCl and 5 ml coupling buffer. The buffer chosen for the ligand coupling reaction should be at pH 8.0 for histidyl residues (e.g., 0.1 M sodium phosphate) or at pH 9–10 for tyrosine or phenolic compounds (e.g., 0.1 M sodium borate).
6. Add to the washed gel 1.0 ml cold coupling buffer and 1.0 ml ligand dissolved in cold coupling buffer. The optimal ligand concentration is roughly dependent on the molecular weight of the compound. For proteins and other large macromolecules, a concentration in the range of 10–20 mg/ml will give good coupling yields and provide a support with a high density of affinity ligand. For small molecules such as tyrosine-containing peptides, a 3–5 mg/ml concentration range will give good results. For ligands that are only sparingly soluble in aqueous environments, the borate coupling buffer may be made with 50% ethanol to aid solubility. The ligand itself may be first dissolved in ethanol and then diluted in coupling buffer to further aid solubility.
7. React overnight at 4°C using a rocker or gentle shaker.
8. Wash the gel extensively with coupling buffer to remove unreacted ligand. If ethanol was included to aid solubility, use the ethanolic buffer. Some decrease in flow rate through the gel may be noticed, especially if severe cross-linking occurred in the matrix during activation and coupling.

2.2.5.2 Mannich Condensation

Activation and Coupling Chemistry

Immobilization chemistry for the coupling of ligands to solid supports is fairly well defined for compounds with suitable functional groups to facilitate such attachment. The types of functional groups generally useful for this operation include those we have already discussed—easily reactive components such as primary amines, carboxylic acids, aldehydes, hydroxyls, or sulfhydryls. Usually, the solid-phase matrix is first activated with a compound that is reactive with one or more of these functionalities. This activated complex then can generate a covalent linkage between the ligand

and the support, thus immobilizing the ligand for subsequent chromatographic applications.

However, for molecules containing no easily reactive functional groups, immobilization can be difficult or impossible using current technologies. Particularly, certain drugs, steroidal compounds, dyes, or other organic molecules often have structures that contain no available "handles" for convenient immobilization. In other cases, functional groups that may be present on a molecule have low reactivity or are sterically hindered from efficient coupling.

Frequently, however, these difficult-to-immobilize compounds do have certain sufficiently active hydrogens that can be condensed with formaldehyde and an amine in the Mannich reaction (for a review, see Adams et al., 1942). Particular hydrogens in ketones, esters, phenols, acetylenes, a-picolines, quinaldines, and a host of other compounds can be aminoalkylated using this reaction.

Formally, the Mannich reaction consists of the condensation of formaldehyde (or sometimes another aldehyde) with ammonia, in the form of its salt, and another compound containing an active hydrogen. Instead of ammonia, however, this reaction can be done with primary or secondary amines, or even with amides. An example of this reaction is illustrated in the condensation of acetophenone, formaldehyde, and a secondary amine salt (the active hydrogens are shown underlined):

$$C_6H_5COC\underline{H}_3 + CH_2O + R_2NH \cdot HCl \longrightarrow C_6H_5COCH_2CH_2NR_2 \cdot HCl + H_2O$$

Some additional examples of compounds with similarly active hydrogens that can participate in the Mannich reaction are shown in Figure 2.44. Noteworthy are those examples in which the molecule would contain no other functional groups that could be used for coupling in typical immobilization schemes.

Whereas the Mannich reaction has been used often in organic solution-phase chemistry, we are unaware of any publication in the literature describing the use of the technique to immobilize a ligand to a chromatographic support.

The use of the Mannich reaction for the preparation of affinity supports is much cleaner than use of the reaction in solution phase conjugations. For instance, when using this kind of reaction in organic synthesis, polymerization is often a problem especially when more than one reactive hydrogen is present. When one of the reactive species is immobilized, however, the reaction is more controlled and inhibits undesirable side reactions. In addition, compounds that contain phenolic residues (often found in drugs) can be coupled without difficulty. Fortunately, the Mannich reaction provides

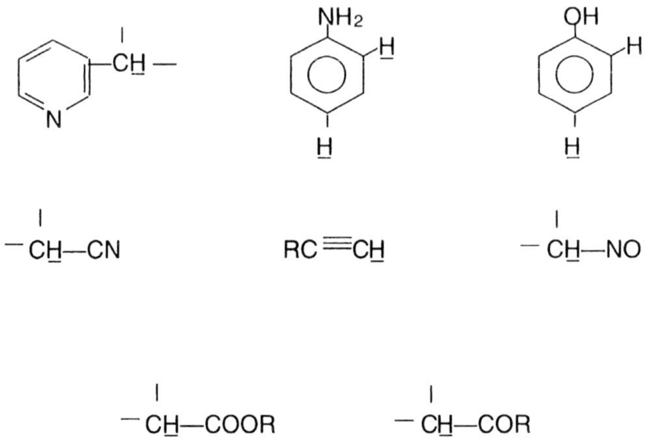

FIGURE 2.44
Active hydrogen-containing compounds that can participate in the Mannich reaction usually have these components present in their structures. The active hydrogens are underlined.

an alternative to the seldom used diazonium coupling method. Diazonium chemistry is often a problem in the preparation of affinity supports because of the disadvantages inherent in the instability of both the diazonium group and the resultant diazo linkage. In contrast, immobilizations done through Mannich condensations result in very stable covalent bonds suitable for the most critical affinity separations.

The immobilization scheme using this method exploits an immobilized diamine as the source of the primary amine. Formaldehyde and the desired ligand to be immobilized (containing an appropriately active hydrogen) are then added to the gel for coupling. Most often amines couple best as their salts, however, both acid- and base-catalyzed reactions have been described in the literature for solution chemistries. Thus, the immobilization protocols can be very flexible and can be customized to suit the ligand being coupled.

Figures 2.45 and 2.46 show the use of the Mannich reaction for the immobilization of phenol red and estradiol, respectively. Notice that in each of these examples, electron releasing groups on the aromatic ring systems activate constituent hydrogens sufficiently to participate in the reaction. (See Figure 2.47 for a list of common ring activators and deactivators.) The rate of observed coupling is directly related to the relative activity of the activated hydrogens. Thus, protocol flexibility should be stressed to allow for different reactivities of various ligands. Often, it is not possible to pre-

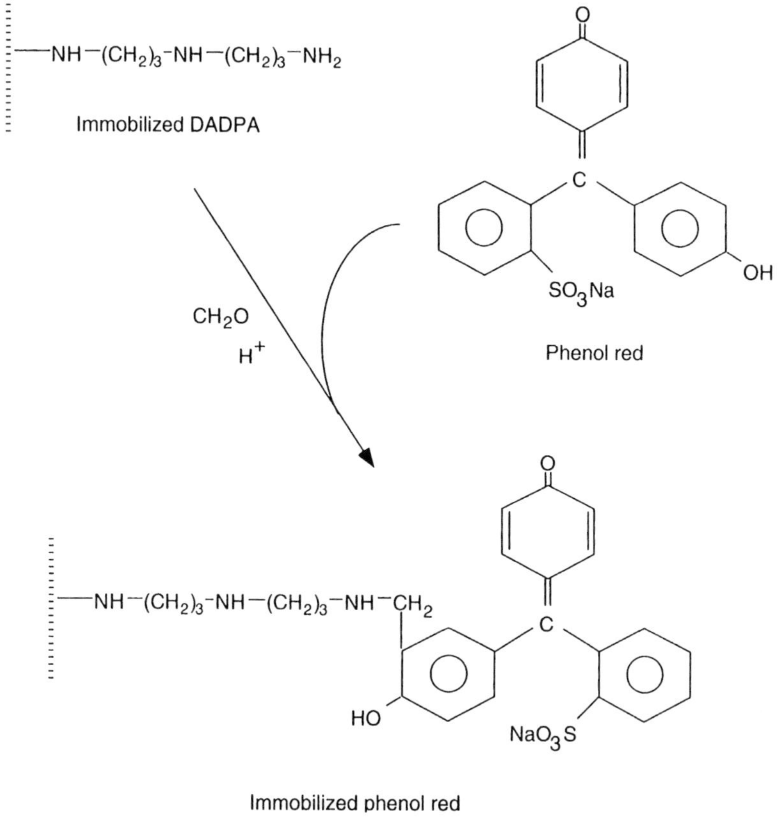

FIGURE 2.45
Phenol red (water soluble; MW 376.36) can be coupled to an amine-containing matrix through condensation with formaldehyde in the Mannich reaction. The active hydrogens present in the phenolic structures of phenol red provide excellent coupling sites.

---

dict the relative reactivity of a ligand without actually attempting to couple it. Trial and error may be the only way to determine with certainty whether a particular compound will react.

A limitation to this method of immobilization should also be realized. Ligands that contain more than one functional group that can potentially participate in the Mannich reaction may undergo solution-phase polymerization and, therefore, couple to a much lower degree or not at all. An example of this is a tyrosine-containing peptide. The phenolic side chain of tyrosine should contain sufficiently active hydrogens to condense with an

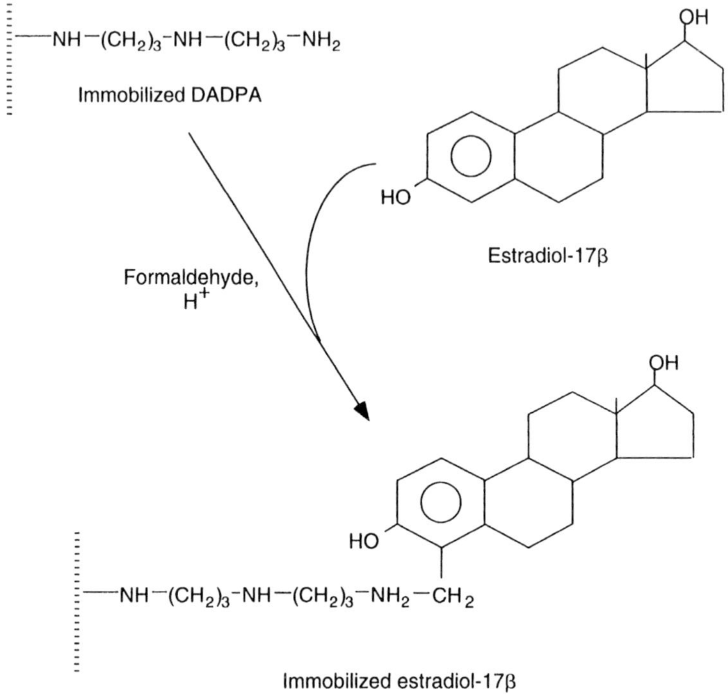

FIGURE 2.46

Estradiol-17β can be immobilized to an amine-containing matrix by condensation with formaldehyde in the Mannich reaction. The molecule is coupled through its active hydrogens *ortho* to the phenolic hydroxyl.

amine and formaldehyde. However, since primary amines are also present in peptides, intra- and intermolecular cross-linking are the more likely reaction. For this reason, no apparent coupling is observed for a tyrosine–lysine dipeptide when reacted with DADPA-agarose. It should be noted that compounds containing aromatic amines (aniline types) appear to couple without difficulty, perhaps because of their excellent ring activating ability and lower amine reactivity.

This technology thus provides an alternative immobilization chemistry that can be useful for coupling compounds with no common functionalities. When functional groups competing with the Mannich reaction do exist (particularly primary amine and formyl groups), then another chemistry should be considered to prevent side reactions.

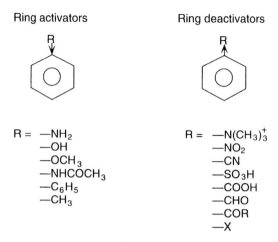

FIGURE 2.47
Common ring activators and deactivators. Hydrogens at positions *ortho* and *para* to a ring activator are often sufficiently active to participate in the Mannich reaction.

ACTIVATION AND LIGAND COUPLING PROTOCOL

The Mannich immobilization procedure uses formaldehyde as the coupling agent, similar to the way a cross-linking reagent is used to react with a functional group on a matrix and with another functionality on a ligand. In this respect, no activation of the matrix occurs in the classical sense. Instead, a support containing a spacer molecule terminating in a primary amine is used as the receiving matrix and a ligand containing an active hydrogen is immobilized.

1. Wash 10 ml primary amine-containing gel with 0.1 M MES, pH 4.7 (coupling buffer). A good choice for a spacer arm in this protocol is DADPA, which provides a terminal primary amine that conjugates well in the Mannich reaction.
2. Dissolve the ligand to be coupled in 10–20 ml coupling buffer. If the ligand is not very soluble in this aqueous environment, a portion of the buffer may be made 50% ethanolic by slowly adding an equal volume of ethanol to the stirred buffer. To further aid solubility, the ligand may be dissolved first in ethanol and then an aliquot added to the 50% ethanol/buffer solution for coupling.
3. Mix the ligand solution with the gel in a sealable vial and add 1.0 ml 37% formaldehyde solution in a fume hood. (**NOTE:** Formaldehyde is a suspected carcinogen.) Seal the vial.

4. React at 37–57°C for a minimum of 24 hr, periodically resuspending the gel by swirling the vial. For compounds with slow reactivity in the Mannich reaction, the temperature may be further increased or the time of reaction extended. Trial and error may be necessary to optimize some immobilizations.
5. After coupling, remove the seal and transfer the gel slurry to a filter funnel for washing.
6. Wash any unreacted ligand from the gel by using the coupling buffer as well as 0.1 M Tris, pH 8.5. This second buffer may also be made up as a 50% ethanolic solution for washing sparingly soluble ligands off the support. Additionally, a brief wash with 100% ethanol is possible to help remove unreacted ligand. If 100% ethanol is used, to prevent buffer precipitation first wash the buffer salts off the gel with water, then apply the ethanol.
7. For storage, a suitable preservative should be added to the gel slurry (0.02% sodium azide) and the affinity support kept at 4°C.

2.2.6 Photoreactive Cross-Linkers

In 1969, Fleet *et al.* introduced the aryl azide group as a photoreactive cross-linking agent. Although unreactive in the dark, the phenyl azide is rapidly converted into an extremely reactive nitrene upon photolysis. The photolysis conversion requires a bright light source usually in the range of 265–275 nm (Ji, 1979). The generated nitrene is short-lived, with a half-life on the order of a millisecond. The derivative will react with covalent bonds to nonspecifically insert in the structure of a target molecule or even react with solvent molecules (Schrock and Schuster, 1984). The nitrene either reacts with another molecule or is quickly destroyed. The nonspecific nature of the photoreactive group coupled with its short half-life usually results in yields of the desired reaction product of less than 10%.

Nevertheless, cross-linking reagents made with a photoreactive group on one end and another reactive functionality on the other end can be used to couple to specific groups on one target molecule and then nonspecifically conjugate to a second target molecule. The stability of the phenyl azide group in the dark allows the photoactivation step to be initiated only when desired.

Heterobifunctional photoreactive cross-linking reagents have been synthesized with a variety of secondary functionalities including hydrazide groups (O'Shannessy and Quarles, 1985; O'Shannessy *et al.*, 1986), NHS esters (Ji and Ji, 1982, Sorenson *et al.*, 1986; Shepard *et al.*, 1988), pyridyl

2.2 Procedures

disulfides (Traut et al., 1989), and formyl groups (Ngo et al., 1981). Examples of some of these reagents are shown in Figure 2.48.

The use of these reagents to immobilize ligands can take one of three approaches. (1) The photoreactive cross-linker first can be conjugated with the ligand in solution using its thermochemical reactive end. For a photoreactive reagent containing an NHS ester group, this would mean initially coupling to the primary amino groups of a protein molecule. Subsequently, the modified protein is isolated and mixed with the desired matrix material. The mixture is then photolyzed and immobilization is accomplished by

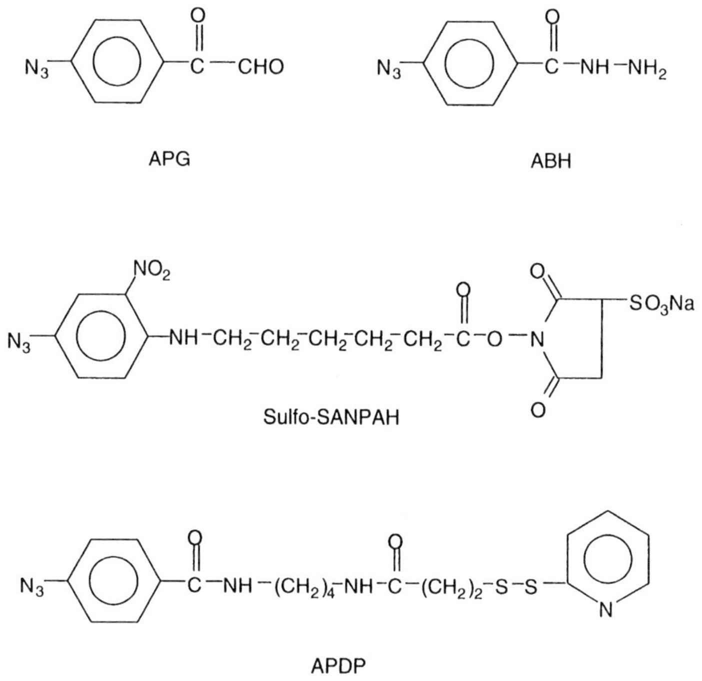

APG

ABH

Sulfo-SANPAH

APDP

FIGURE 2.48
Heterobifunctional photoreactive cross-linking reagents that contain custom thermoreactive groups able to couple to various functionalities are commercially available (Pierce). APG (p-azidophenyl glyoxal) can selectively react with arginine residues in proteins. ABH (azidobenzoyl hydrazide) can couple to oxidized sugar residues in glycoproteins. Sulfo-SANPAH [sulfosuccinimidyl-6-(4'-azido-2'-nitrophenylamino)hexanoate] contains an active ester end that can react with primary and secondary amine groups. APDP [N-[4-(p-azidosalicylamido)butyl]-3'(2'-pyridyldithio)propionamide] has a pyridyl disulfide group that can reversibly couple to sulfhydryl-containing ligands. The photoreactive end of these cross-linking agents can insert nonselectively into various matrix materials, particularly supports that contain no convenient coupling groups.

nonselective coupling of the reactive nitrene with the support. [See Section 3.4.2.4, Chapter 3, for an example of this type of immobilization (Fig. 2.49).] (2) In a second strategy, the NHS ester end of the cross-linker is first reacted with an amine-containing matrix to yield immobilized photoreactive groups. A ligand solution is then mixed with this derivatized matrix and photolyzed to nonselectively couple the ligand to the support. (3) Alternatively, photoreactive cross-linking reagents can be used to functionalize a matrix material that contains no convenient chemical groups for use in traditional activation procedures. In this sense, the photoreactive end is first reacted with the matrix by photolysis, resulting in nonspecific insertion of the phenyl azide into the support material. Next, the ligand is covalently coupled to the attached cross-linker using its other thermoreactive end.

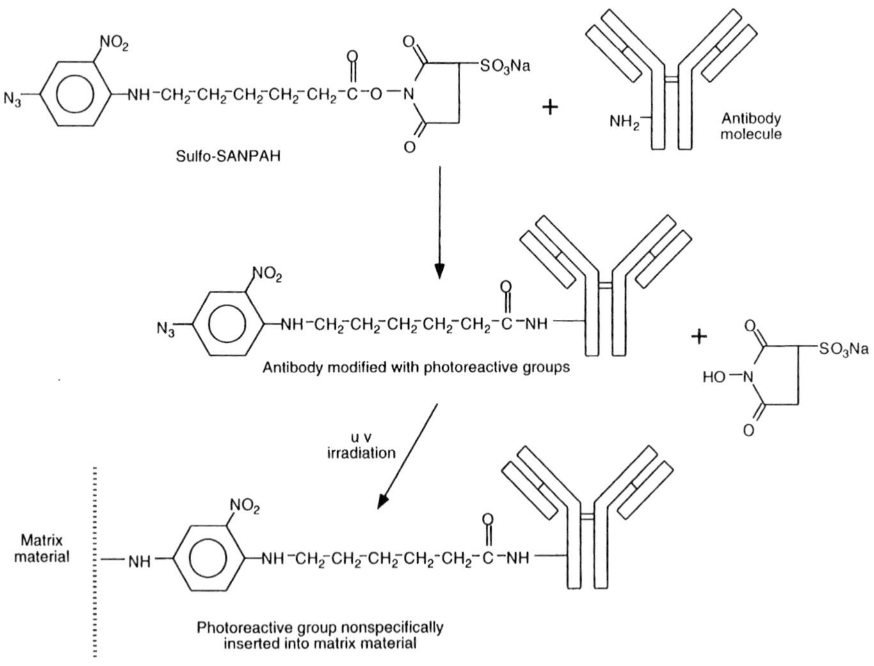

FIGURE 2.49

The use of photoreactive cross-linking agents to immobilize ligands to matrix materials is illustrated by the use of Sulfo-SANPAH. The NHS ester end of the cross-linking agent is first reacted with an antibody molecule, coupling to the available amine groups of the protein. Next, the modified immunoglobulin is mixed with the matrix and photolyzed with uv light for several minutes. The photoreactive group will nonspecifically insert into the matrix structure.

This approach works well for the modification of polystyrene microplates and provides a route for obtaining functionalities in good yield on a surface that is rather difficult to derivatize.

All these methods work, but the resulting yield of immobilized ligand is rather low compared with the nonphotoreactive derivatization techniques discussed previously. The utility of these methods lies in the ability of the photoreactive group to nonspecifically insert and couple to target substances. This characteristic is particularly useful when the matrix or ligand contains no available functionalities for immobilization that would normally enable the use of more traditional chemistries. Thus, difficult-to-derivatize matrix materials such as polystyrene can be modified using a photoreactive cross-linker, while the thermochemical end can be used to immobilize ligands (Guire, 1978, 1988, 1990; Guire and Chudzik, 1989).

PHOTOCHEMICAL MODIFICATION PROTOCOL

A general method of immobilization using these techniques is difficult to write, since the procedures will vary depending on the photoreactive cross-linker and matrix material employed. For each specific immobilization problem, solubility considerations must be worked out for the cross-linking reagent, as must solvent stability concerns for the matrix (especially when working with polystyrene microplates). The basic strategy for the photochemical modification of a matrix with a cross-linker and the subsequent coupling of a ligand through a thermochemical group can be outlined as follows.

1. Select a photoreactive cross-linker (Pierce Chemical, Rockford, Illinois) with another reactive group suitable for coupling to functionalities on the ligand to be immobilized.
2. Dissolve the cross-linker in a buffer or solvent medium, the composition of which depends on the unique solubility properties of the compound. When working with thermochemical groups that hydrolyze in an aqueous environment, use organic solvents such as DMSO, acetonitrile (OK for polystyrene), or DMF whenever possible. For aqueous soluble cross-linkers, a high salt environment will help concentrate the photoreactive end near hydrophobic surfaces (such as polystyrene microplate wells). A cross-linker concentration of 3–5 mg/ml is sufficient for the initial photochemical reaction with a matrix material.
3. Mix the photoreactive cross-linker solution with the matrix and photolyze with light at a wavelength of 265–275 nm. Several minutes of intense light exposure may be necessary to complete the reaction.
4. Wash the matrix free of unreacted cross-linker using the reaction buffer.

5. Add to the matrix the ligand solution made in a buffer optimal for the coupling reaction involving the thermochemical reactive group being used. Reference to the buffers recommended in the previous sections on activation chemistries will identify the best buffer composition to use.
6. Wash the matrix free of uncoupled ligand using the reaction buffer, water, 1 M NaCl, and water.

3

IMMOBILIZATION OF LIGANDS

The following sections contain protocols for the immobilization of some of the most frequently used ligands and spacer molecules. Several matrix materials and coupling chemistries are presented to show the broadest selection of immobilization techniques possible. For each type of ligand, alternative supports or chemistries may be substituted for the ones illustrated with minimal or no change in usefulness of the affinity matrix. Cross-referencing to the corresponding sections in Chapters 1 and 2 will provide additional background for understanding the methods described here.

3.1 SMALL LIGANDS

3.1.1 Spacer Arms

Use of Spacers

Spacer arms or leashes are low-molecular-weight molecules that are used as intermediary linkers between a support material and an affinity ligand. Usually spacers consist of linear hydrocarbon chains with functionalities on both ends for easy coupling to the support and ligand. First, one end of the spacer is attached chemically to the matrix using traditional immobilization chemistries; the other end is connected subsequently to the ligand using a secondary coupling procedure. The result is an immobilized ligand that sticks out from the matrix backbone by a distance equal to the length of the spacer arm chosen.

Depending on the characteristics of a particular affinity binding site, the use of a spacer may be crucial for successful interaction between a ligand and a complementary protein or other macromolecule (Cuatrecasas, 1970;

Steers *et al.*, 1971). Often, ligand binding sites are buried or in a pocket just below the surface of a protein (e.g., avidin/biotin interactions). A ligand that is attached directly to a polymeric support material may not protrude far enough from the matrix surface to reach the level of the binding site on an approaching protein molecule (Fig. 3.1). The result may be a weakened interaction or no binding at all (Lowe *et al.*, 1973).

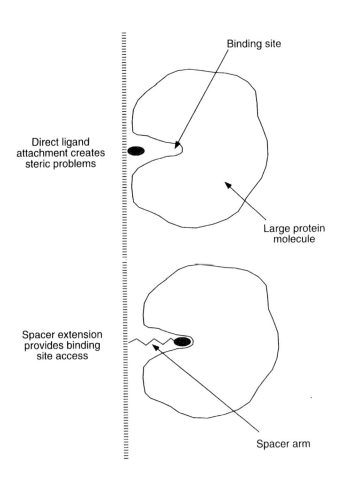

FIGURE 3.1

The principle advantage of using a spacer arm is that it provided ligand accessibility to the binding site of a target molecule. When the target molecule is a protein with a binding site somewhat beneath its outer surface, a spacer is essential to extend the ligand out far enough from the matrix to allow interaction.

3.1 Small Ligands

With rigid support materials, a spacer molecule may also provide greater flexibility, allowing the immobilized ligand to move into position to establish the correct binding orientation with a protein. The degrees of freedom that a hydrocarbon extender can provide are much greater than the movement possible within the polymeric backbone of a matrix (Cuatrecasas, 1970).

The choice of spacer molecule can affect the relative hydrophilicity of the immediate environment of an immobilized ligand. Molecules containing long hydrocarbon chains may increase the potential for nonspecific hydrophobic interactions, especially when the affinity ligand is small and of low molecular weight (O'Carra et al., 1973). Selecting spacers that have more polar constituents, such as secondary amines, amide linkages, ether groups, or hydroxyls will help keep hydrophobic effects at a minimum.

It is also important to consider the ionic effects a spacer molecule may impart to a gel. Spacers with terminal primary amine groups should be completely coupled with ligand or blocked by a nonrelevant molecule (e.g., acetic anhydride; see Section 3.1.1.9) to eliminate the potential for creating a positive charge on the support. With small ligands, these residual charges can form a secondary environment that may cause considerable nonspecific interactions with proteins. The same holds true for spacers with terminal carboxylic groups. In general, a negatively charged spacer will cause less nonspecific protein binding than a positively charged one, but blocking excess remaining groups is still a good idea. A good blocking agent for use with carboxylic residues is ethanolamine, which leaves a terminal hydroxyl group (see Section 3.1.1.9).

A few words of caution should be given. It is better to cap off excess spacer molecules with ligand than with some other nonrelevant molecule, since a secondary capping component may itself cause additional nonspecificity. Controlling the amount of excess spacer can be accomplished by limiting the degree of spacer substitution on the gel in the first place. Thus, at the point of ligand coupling, there is a better chance of establishing complete substitution with no remaining spacers to cause nonspecific binding. In addition, care should be taken in capping excess spacer under those circumstances in which the ligand itself may also be modified. In other words, do not cap off excess amine spacers if your ligand contains amine groups.

When immobilizing large molecular weight molecules such as proteins, the use of a spacer arm may not be as critical as it is with small ligands. The three-dimensional size of a protein molecule is huge in comparison to even the longest of spacers. A spacer could only extend the protein from the matrix surface by a small percentage of the overall width of the protein. In this case, the result would provide little if any advantage over direct attachment to the matrix without a spacer.

The use of spacers can be advantageous with macromolecules, however, when site-directed immobilization is desired (Section 3.4.2.3). Proteins usually have a number of potential immobilization sites, corresponding to particular functionalities on the molecules. These groups can be polypeptide derived, for example, amines, carboxylic acids, active hydrogens on aromatic side chains, and sulfhydryls, or they may be associated with posttranslational modifications, for example, carbohydrate chains (hydroxyls and chemically generated formyl groups). The coupling chemistry can be designed to attach via only one of these functionalities to direct the orientation of the protein to favor retention of activity. Usually, site-directed immobilization is aimed at functional groups of relatively low abundance (sulfhydryls, carbohydrates, or active hydrogens) within the protein molecule and in known locations away from active sites. These specific activation and coupling methods often require additional functionalities to be added that are not present in the base matrix composition of the support. Specific immobilization chemistries can be constructed from the appropriate choice of spacer molecules. The site-directed chemistry can be built on the end of the spacer, and its length can aid in reaching the appropriate functionality on the protein. For examples of this concept, see the discussion on hydrazide and iodoacetyl coupling methods in Chapter 2.

In many respects, the base matrix and the choice of primary activation chemistry are only part of the total picture in constructing an optimal affinity support. Even before specific ligands are attached, spacer molecules often may be vital pieces in making the final matrix successful in an affinity purification process.

Types of Spacers

At least in theory, any small molecule can function as a intermediary linker between a matrix and a ligand. However, most spacers are built from individual molecules no more than 10 atoms in length. The best choices have appropriate coupling functionalities on either end and an overall hydrophilic character. In this sense, the choice of a spacer molecule can be made simply by browsing through an organic chemical catalog and choosing the most likely candidates.

Over the years, this process of selection has narrowed the variety of spacers to a few that are used recurringly. This does not mean that other molecules are not appropriate for use in a particular application, but that these spacers seem to be used most frequently by the practitioners of the art in preparing affinity supports. Figure 3.2 shows the structures and names of the most popular spacer molecules.

3.1 Small Ligands

$NH_2-CH_2-CH_2-CH_2-NH-CH_2-CH_2-CH_2-NH_2$

Diaminodipropylamine (DADPA)
3,3'-Iminobispropylamine

$HOOC-CH_2-CH_2-COOH$

Succinic acid
(used as the anhydride)

$NH_2-CH_2-CH_2-CH_2-CH_2-CH_2-COOH$

6-Aminocaproic acid (6-AC)

$NH_2-CH_2-\overset{\overset{OH}{|}}{CH}-CH_2-NH_2$

1,3-Diamino-2-propanol

$NH_2-CH_2-CH_2-CH_2-CH_2-CH_2-CH_2-NH_2$

1,6-Diaminohexane (DAH)

$NH_2-CH_2-CH_2-NH_2$

Ethylenediamine (EDA)

FIGURE 3.2
Some common spacer molecules used to extend a ligand out from a matrix surface.

The following sections describe the initial immobilization and coupling protocols necessary to use these spacers in building affinity supports. Most of these spacers can be found throughout this chapter in our presentation of real examples of spacer—ligand immobilizations.

3.1.1.1 Diaminodipropylamine

Diaminodipropylamine (DADPA) (found in some catalogs as 3,3'-iminobispropylamine) is a nine-atom spacer with primary amines at either end and a secondary amine linkage in the middle (Harris *et al.*, 1973; Sica *et al.*, 1973). The secondary amine group provides a certain amount of internal hydrophilicity, while creating a 120° bend in the molecule. The compound is a liquid at room temperature (d = 0.938, mp, −14°C) and gives off a pungent amine odor. Since it is a volatile substance and a highly toxic corrosive, it is best to handle pure DADPA and all nonneutralized solutions of it in a fume hood. The compound is freely soluble in aqueous solutions and in many organics [e.g., acetone, dioxane, dimethyl sulfoxide (DMSO), dimethylformamide (DMF), and ethanol].

When preparing a buffered coupling solution of DADPA, the pH must be adjusted with HCl to neutralize the extremely basic character of the amines. To prevent a significant release of heat on neutralization, the solution should be stirred on ice before addition of acid (in this case, 6 N HCl). Alternatively, we often add the appropriate amount of DADPA directly to chopped ice made from deionized water, and then begin to adjust the pH with concentrated HCl. The heat of reaction will melt the ice at just about the point that the correct pH value is reached. The neutralized solution can then be buffered by adding the appropriate buffer salts and readjusting the pH.

Coupling any spacer molecule with amine groups at both ends to an amine-reactive support requires adding the spacer in significant molar excess to avoid cross-linking the matrix. In these protocols, the molar amount of amines present from the spacer is at least 40-fold higher than the theoretical amount of active groups present in the gel. This promotes "one-on" hits instead of having both ends of the spacer couple to the active sites on the gel. Thus, the result will be the desired free primary amine group protruding from the support.

IMMOBILIZATION PROTOCOL

Two protocols are given for the immobilization of DADPA. The first one uses the reductive amination procedure for periodate-oxidized agarose and is done in an aqueous solution. This process yields a secondary amine linkage between the matrix and spacer that is very stable to leakage. The second method uses carbonyl diimidazole (CDI)-activated agarose and is performed in acetone. This chemistry gives a carbamate linkage, one of the most stable known. Figure 3.3 shows these reactions and the resultant linkages. See Chapter 2 for methods on how to prepare the activated gels required for these procedures.

a. Immobilization Using Reductive Amination

1. Wash 100 ml periodate-oxidized Sepharose CL-4B into coupling buffer (0.1 M sodium phosphate, pH 7.0). Filter off excess buffer solution until the gel is left as a wet cake.
2. Working in a fume hood, add 20 gm (21.3 ml) DADPA (Aldrich) to about 100 ml chopped ice made from deionized water. (Wear safety glasses and gloves for this procedure.) Slowly add 8–10 ml concentrated HCl dropwise, while stirring the mixture. The DADPA mixture will fume slightly and become warm from the acid addition. The ice prevents extensive fuming and developing of a hot solution. At this point, the ice should begin to melt enough to add a stir bar and an electrode for mea-

3.1 Small Ligands 143

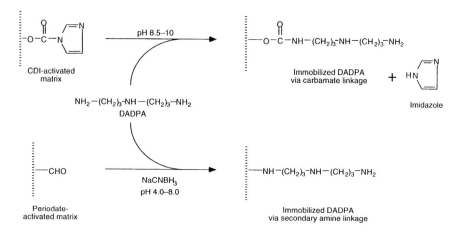

FIGURE 3.3
Diaminodipropylamine (DADPA) can be coupled to a matrix by amine reactive activation chemistries. A carbonyl diimidazole (CDI)-activated support will react with DADPA to give an uncharged carbamate linkage. Periodate-oxidized matrices providing formyl groups can be reacted with DADPA via reductive amination to generate secondary amine linkages. CDI coupling can result in the higher density of space substitution, if this is desired.

---

suring the pH. Continue to add HCl with stirring until the pH is about 7.0. At the completion of the neutralization, the ice should be melted fully. If the volume is less than 100 ml, bring it up to that mark with water. Finally, add sodium phosphate buffer salts to this solution to prepare a 0.1 M concentration. Readjust the pH to 7.0.
3. Add the washed gel to the DADPA solution and stir with a paddle in a fume hood.
4. Add 1.2 gm sodium cyanoborohydride (toxic) and continue stirring for at least 4 hr.
5. Extensively wash the DADPA-/ /-Sepharose CL-4B with water, 1 M NaCl, and again with water. Use the TNBS test (see Section 4.1.3, Chapter 4) to examine the completeness of washing by checking for the presence of amines in the wash solution. This test may also be used to appraise the success of DADPA coupling.

b. Immobilization Using CDI Activation

1. In a fume hood, dissolve 20 gm DADPA (21.3 ml) in 100 ml dry acetone.
2. Drain 100 ml CDI-activated Sepharose CL-6B of excess acetone, and then add the gel to the DADPA solution.

3. Stir the reaction mixture with a paddle for at least 3 hr at room temperature.
4. Wash the gel with 1 L acetone to remove uncoupled amines, then sequentially wash into water. Continue to wash with water, 1 M NaCl, and water to completely remove excess DADPA. Check for the success of spacer coupling and completeness of washing by using the TNBS test (see Section 4.1.3, Chapter 4).

Gels substituted with DADPA contain free primary amines. These supports are now suitable either for coupling with a carboxyl-containing ligand or for further modification with another spacer molecule that also contains a carboxylate group.

3.1.1.2 1,6-Diaminohexane

The use of diaminohexane (DAH) as a spacer molecule is similar to the use of DADPA just described because a gel substituted with this group will provide primary amine functionalities for further coupling (O'Carra and Barry, 1972). Since DAH contains a linear uninterrupted 6-carbon chain, its hydrophobic character is considerably greater than that of DADPA. Small ligands immobilized on the end of this spacer may have a competing site of interaction because of this hydrophobic character. Larger ligands such as proteins probably will not be affected by the hydrophobicity, since the spacer is small compared with the protein. Despite these facts, many small ligand affinity supports have been prepared successfully using this compound.

DAH is a solid at room temperature (mp, 42–45°C), highly corrosive, and hygroscopic. Yellow to brown colored crystals of DAH as obtained commercially are satisfactory for use without further purification. Aqueous solutions of this diamine require neutralization similar to DADPA, as described. This is best done by adding the solid to ice cold water or buffer, while stirring, in a beaker that is placed in an ice bath. In a fume hood, bring the pH down to the optimal coupling point using concentrated HCl, while continuing to cool the solution. The dihydrochloride form of this spacer is often more convenient to use, since the neutralization process is unnecessary.

IMMOBILIZATION PROTOCOL

The following protocol requires the preparation of tresyl-activated Trisacryl GF-2000LS. (See Chapter 2 for tresyl chemistry as well as Sec-

3.1 Small Ligands

tion 1.3.1.2, Chapter 1). Figure 3.4 shows the reactions involved in preparing this support.

1. In a fume hood, dissolve 10 gm DAH in a minimum quantity of coupling buffer (0.2 M sodium phosphate, pH 7.5) and chill in an ice bath. Adjust the pH to 7.5 with concentrated HCl while maintaining the solution at 0–4°C. If the final solution volume is less than 100 ml, adjust to this mark with additional coupling buffer. If it is over this amount by a little, leave the volume unchanged. Alternatively, dissolve 16.2 gm DAH dihydrochloride in 100 ml coupling buffer, and readjust the pH to 7.5 if necessary.
2. Wash 100 ml tresyl-activated Trisacryl GF-2000LS into coupling buffer at 4°C and drain off excess buffer.
3. Add the gel as a wet cake to the DAH solution and stir at 4°C for 5 hr. The reaction can also be done at room temperature in 2 hr.
4. Wash the support with coupling buffer, 1 M NaCl, and water to remove unreacted DAH. The TNBS test may be used to check for completeness of the washing step as well as for the success of coupling (see Section 4.1.3, Chapter 4).

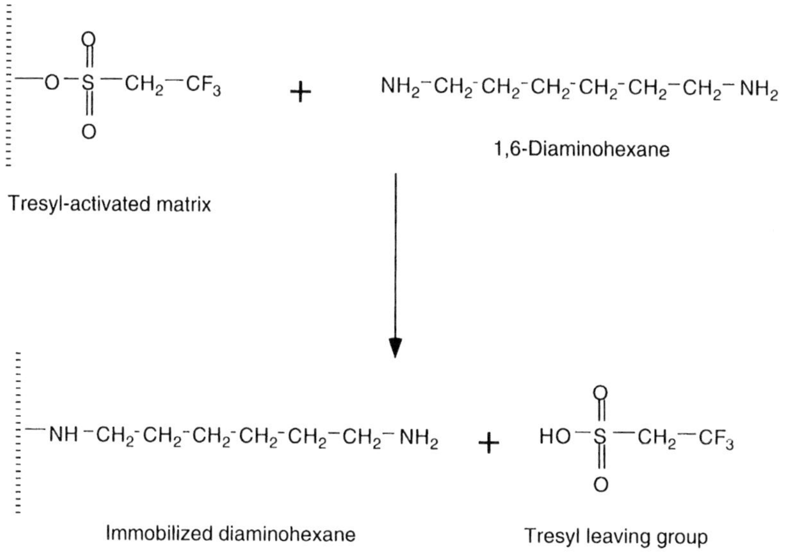

FIGURE 3.4

Tresyl chloride activation of an hydroxylic matrix produces reactive sulfonates that can immobilize amine-containing ligands or spacers in high yield.

3.1.1.3 6-Aminocaproic Acid

A popular spacer used to create a terminal carboxylic group on a matrix is 6-aminocaproic acid (6-AC) (Schmer, 1972). This compound provides a primary amine coupling functionality on one end and a carboxylic group on the other. Amine-reactive supports are used to first couple the spacer to the matrix, and secondary chemistries [i. e., 1-ethyl-3-(3-dimethylaminopropyl) carbodiimide (EDC)] can be used to immobilize an amine-containing ligand or further lengthen the spacer on the other end.

Like DAH, 6-AC may create some mild hydrophobic interactions because of its 5-carbon linear structure. In practice, however, these effects are not usually severe, especially when using hydrophilic ligands. In addition, after ligand coupling some carboxylate groups often remain that have not been blocked by ligand. These negatively charged spacers will effectively negate the hydrophobic environment of the hydrocarbon chain.

6-AC is a solid that can be added to buffers with a minimal effect on pH. Unlike the diamines discussed earlier, no unpleasant neutralization step is required with this compound.

IMMOBILIZATION PROTOCOL

The following procedure uses the cyanogen bromide (CNBr) activation method to couple 6-AC to Toyopearl HW-65f. This procedure results in an isourea bond formed between the matrix and the spacer (Fig. 3.5). Chapter 2 should be consulted to prepare the activated matrix. The use of other supports and different coupling protocols that react with primary amines will also work well for immobilizing 6-AC.

1. Dissolve 10 gm 6-AC in 100 ml cold 0.1 M sodium phosphate, pH 7.5. Readjust the pH if necessary.
2. Add 100 ml freshly prepared CNBr-activated Toyopearl HW-65f support to the spacer solution and stir with a paddle at 4°C.
3. Allow the coupling reaction to continue overnight at 4°C.
4. Using a Buchner funnel with a glass fiber filter pad, wash the gel extensively with water, 1 M NaCl, and again with water to remove unreacted 6-AC. A negative test for the presence of amines in the washes using TNBS is a good check for completeness of the washing process (see Section 4.1.3, Chapter 4). Note that TNBS will not react with the spacer on the gel since the amine end has coupled to the support.

The support is now ready for coupling an amine-containing ligand or for further extension using another amine-containing spacer molecule.

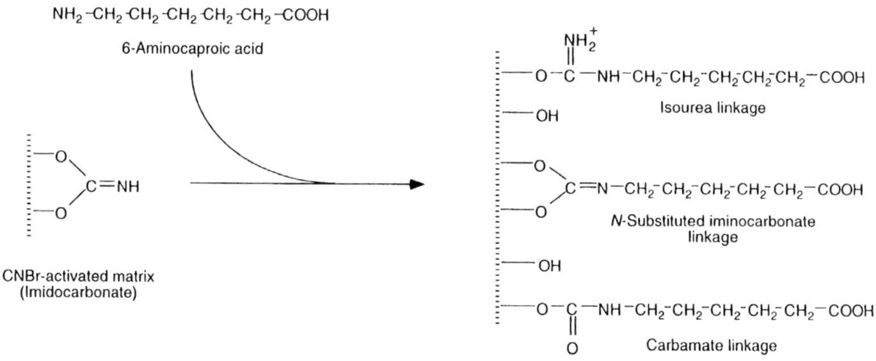

3.1 Small Ligands

FIGURE 3.5
The coupling of 6-aminocaproic acid (6-AC) to a CNBr-activated support can result in several different linkages. The carbamate and iminocarbonate linkages are usually dominated by the charged isourea bond. The positive charge associated with this group at basic pH may introduce nonspecific character into any resultant affinity support, especially when using low molecular weight ligands.

3.1.1.4 Ethylene Diamine

Ethylene diamine (EDA) is a convenient spacer molecule that is often used to convert carboxylate terminals into amines. The short two-carbon length of EDA does not cause any steric problems, and using it assures hydrophilicity because of its proximal amines. The molecule also is used in the transamidination reaction of polyacrylamide to create primary amine functionalities on the gel (see protocol that follows). Figure 3.6 shows the carbodiimide-mediated reaction of EDA with the immobilized spacer 6-AC.

EDA is a thick liquid (d = 0.899) that is flammable and highly corrosive. All operations using this compound should be done in a fume hood. Aqueous solutions of EDA must be neutralized prior to an immobilization reaction. To prevent fuming, pH adjustments are done on ice. Alternatively, the recommended amount of diamine can be added to ice chips and the pH adjusted with concentrated HCl in a manner similar to that described for DADPA. The use of the dihydrochloride form of EDA (Aldrich, Milwaukee, Wisconsin) eliminates the need for neutralization altogether.

IMMOBILIZATION PROTOCOL

Two methods are outlined here for use of EDA. In the first procedure, immobilized 6-AC is further modified with EDA to give a spacer gel with a

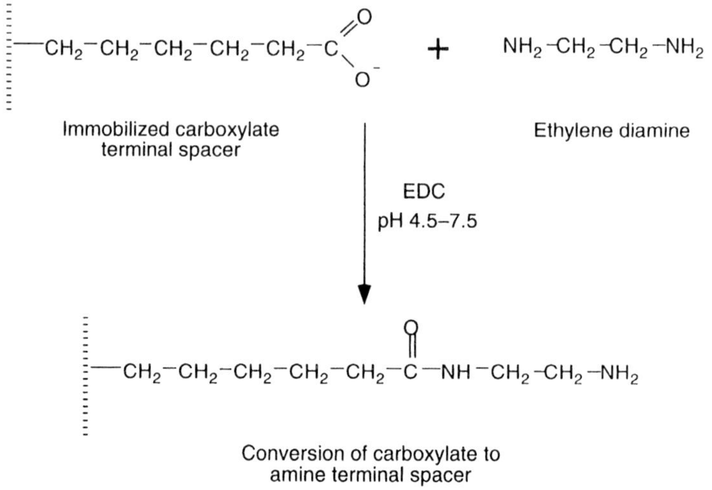

—CH$_2$—CH$_2$—CH$_2$—CH$_2$—CH$_2$—COO$^-$ + NH$_2$—CH$_2$—CH$_2$—NH$_2$

Immobilized carboxylate terminal spacer   Ethylene diamine

EDC
pH 4.5–7.5

—CH$_2$—CH$_2$—CH$_2$—CH$_2$—CH$_2$—C(=O)—NH—CH$_2$—CH$_2$—NH$_2$

Conversion of carboxylate to amine terminal spacer

FIGURE 3.6

Ethylene diamine can be used to convert a carboxylate terminal spacer into an amine terminal spacer. The water-soluble carbodiimide EDC facilitates the formation of an amide bond linkage.

---

terminal amino group. This is accomplished using the water-soluble carbodiimide EDC to form an amide bond between the carboxylate group of 6-AC and one of the amines of EDA. In the second protocol, EDA is used to modify polyacrylamide beads by transamidination to provide functional groups on the gel for subsequent ligand immobilization.

a. EDA Coupling to Immobilized 6-Aminocaproic Acid

1. Wash 100 ml 6-AC–/ /–Toyopearl HW65f support (made according to the protocol just described) with water and then with 0.1 M MES, pH 4.7 (coupling buffer). Drain the support of excess buffer.
2. Dissolve 40 gm EDA dihydrochloride in 100 ml coupling buffer. Readjust the pH if necessary.
3. Add the washed gel cake to the EDA solution, and stir the slurry in a fume hood with a paddle stirrer.
4. Add 3 gm EDC hydrocholoride (Pierce; Sigma; Aldrich). React for 2 hr at room temperature.

5. Wash the gel extensively with water, 1 M NaCl, and again with water. Test for completeness of washing and for success of EDA derivatization using the TNBS test (see Section 4.1.3, Chapter 4).

The gel is now ready for the coupling of a carboxylic acid-containing ligand (which may be done using the EDC reaction again).

b. EDA Modification of Polyacrylamide Beads by Transamidination

The amide bonds in the polymeric backbone of polyacrylamide beads may be broken and, under the right conditions, substituted with an amine-containing compound (Inman and Dintzis, 1969). Polyacrylamide beads will conveniently swell in a solution of 100% EDA. When this slurry is heated, transamidination will occur and EDA molecules will be inserted into the polymer, forming stable amide bonds with the matrix. The following protocol is useful with any of the commercially available polyacrylamide supports. This technique works equally well for insertion of hydrazide functionalities using hydrazine.

1. In a fume hood, heat 500 ml EDA to 90°C using a three-necked round-bottom flask and a heating mantle. Stirring should be done with a paddle to prevent subsequent gel damage. Insert a thermometer into one of the necks of the flask to measure temperature.
2. Slowly add to the vigorously stirring EDA solution 25 gm dry polyacrylamide beads. The addition should be in small portions to avoid clumping of the support as it swells.
3. Continue stirring at 90°C for at least 5 hr. The rate of transamidination is such that about 0.3 mmol/gm amide is converted per hour of the reaction. The maximal substitution level of EDA is about 1.2 mmol/gm beads.
4. At the end of the reaction period, remove the heating mantle and allow the solution to come to room temperature. Once it has cooled, an equal volume of crushed ice is added to keep the solution chilled as it comes into contact with aqueous solutions.
5. Transfer the slurry to a Buchner funnel with a glass fiber filter pad (preferably in a fume hood). Wash the gel free of excess EDA with cold 0.2 M NaCl, 0.001 N HCl. As the amines are washed out and the gel is neutralized, the solutions may have a tendency to become warm. Add more ice as needed to prevent a severe heat release. Once the gel has become neutral or acidic, continue washing with 0.2 M NaCl. Check for

completeness of washing and for success of EDA coupling using the TNBS test (see Section 4.1.3, Chapter 4).

The aminoethylated polyacrylamide support is now ready for further substitution, either with ligand or with a carboxylate-containing spacer arm.

3.1.1.5 Amino Acids as Spacers

Amino acids are often convenient choices for use in spacer arm applications because of their amine- and carboxylate-containing backbones (Sica *et al.*, 1973). Care should be taken in choosing the right amino acid, however, because some side-chain structures could interfere with affinity binding. Particularly, hydrophobic structures such as those on tyrosine, phenylalanine, and tryptophan should be avoided. Other side-chain groups may be advantageous. For instance, coupling a lysine spacer to a matrix through its carboxylic group provides two amine functionalities per molecule for further modification. Similarly, aspartate or glutamate provides two carboxylate groups if either of these compounds is first immobilized through their amino group.

Perhaps the most useful amino acids are relatively inexpensive and small: glycine and alanine. The β-alanine analog (actually, 3-aminopropionic acid) is especially suitable as a spacer. In our experience, β-alanine couples much more efficiently via reductive amination procedures than does glycine or alanine. Detrimental side-chain interactions are avoided with these molecules, since they consist of only a hydrogen or, at most, a methyl group. All three amino acids are appropriate for coupling to an amine-reactive matrix to provide carboxylate groups for further derivatization. This may be an important consideration in instances in which an amide bond between the matrix and ligand is desired. For instance, some aromatic amines couple with greater efficiency in an EDC-mediated reaction with a carboxylic group to form an amide linkage than they do if reacted directly to an activated support.

Polymers of amino acids have also been used for spacer applications. Poly-L-lysine and polymers of glutamic and aspartic acids have found application for adding multiple amines or carboxylic groups to a matrix (Villarejo and Zabin, 1973). These polymers will extend from the matrix surface in chains but will often be attached to the support at more than one point. This potential for multisite attachment is probably the reason for the frequent low yields of active group substitution using these molecules. In other words, a greater density of primary amines can be generated on the

3.1 Small Ligands

surface of a support by immobilizing lysine than poly-L-olyine. We have found that individual molecules as spacers are more successful than polymers when adding custom functionalities to a support.

Other amino acids such as cysteine and closely related compounds such as homocysteine can be used as spacer molecules to provide sulfhydryl functionalities on a matrix. These compounds can be used as the basis for forming a disulfide-reducing support or even a covalent chromatography matrix that will reversibly bind sulfhydryl-containing proteins.

Figure 3.7 shows the structures of the most useful amino acid spacers.

The following example illustrates the immobilization of the amino acid β-alanine to glycidol-modified Trisacryl GF-2000LS that has been periodate treated to generate reactive formyl groups (Fig. 3.8). See Chapters 1 and 2 for a description of the Trisacryl support and its activation and coupling procedures, respectively.

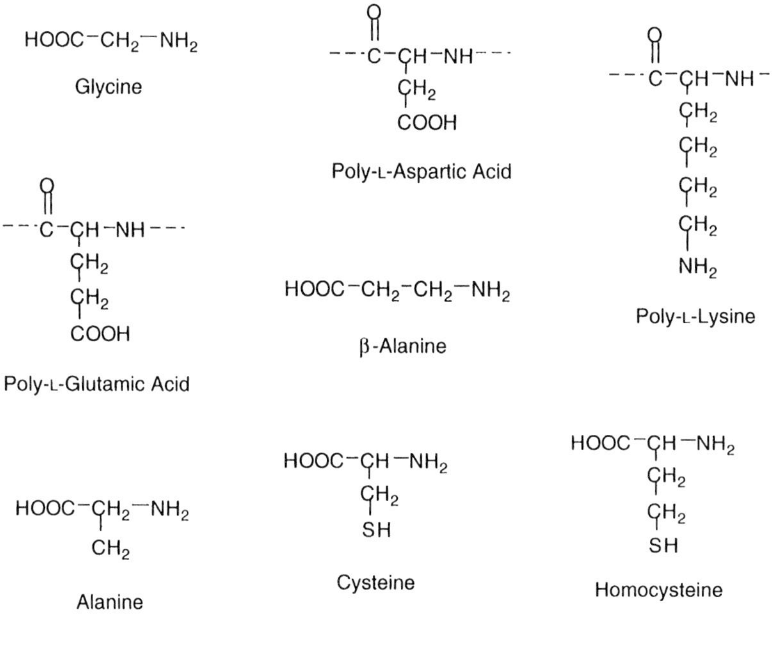

HOOC−CH$_2$−NH$_2$
Glycine

−−−C(=O)−CH−NH−−−
    |
    CH$_2$
    |
    COOH
Poly-L-Aspartic Acid

−−−C(=O)−CH−NH−−−
    |
    CH$_2$
    |
    CH$_2$
    |
    CH$_2$
    |
    CH$_2$
    |
    NH$_2$
Poly-L-Lysine

−−−C(=O)−CH−NH−−−
    |
    CH$_2$
    |
    CH$_2$
    |
    COOH
Poly-L-Glutamic Acid

HOOC−CH$_2$−CH$_2$−NH$_2$
β-Alanine

HOOC−CH−NH$_2$
    |
    CH$_2$
Alanine

HOOC−CH−NH$_2$
    |
    CH$_2$
    |
    SH
Cysteine

HOOC−CH−NH$_2$
    |
    CH$_2$
    |
    CH$_2$
    |
    SH
Homocysteine

FIGURE 3.7
Amino acids commonly used as spacer molecules.

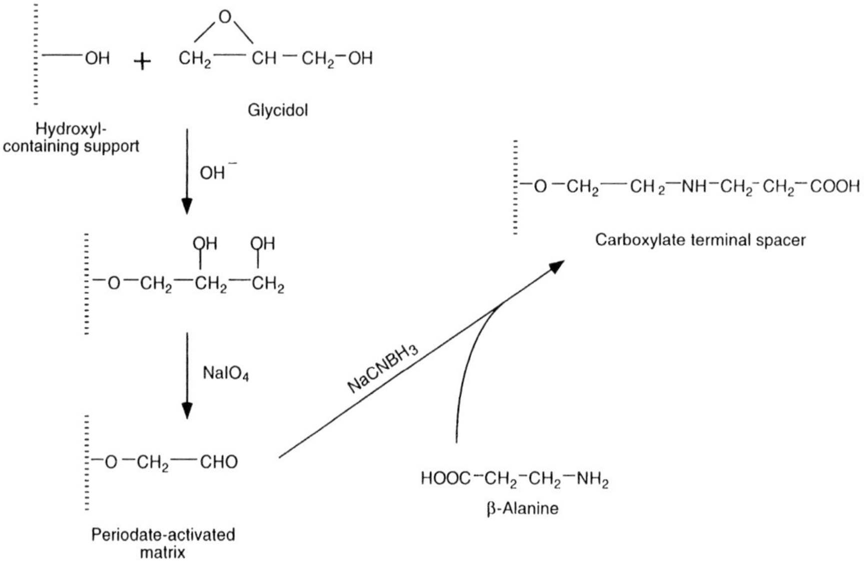

FIGURE 3.8
The epoxide compound glycidol can be used to introduce adjacent hydroxyls into a hydroxyl-containing matrix. Subsequent oxidation with sodium *meta*-periodate generates reactive formyl groups for ligand coupling via reductive amination with sodium cyanoborohydride. The spacer β-alanine can be immobilized with excellent yields using this procedure.

IMMOBILIZATION PROTOCOL

1. Wash 100 ml periodate-oxidized glycidol-treated Trisacryl GF-2000LS with water and then with coupling buffer (0.1 M sodium phosphate, pH 7.0).
2. Dissolve 1 gm β-alanine (Aldrich) in 100 ml coupling buffer, and readjust the pH to 7.0 if necessary.
3. Add the gel as a wet cake to the β-alanine solution, and stir the slurry with a paddle stirrer in a fume hood.
4. Add 1.2 gm sodium cyanoborohydride (**Caution:** Toxic) to the stirring gel slurry.
5. React for at least 4 hr at room temperature.
6. Wash the gel with coupling buffer, 1 M NaCl, and water to remove unreacted β-alanine and cyanoborohydride. Dispose of cyanide-containing washes according to EPA guidelines. The TNBS test may be used to check the washes for the presence of β-alanine and the completeness of the washing process (see Section 4.1.3, Chapter 4). TNBS will not give

a positive orange color with the matrix, however, since the spacer molecule terminates in a carboxyl group.

This support is now ready for coupling to an amine-containing ligand using, for example, the EDC procedure (chapter 2, Section 2.2.1.6).

3.1.1.6 Aminated Epoxides

The compounds useful in producing an epoxy-activated support, namely epichlorohydrin and 1,4-butanediol diglycidyl ether (or other similar bisoxiranes), inherently provide a spacer coming off the matrix because of their length. However, these epoxy compounds can also be used in producing an amine-terminating spacer if the oxirane groups are reacted and capped with ammonium ions. The result is a very hydrophilic spacer arm containing secondary hydroxyls from the opened epoxides, a relatively stable ether linkage to the matrix, and a terminal primary amine for further derivatization (Fig. 3.9).

IMMOBILIZATION PROTOCOL

The following procedure uses a hydroxyl-containing matrix, in this case Sepharose CL-6B, that has been activated with 1,4-butanediol diglycidyl ether. See Chapter 2 for the protocol on how to epoxy activate a support.

1. Wash 100 ml fresh epoxy-activated Sepharose with 100 ml 1 M NH$_4$OH and then suspend in 100 ml ammonia solution.
2. Transfer the gel suspension to a three-necked round-bottom flask and heat with a heating mantle to 40°C. Maintain mixing using a paddle stirrer and check temperature with a thermometer inserted into the slurry.
3. React for 3 hr at 40°C.
4. Cool the reaction mixture, then wash the aminated spacer gel with water, 1 M NaCl, and again with water. Test for the presence of amines on the matrix using the TNBS test (see Section 4.1.3, Chapter 4).

This gel is now useful for coupling ligands containing carboxylic groups (via EDC) or other compounds with amine-reactive chemistries.

3.1.1.7 Use of Anhydrides to Extend Spacer Length

Small cyclic anhydrides are often used to extend the length of a spacer arm by reacting with a terminal amino group on a matrix (Cuatrecasas, 1970). The most frequently used extenders are succinic anhydride and glutaric anhydride, containing a 4- and 5-carbon chain, respectively.

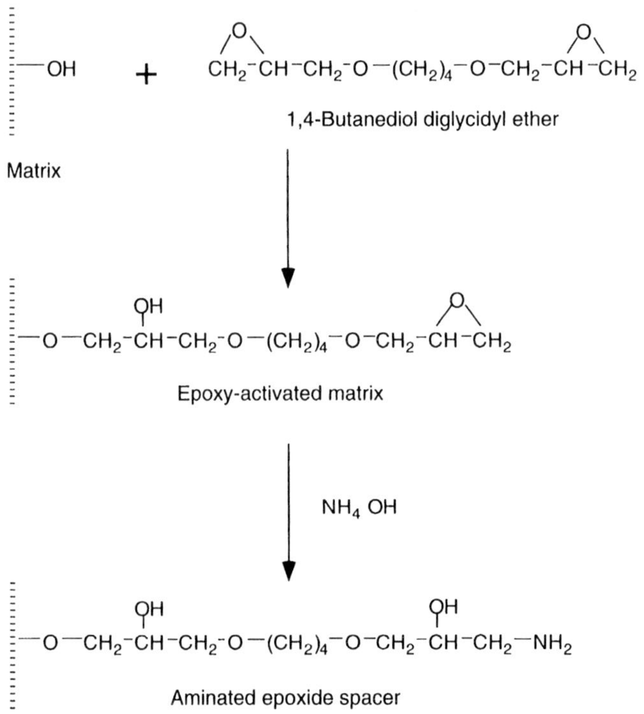

```
⋮—OH   +   CH₂-CH-CH₂-O—(CH₂)₄—O-CH₂-CH-CH₂
⋮                \O/                        \O/
```
1,4-Butanediol diglycidyl ether

Matrix

```
       OH
⋮—O—CH₂-CH-CH₂-O—(CH₂)₄—O-CH₂-CH-CH₂
                                  \O/
```
Epoxy-activated matrix

NH₄OH

```
       OH                        OH
⋮—O—CH₂-CH-CH₂-O—(CH₂)₄—O-CH₂-CH-CH₂—NH₂
```
Aminated epoxide spacer

FIGURE 3.9
Amine terminal spacers can be converted to carboxylate terminal spacers through the use of succinic or glutaric anhydride. The anhydrides react rapidly and usually completely block any amine terminals. When DADPA is used as the initial amine spacer, the carboxylate derivatives provide excellent spacer length and a hydrophilic environment.

Immobilized DADPA that has been succinylated with succinic anhydride provides a long 13-atom hydrophilic spacer that terminates in a carboxylic group. We have found that this extended spacer arm is especially useful for coupling small amine-containing ligands.

IMMOBILIZATION PROTOCOL

The following protocol makes use of immobilized DADPA on Sepharose CL-6B as the amine-terminating spacer gel that is then modified with an anhydride (Fig. 3.10). See Section 3.1.1.2 for the production of DADPA–/ /–agarose.

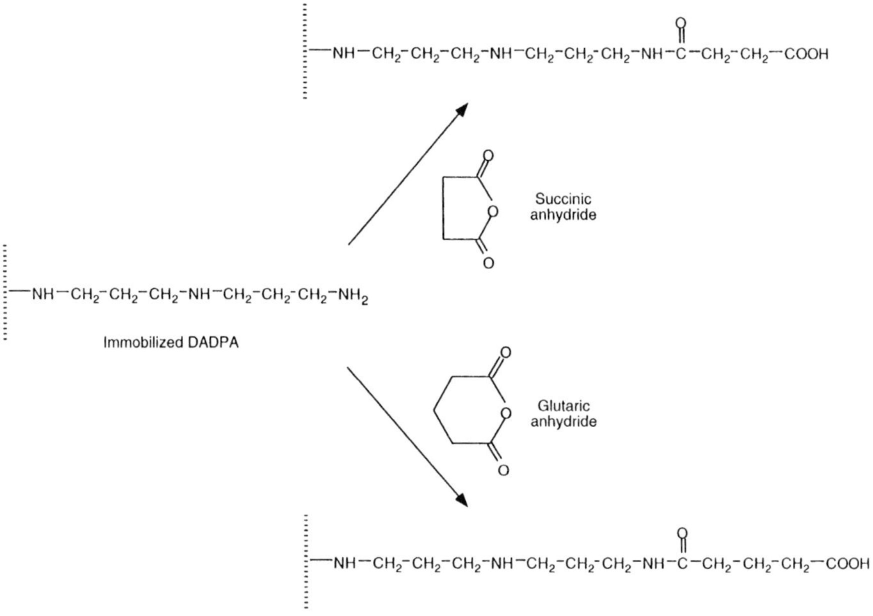

3.1 Small Ligands

—NH—CH$_2$-CH$_2$-CH$_2$-NH—CH$_2$-CH$_2$-CH$_2$-NH—C(=O)—CH$_2$-CH$_2$—COOH

Succinic anhydride

—NH—CH$_2$-CH$_2$-CH$_2$-NH—CH$_2$-CH$_2$-CH$_2$-NH$_2$

Immobilized DADPA

Glutaric anhydride

—NH—CH$_2$-CH$_2$-CH$_2$-NH—CH$_2$-CH$_2$-CH$_2$-NH—C(=O)—CH$_2$-CH$_2$-CH$_2$—COOH

FIGURE 3.10
Amine terminal spacers can be converted to carboxylate terminal spacers through the use of succinic or glutaric anhydride. The anhydrides react rapidly and usually completely block any amine terminals. When DADPA is used as the initial amine spacer, the carboxylate derivatives provide excellent spacer length and a hydrophilic environment.

---

1. Wash 100 ml DADPA–/ /–Sepharose CL-6B with water, and then suspend the gel in an equal volume of water.
2. Stir the gel with a paddle while slowly adding 10 gm succinic anhydride.
3. React for 1 hr at room temperature.

NOTE: Some protocols recommend keeping the pH at about 6.0 with NaOH during the course of reaction. We have found this to be unnecessary, since the succinylation reaction proceeds rapidly and with high yield without NaOH addition.

4. Wash the succinylated gel extensively with water, 1 M NaCl, and again with water to remove unreacted succinic acid. A negative test with TNBS (Section 4.1.3, Chapter 4) indicates that all the amines of DADPA were successfully blocked with succinic acid.

This gel is now ready for coupling to an amine-containing ligand. The protocol for modifying with glutaric anhydride is virtually identical.

3.1.1.8 Blocking or Capping Agents

Small molecules are often used to block or cap an activated support or spacer molecule. After ligand immobilization, these terminators couple and block residual active sites and essentially eliminate them from potentially irreversibly binding protein during an affinity operation. When using chemistries that are relatively stable in aqueous solutions, a blocking step after ligand immobilization may be crucial to preparing a support with no non-specific binding properties. This is particularly true when the ligand that is immobilized is large, for example, a protein. In this case, all the active groups on the matrix may not have been exposed to protein because of steric hinderance. Small molecules must enter afterward and completely block these untouched active groups.

An alternative to blocking is to leave the support in coupling buffer long enough to hydrolyze any remaining active sites. Nearly all activation methods will eventually (within 1–2 days) lose activity due to hydrolysis and become free of nonspecific coupling potential.

Used with spacer molecules, capping agents can remove residual charges from a matrix after ligand immobilization. Incomplete coupling of ligand to a carboxylic- or amine-terminating spacer can lend significant ion-exchange character to the resulting gel. In some cases, small ligands can be added in sufficient molar excess to insure nearly complete spacer use. In other situations, the desired ligand density may not be as great as the initial spacer density on the support. Capping the remaining spacer functionalities can thus avoid secondary ionic interactions.

One word of caution should be given for any capping operation. If the ligand contains the same functionality as the spacer (for instance, a carboxylate-terminating spacer and a ligand that has both an amine for coupling to the spacer and a carboxylic acid group), then a blocking step will also modify the ligand. Since any such ligand modification may change the resulting affinity interaction, care should be taken when deciding whether to block.

Some of the most common blocking or capping molecules are shown in Figure 3.11. Tris, ethanolamine, and glycine are used to block active sites on amine-reactive supports. Ethanolamine may also be used to cap excess carboxylate-terminal spacers, whereas acetic anhydride works well for amine-terminal spacer molecules. Cysteine, β-mercaptoethanol, and β-mercaptoethylamine may be used to block sulfhydryl reactive matrices.

Short small hydrophilic compounds are the best choices for blocking or capping reagents. Avoid molecules that contain secondary functionalities,

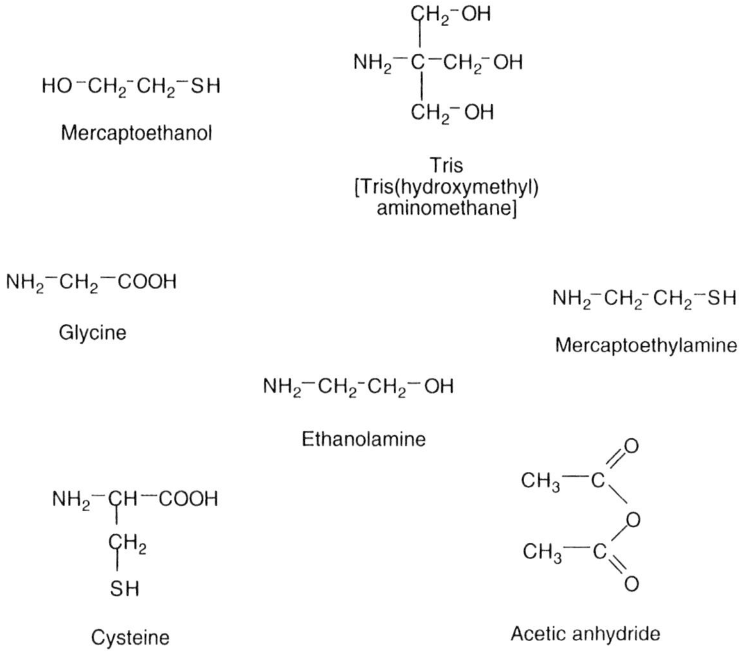

HO−CH$_2$−CH$_2$−SH

Mercaptoethanol

NH$_2$−C(CH$_2$−OH)(CH$_2$−OH)(CH$_2$−OH)

Tris
[Tris(hydroxymethyl)
aminomethane]

NH$_2$−CH$_2$−COOH

Glycine

NH$_2$−CH$_2$−CH$_2$−SH

Mercaptoethylamine

NH$_2$−CH$_2$−CH$_2$−OH

Ethanolamine

NH$_2$−CH(COOH)−CH$_2$−SH

Cysteine

(CH$_3$−C(=O)−O−C(=O)−CH$_3$)

Acetic anhydride

FIGURE 3.11

Common blocking or capping agents. Glycine, ethanolamine, and Tris are usually used to block remaining active groups on amine-reactive supports. Acetic anhydride is used as a blocking agent to eliminate any remaining amine groups on an amine terminal spacer. Cysteine, mercaptoethanol, and mercaptoethylamine are used to block remaining active groups on sulfhydryl-reactive matrices.

---

hydrophobic portions, or long hydrocarbon chains. A blocking agent that terminates in one or more hydroxyl groups is often the best choice.

BLOCKING PROTOCOL

The following two protocols give examples of how to block the active sites remaining on a gel after ligand immobilization and the procedure for capping excess spacer molecules. In the first protocol, an amine-reactive matrix is blocked with a buffered Tris solution after ligand immobilization. The second protocol uses acetic anhydride to cap excess amine-terminal spacer molecules after ligand coupling.

a. Use of Tris to Block Excess Amine-Reactive Sites

1. After immobilization of ligand to an amine-reactive support, wash the gel with water and then suspend it in an equal volume of 1 M Tris at a pH that is optimal for the coupling reaction. For most activation chemistries that react with amine ligands, this will be in the range of pH 7.5–8.5. Refer to Chapter 2 for the correct coupling pH value for the method you are using.
2. React for at least 2 hr at room temperature while stirring the suspension with a paddle.
3. Wash the blocked support with water, 1 M NaCl, and again with water.

b. Use of Acetic Anhydride to Cap Excess Amine-Terminal Spacers

Acetic anhydride is a flammable, volatile, and corrosive substance that should be handled in a fume hood. Avoid contact with skin or eyes, and do not breathe vapors.

1. After the immobilization of ligand on a support containing an amine-terminal spacer, wash off uncoupled ligand and suspend the gel in an equal volume of 0.2 M sodium acetate.
2. Add an aliquot of cold acetic anhydride equal to one-half the gel volume and react at 0°C for 30 min. This is followed by a second one-half volume addition of acetic anhydride for 30 min at room temperature.
3. Filter the gel of reactants, and then wash it sequentially with water, 0.1 N NaOH, and water. The gel at this stage gives a negative color test with TNBS, indicating that no free amino groups are present (Section 4.1.3, Chapter 4).

3.1.2 Sugars and Glycosaminoglycans

3.1.2.1 Preparation of Melibiose–/ /–Polyacrylamide by Reductive Amination

Immobilized melibiose is prepared on polyacrylamide beads through a multistep process incorporating the spacer molecule ethylenediamine. The method involves first creating an aminoethyl derivative by transamidination, and then reductively aminating the reducing end of melibiose onto the amine spacer of the gel (Fig. 3.12). Such an affinity support made from melibiose (Baues and Gray, 1977) has been used in the preparation of lectins (carbohydrate-binding proteins) that have a binding specificity for melibiose or one of its monomers, galactoside (e.g., *Bandieria simplicifolia* lectin; see Chapter 5, Section 5.1.4.4).

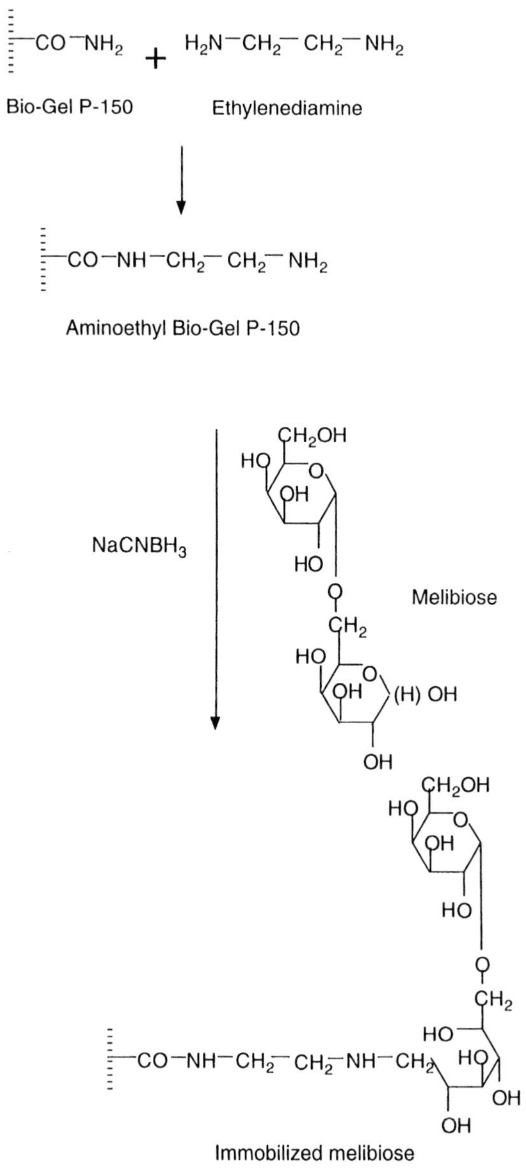

3.1 Small Ligands

FIGURE 3.12
Reaction sequence for the preparation of immobilized melibiose on an acrylamide matrix.

IMMOBILIZATION PROTOCOL

The first step involving the preparation of the spacer intermediate, aminoethyl BioGel P-200, is done as described in Section 3.1.1.4 (Inman and Dintzis, 1969).

1. Heat two liters of ethylenediamine (99%; Aldrich, Milwaukee, Wisconsin) to 90°C and stir with a paddle, while gradually adding 100 gm BioGel P-200 (100–200 mesh; BioRad Laboratories, Richmond, California).
2. Stir the reaction mixture for 5 hr at 90°C.
3. To the reaction mixture, add 1 L crushed ice, filter the gel, and wash with 1 M NaCl until the washings give a negative color with TNBS (Section 4.1.3, Chapter 4). Aminoethyl BioGel P-200 prepared in this fashion contains 40 μmol amine functionality per ml gel as determined by the ninhydrin color reaction (as described in Chapter 4).

Next, melibiose is coupled to Aminoethyl BioGel P-200 as described by Baues and Gray (1977), with some modifications.

1. Suspend aminoethyl BioGel P-200 (900 ml settled gel) in 0.2 M potassium phosphate, pH 9.0 (900 ml).
2. To the stirred gel suspension in a fume hood, add 27 gm melibiose (Pfanstiehl Laboratories, Waukegan, Illinois) and 13.5 gm sodium cyanoborohydride (Aldrich). (**Caution:** Cyanide fumes.)
3. Stir the reaction mixture for 15 min, seal in a flask with Parafilm, and leave at room temperature for 3 weeks. (This is a slow reaction because of the small percentage of time the sugar is in its formyl or open form.) Since the reduction step slowly will generate hydrogen bubbles, the flask should not be sealed with anything that will not allow gas volume expansion on the inside. Periodic mixing should be done over the course of the reaction.
4. Filter the reaction mixture, and wash the gel with water (4 L) and 0.2 M sodium acetate (4 L).
5. To acetylate the excess amino groups, suspend the gel in 1 L 0.2 M sodium acetate and treat with 500 ml acetic anhydride at 0°C for 30 min, followed by a second addition of 500 ml acetic anhydride for 30 min at room temperature.
6. Filter the gel and wash sequentially with water (4 L), 0.1 N NaOH (4 L), water (4 L), and 0.9% NaCl containing 0.02% sodium azide (4 L). The gel at this stage gives a negative color test with TNBS, indicating that no free amino groups are present (Section 4.1.3, Chapter 4).

Immobilized melibiose can be stored in 0.9% NaCl containing 0.02% sodium azide at 4°C. The amount of melibiose coupled to the gel can be determined by the phenol-sulfuric acid method (Dubois *et al.*, 1956). Typically, 35–40 μmol melibiose is coupled per ml gel.

NOTE: Lactose, maltose, cellibiose, and other carbohydrates with reducing ends also can be immobilized using this method.

3.1.2.2 Preparation of Lactose–/ /–Sepharose 4B by the Bisoxirane Method

Lactose may be immobilized through its hydroxyl groups using an epoxide coupling procedure. Activation of agarose (Sepharose 4B) with the bisoxirane 1,4-butanediol diglycidyl ether gives a support able to react with carbohydrate hydroxyls (Fig. 3.13). This method is also appropriate for immobilizing many other sugars and polysaccharides, especially those that have no other convenient functionalities useful in coupling (Uy and Wold, 1977). Immobilized lactose has been used to purify lactose-binding lectins (carbohydrate-binding proteins) or proteins that bind its monomers, galactose, or glucose.

IMMOBILIZATION PROTOCOL

The following protocol uses a bisoxirane in the activation step. All operations involved with activating the gel should be done in a fume hood to prevent exposure to this toxic compound.

1. Activate Sepharose 4B (100 ml settled gel) with 1,4-butanediol diglycidyl ether as described in Chapter 2 and wash to remove excess bisoxirane.
2. Wash the epoxy-activated Sepharose 4B with 1 L water, suction dry to a moist cake, and add to a lactose-containing solution (15 gm lactose dissolved in 100 ml 0.1 N NaOH).
3. Place the reaction mixture in a round-bottom flask in a fume hood and heat with stirring to 40°C.
4. Stir the reaction at 40°C for 24 hr.
5. Wash the gel successively with 2 L each of water, 0.1 M sodium borate buffer, pH 8.0, and water. Lactose–/ /–Sepharose 4B can be stored in 0.02% sodium azide at 4°C.

3.1.2.3 Preparation of *N*-Acetylglucosamine–/ /–Sepharose 4B by the Bisoxirane Method

N-Acetylglucosamine is a monosaccharide made of a glucosyl backbone on which the 2-carbon hydroxyl group has been replaced by an acetylated

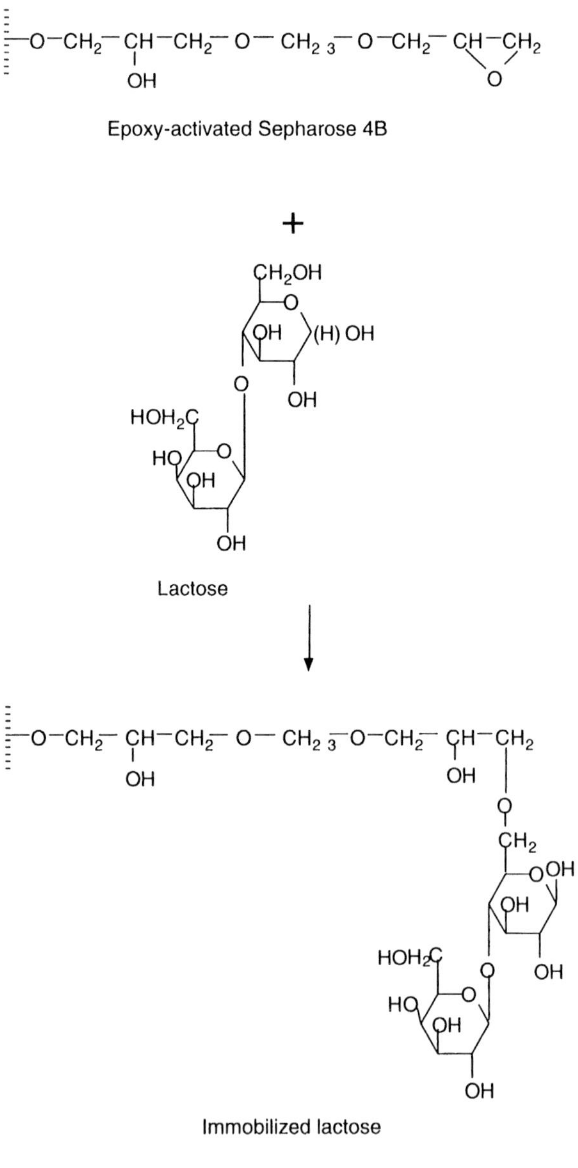

FIGURE 3.13

Reaction sequence for the preparation of immobilized lactose on agarose.

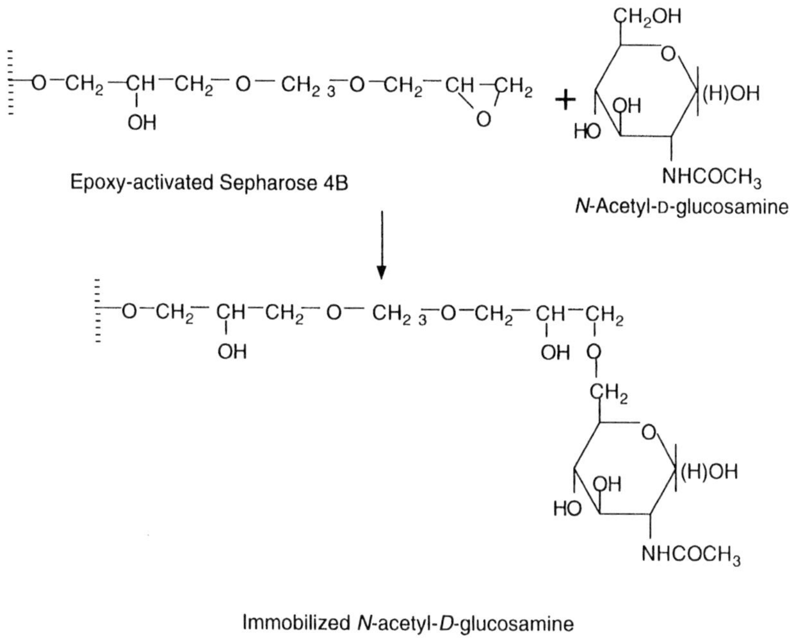

FIGURE 3.14
Reaction sequence for the preparation of immobilized N-acetyl-D-glucosamine on agarose.

amino group. It is one of two sugars found in the cell wall polysaccharides of bacterial cells.

A matrix that has been activated by the bisoxirane method will couple to this sugar through its hydroxyl groups (Fig. 3.14). An affinity support prepared in this manner will bind lectins with specific binding sites able to interact with this sugar. Wheat germ agglutinin is an example of a protein that will bind terminal N-acetylglucosamine residues (Vretblad, 1976).

IMMOBILIZATION PROTOCOL

The following protocol makes use of the bisoxirane activation method. All operations using this compound should be done in a fume hood.

1. Activate Sepharose 4B (100 ml settled gel) with 1,4-butanediol diglycidyl ether as described in Chapter 2 and wash free of uncoupled bisoxirane.

2. Wash the epoxy-activated Sepharose 4B with 1 L water, suction dry to a moist cake, and add to an N-acetylglucosamine solution (5 gm N-acetylglucosamine dissolved in 100 ml 0.1 N NaOH).
3. Place the reaction mixture in a round-bottom flask and heat with stirring in a fume hood to 40°C.
4. Stir the reaction at 40°C for 24 hr.
5. Wash the gel successively with 1 L each of water, 0.1 M sodium borate buffer, pH 8.0, and water. N-acetylglucosamine–/ /–Sepharose 4B can be stored in 0.02% sodium azide at 4°C.

NOTE: N-Acetylgalactosamine–/ /–Sepharose 4B also can be prepared by this method.

3.1.2.4 Preparation of D-Mannose–/ /–Sepharose 4B by the Divinylsulfone Activation Method

D-Mannose is a monosaccharide that may be immobilized by a number of methods. For instance, both the reductive amination and bisoxirane procedures described for N-acetylglucosamine and melibiose will work for this sugar.

In this protocol, another powerful way of immobilizing hydroxyl-containing compounds is presented (Fornstedt and Porath, 1975). Agarose is first activated with the bifunctional reagent divinylsulfone (DVS) and then coupled under basic conditions to the hydroxyls of D-mannose (Fig. 3.15).

Depending on the way a sugar or polysaccharide is immobilized, it may have more or less affinity for the protein that normally binds it. Sometimes several activation and coupling methods need to be tried before the immobilized sugar is fixed in the correct orientation to interact with the binding site on the lectin. Reductive amination, epoxy, and DVS activation provide three excellent alternatives for obtaining an optimal affinity support.

IMMOBILIZATION PROTOCOL

The following method uses DVS to activate an agarose gel. Since this compound is highly toxic, all operations involved in activation and coupling should be done in a fume hood.

1. Activate Sepharose 4B (100 ml settled gel) with DVS as described in Chapter 2.
2. Wash the DVS-activated Sepharose 4B with 1 L water, suction dry to a moist cake, and add to a D-mannose solution (20% solution of D-mannose in 0.5 M sodium carbonate).

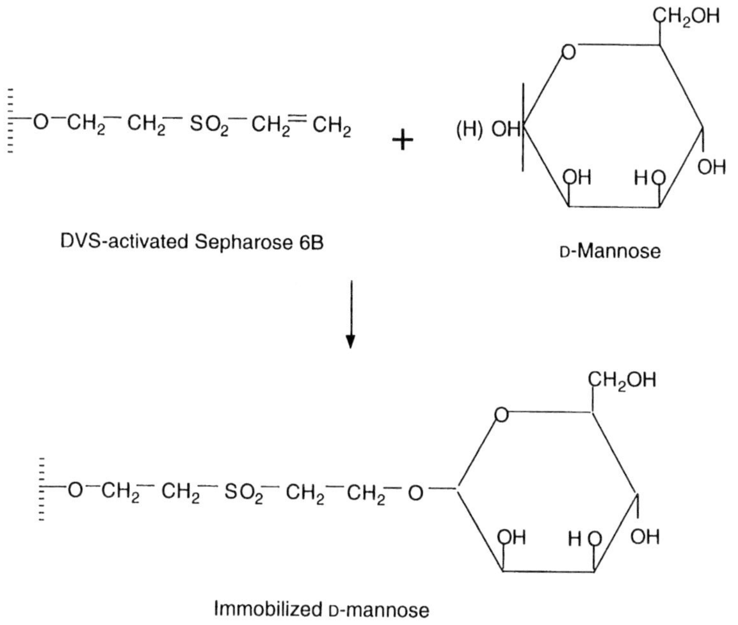

3.1 Small Ligands

FIGURE 3.15
Reaction sequence for the preparation of immobilized D-mannose using DVS-activated agarose.

---

3. Stir the reaction mixture at room temperature for 24 hr.
4. Filter and wash the gel successively with 2 L each of water and 0.5 M sodium bicarbonate.
5. In a fume hood, suspend the gel in 100 ml 0.5 M sodium bicarbonate containing 5 ml 2-mercaptoethanol.
6. Mix at room temperature for 2 hr to block the excess vinyl reactive groups.
7. Finally, wash the gel with 2 L water. D-Mannose–/ /–Sepharose 4B can be stored in 0.02% sodium azide at 4°C.

3.1.2.5 Preparation of Heparin–/ /–Sepharose 4B

Heparin is a highly sulfated mucopolysaccharide that is composed of 1,4-linked D-glucuronic acid and D-glucosamine residues. The molecular weight of the polymer varies from 6000 to 20,000, depending on the method of preparation and the source. Different heparin samples will have

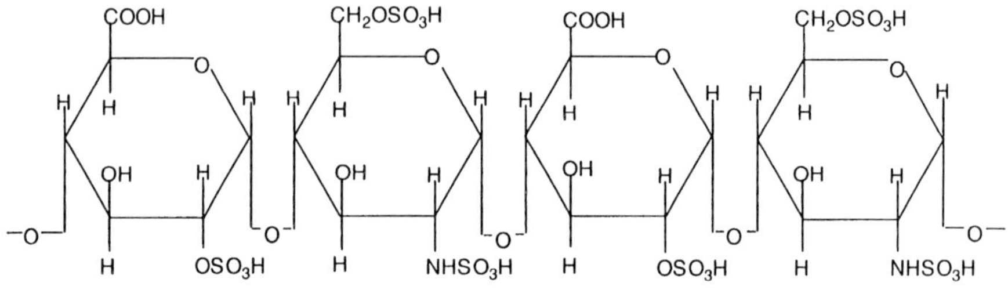

FIGURE 3.16
Partial chemical structure of heparin.

varying levels of N-sulfation within hexosamine residues, and varying extents of O-sulfation at carbon 2 of hexuronic acid and carbon 3 of hexosamine (Fig. 3.16).

Immobilized heparin has been used to purify a number of proteins, particularly certain coagulation factors, anti-heparin binding proteins, basement membrane proteins, enzymes, and lectins (for review see Farooqui, 1980). Heparin is a powerful general purpose affinity support that allows the purification of many proteins by differential elution using a salt gradient.

IMMOBILIZATION PROTOCOL

The following method uses CNBr to activate the heparin polymer for coupling to a matrix. The covalent linkages probably result from cross-links between the heparin polymer and the hydroxyls on the support. The free (unsulfated) amines on the glucosamine residues of heparin most likely also participate in the coupling. Because of cyanide fumes, all operations should be done in a well-ventilated hood. This protocol describes the immobilization of heparin on an agarose support, but the same general method is suitable for other matrices.

1. Add heparin solution (1.7 gm; 120 units/mg) to 400 ml ice-cold water in a 1-L beaker. Keep the beaker with the heparin solution in an ice bath and stir with a paddle in a fume hood.
2. Add CNBr (6 gm) to the heparin solution. When all the CNBr has dissolved, add Sepharose 4B (100 ml settled gel) that has been washed previously with 1 L water and suction dried to a moist cake.

3. Maintain the pH of the reaction mixture at 11.0 for 25 min by adding, drop by drop, 6 N NaOH.
4. Stir the reaction mixture at room temperature overnight.
5. Wash the gel with 2 L water, suspend in 100 ml 1.0 M ethanolamine, pH 9.0, and stir for 1 hr at room temperature to block the excess reactive groups.
6. Finally, wash the gel successively with 2 L each of water, 0.1 M sodium acetate buffer, pH 4.7, 0.5 M sodium bicarbonate, and water. Heparin–/ / –Sepharose 4B can be stored in 0.02% sodium azide at 4°C.

NOTE: Dextran sulfate–/ /–Sepharose 4B also can be prepared according to this procedure.

3.1.3 Inhibitors and Various Biospecific Binders

3.1.3.1 Preparation of Immobilized *p*-Aminobenzamidine

The ligand *p*-aminobenzamidine (*p*-AB) has been used extensively for the purification of urokinases, serine proteases, plasminogen activators, and other similar enzymes that contain a benzamidine (or arginine) binding site (Mares-Gula and Shaw, 1965; Markwardt *et al.*, 1968; Hixson and Nishikawa, 1973; Jany *et al.*, 1976; Holleman and Weiss, 1978; Honda *et al.*, 1986; Ito *et al.*, 1987; Steil *et al.*, 1989; Male *et al.*, 1990). Enzymes usually will bind tightly at physiological pH and elute under mild acidic conditions or by using benzamidine as a counter ligand.

Immobilized *p*-AB is best prepared by using a spacer arm to extend the ligand some distance from the matrix. This allows approaching enzymes to wrap around the protruding ligand, reaching binding sites that may be somewhat buried below the surface of the protein. We have found that two spacers work particularly well for this ligand: 6-AC and succinylated DADPA. Both spacers and the resulting affinity gels have their own unique properties for certain purification applications.

In the first protocol given here, *p*-AB is coupled to an agarose gel containing the spacer arm 6-AC. The terminal carboxylate group of the spacer may be coupled to the amine of *p*-AB by using the carbodiimide (EDC) conjugation procedure. This support works very well for the majority of purification problems. However, the slight hydrophobic nature of the 6-AC spacer will cause some problems in applications in which a high degree of purity (with no contamination from other proteins) is desired. Figure 3.17 shows the structures of these immobilized preparations. See Section 3.1.1 for details on how to prepare the intermediate spacer arm supports.

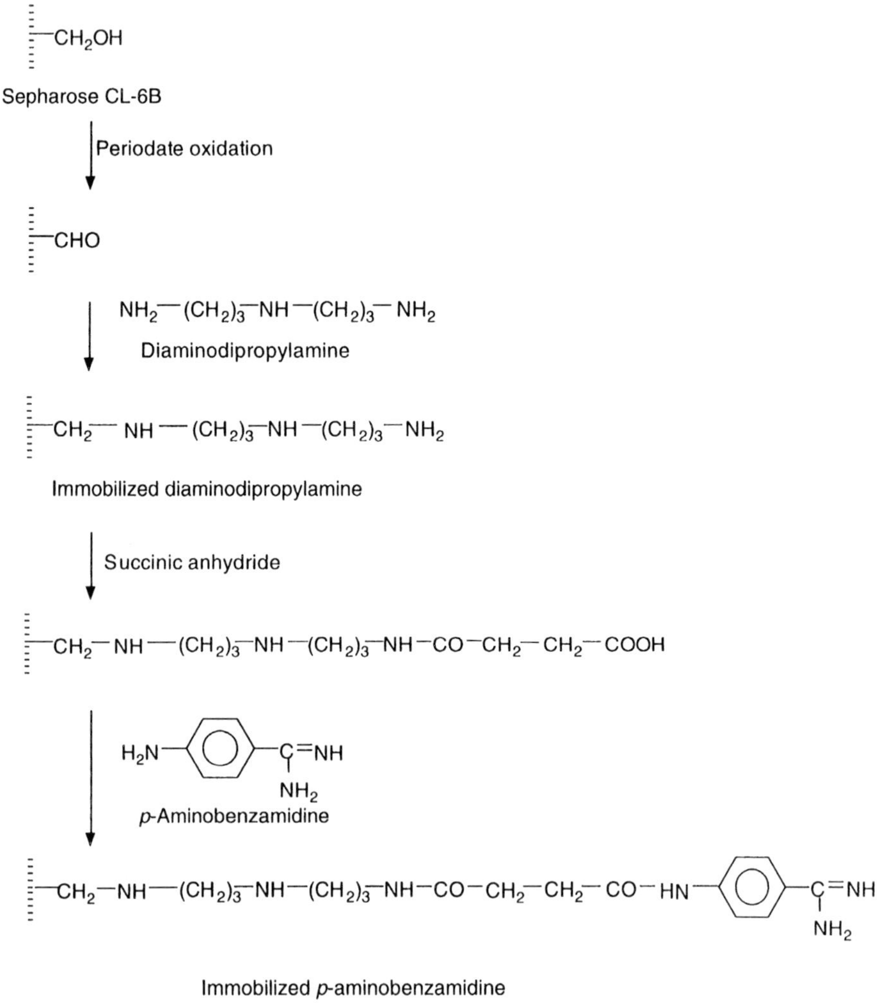

168   3. Immobilization of Ligands

```
┊─CH₂OH
┊
Sepharose CL-6B
        │ Periodate oxidation
        ▼
┊─CHO
┊
        │   NH₂─(CH₂)₃─NH─(CH₂)₃─NH₂
        │       Diaminodipropylamine
        ▼
┊─CH₂─NH─(CH₂)₃─NH─(CH₂)₃─NH₂
┊
    Immobilized diaminodipropylamine
        │ Succinic anhydride
        ▼
┊─CH₂─NH─(CH₂)₃─NH─(CH₂)₃─NH─CO─CH₂─CH₂─COOH
┊
        │   H₂N─⟨C₆H₄⟩─C(=NH)NH₂
        │       p-Aminobenzamidine
        ▼
┊─CH₂─NH─(CH₂)₃─NH─(CH₂)₃─NH─CO─CH₂─CH₂─CO─HN─⟨C₆H₄⟩─C(=NH)NH₂
┊
    Immobilized p-aminobenzamidine
```

FIGURE 3.17
Reaction sequence for the preparation of immobilized *para*-aminobenzamidine.

IMMOBILIZATION PROTOCOL

Immobilized DADPA is prepared as described in Section 3.1.1.1, immobilized 6-AC as in Section 3.1.1.3, and succinylated DADPA as in Section 3.1.1.8.

3.1 Small Ligands

a. Using Succinylated DADPA as Spacer

1. Wash immobilized succinylated DADPA (100 ml settled gel) with 1 L water, suction dry to a moist cake, and transfer to a 500-ml beaker.
2. Suspend the gel in 100 ml 0.1 M MES buffer, pH 4.7, and stir.
3. Add 1 gm p-aminobenzamidine·HCl (Aldrich, Milwaukee, Wisconsin) and 3 gm EDC (Pierce, Rockford, Illinois) to the gel suspension, and maintain the pH of the reaction mixture at 4.7 for 1 hr by adding either 1 N NaOH or 1 N HCl. In MES buffer, the pH drift will be minimal.
4. Stir the reaction mixture overnight at room temperature.
5. Finally, wash the gel successively with 2 L each of water, 0.1 M sodium acetate, pH 4.7, 0.5 M sodium bicarbonate, and water.

Immobilized p-AB can be stored in 0.02% sodium azide at 4°C.

b. Using 6-AC as Spacer

1. Wash immobilized 6-AC with water and then with 0.1 M MES, pH 4.7.
2. Couple p-aminobenzamidine to this gel using the EDC protocol described earlier for the succinylated DADPA spacer.

3.1.3.2 Preparation of Immobilized D-Biotin

Biotin, or vitamin H (MW 244.31), is a small naturally occurring cofactor that is present in every living cell in very minute amounts (usually less than 0.0001%). The biotin molecule normally exists bound to proteins (such as pyruvate carboxylase) through its valeric acid carboxylic group by an amide bond to lysine side-chain amines. *In vivo*, it is a carrier of activated CO_2 through the N-1 nitrogen atom of the biotin ring.

One of the most useful interactions in immunochemistry involves the specific binding of biotin to the egg white protein, avidin, or the similar bacterial protein, streptavidin. Avidin is a tetramer containing four identical subunits of molecular weight 15,000. Each subunit contains one high-affinity binding site for biotin with a dissociation constant of approximately 10^{-15} M. *Streptavidin* has an almost identical tetrameric structure and binding affinity for biotin. The binding is undisturbed by extremes of pH, buffer salts, or even chaotropic agents such as guanidine hydrochloride (up to 3 M). The strength of the avidin–biotin interaction has provided researchers with a unique tool for immunoassays, receptor studies, immunocytochemical staining, and protein isolation. (For a book on this subject, see Wilchek and Bayer, 1990.)

Immobilized biotin may be used to recover or purify biotinylated protein–avidin complexes from solution. It may also be used in the purification of avidin from egg whites or streptavidin from bacterial fermentations. The only disadvantage in using this ligand for affinity isolations is the severe elution conditions necessary to release the bound avidin or streptavidin. Elution will occur with 6 M guanidine hydrochloride, pH 1.5, an environment sometimes too extreme for recovery of many proteins with high activity. The biotin analog 2-iminobiotin is often a good choice because of the milder elution conditions required to break its interaction with avidin.

Immobilized D-biotin is prepared by coupling its valeric acid side chain to an amine-containing spacer arm. In the protocol given here, an immobilized DADPA–/ /–Sepharose gel is used as the intermediate support. The water-soluble carbodiimide reaction using EDC is then performed to couple biotin to the terminal amines on the matrix (Fig. 3.18).

IMMOBILIZATION PROTOCOL

1. Wash 100 ml immobilized DADPA on Sepharose CL-6B (prepared by the periodate oxidation/reductive amination method) with 1 L water and suction dry to a moist cake.
2. Prepare a D-biotin solution by dissolving 120 mg D-biotin in 100 ml MES buffer, pH 5.8.
3. Add the moist gel cake to the biotin solution in a beaker. While stirring with a paddle, add EDC (1 gm) to the gel suspension and maintain the pH of the reaction mixture at 5.8 for 1 hr by adding either 1 N NaOH or 1 N HCl drop by drop.
4. Stir the reaction mixture at room temperature overnight.
5. Finally, wash the gel successively with 2 L each of water, 0.1 M sodium acetate, pH 4.7, 0.5 M sodium bicarbonate, and water. Immobilized D-biotin can be stored in 0.02% sodium azide at 4°C.

3.1.3.3 Preparation of Immobilized 2-Iminobiotin

2-Iminobiotin is an analog derivative of biotin in which the cyclic ureido group is replaced by a cyclic guanidino group (Fig. 3.19). The compound may be synthesized from biotin using a two-step process involving alkaline hydrolysis followed by reaction with cyanogen bromide (Hofmann *et al.*, 1941; Hofmann and Axelrod, 1950). The guanidino ring is protonated and positively charged at pH values below about 11 (pK_a 11–12) and uncharged at or above this value. In its uncharged state, the compound is able to interact with avidin or streptavidin with a binding constant nearly as strong as D-biotin (KD 3.5×10^{-11} M). However, when protonated, 2-iminobiotin loses much of its ability to bind to these proteins, its dissociation constant falls to less than 10^{-3} M.

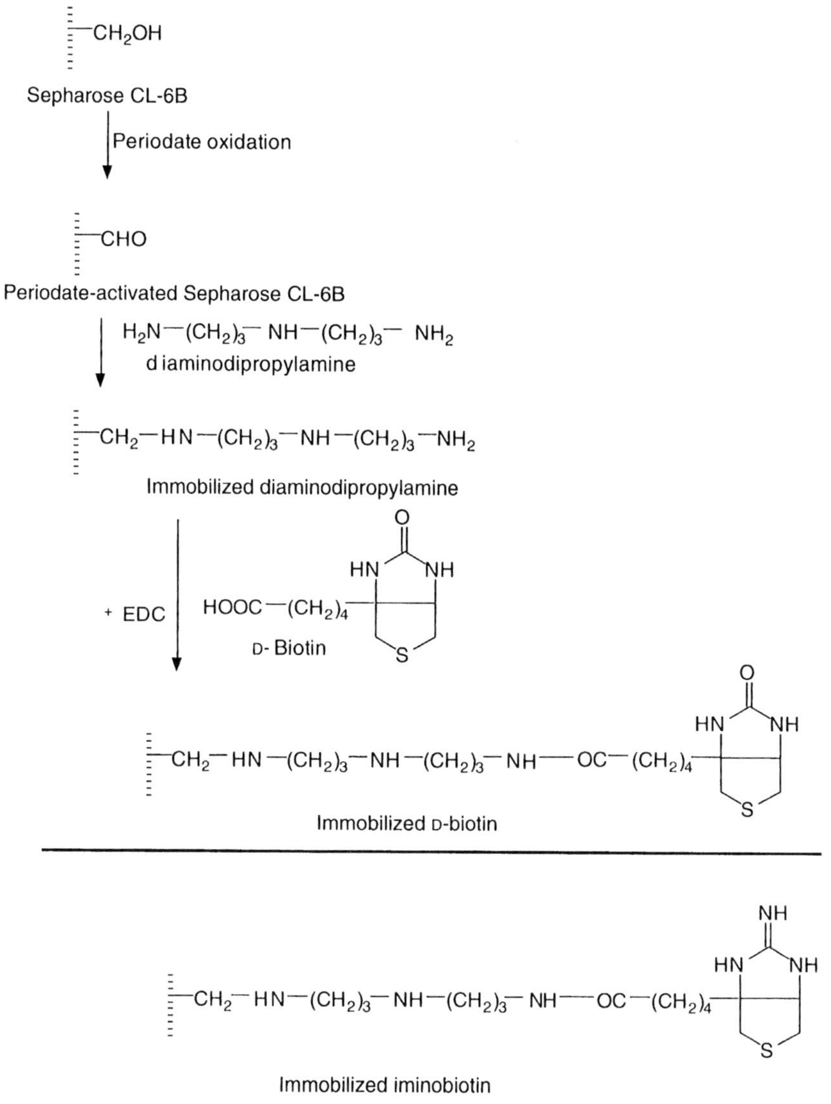

3.1 Small Ligands

FIGURE 3.18
Reaction sequence for the preparation of immobilized D-biotin and immobilized iminobiotin.

Therefore, an affinity column made from 2-iminobiotin will bind avidin or streptavidin tightly at pH 11 and fully release the proteins at pH 4.0. This mild elution step, compared with that required for D-biotin affinity sup-

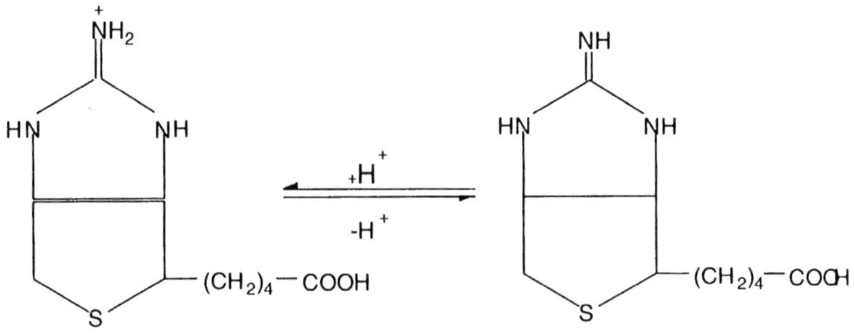

FIGURE 3.19
The chemical structure of iminobiotin.

ports, makes immobilized 2-iminobiotin ideal for purifying either avidin or streptavidin with excellent retention of activity.

IMMOBILIZATION PROTOCOL

The method for the preparation of immobilized iminobiotin is essentially same as the method just given for the preparation of immobilized D-biotin, except that iminobiotin is used instead of D-biotin.

3.1.3.4 Preparation of Immobilized Pepstatin A

Carboxyl proteases, also called acid proteases because they are active at acid pH, can be inhibited by extremely low concentrations (10^{-10} M) of the hexapeptide transition-state analog, pepstatin A. Pepstatin A (Fig. 3.20) contains two statine residues (4-amino-3-hydroxy-6-methylheptanoic acid). The carboxylate groups of these residues are convenient handles for immobilization to an amine-containing support. The resultant affinity support has been shown to be useful for the purification of such acid proteases as pepsin, rennin, cathepsin D, and various milk clotting enzymes (Dzau *et al.*, 1979; Afting and Becker, 1981; Kabayashi *et al.*, 1982; Helseth and Veis, 1984).

IMMOBILIZATION PROTOCOL

1. Wash immobilized DADPA (100 ml settled gel; prepared as described in Section 3.1.1.1) with 1 L each of water and 50% ethanol, suction dry to a moist cake, and transfer to a 2-L beaker.

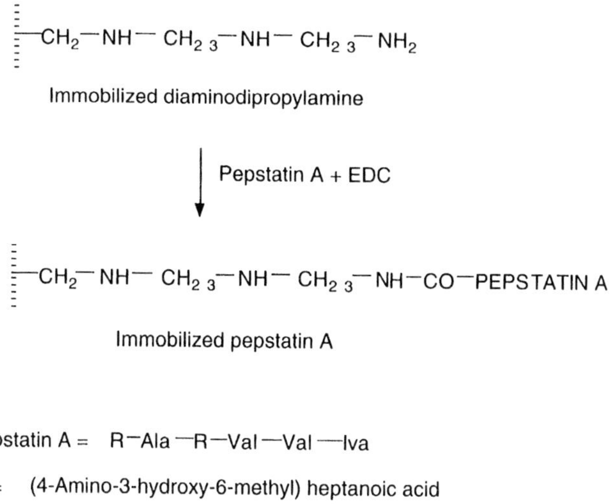

3.1 Small Ligands

```
┊―CH₂―NH―CH₂ ₃―NH―CH₂ ₃―NH₂
┊
    Immobilized diaminodipropylamine

              │ Pepstatin A + EDC
              ▼

┊―CH₂―NH―CH₂ ₃―NH―CH₂ ₃―NH―CO―PEPSTATIN A
┊
    Immobilized pepstatin A
```

Pepstatin A = R―Ala―R―Val―Val―Iva

R = (4-Amino-3-hydroxy-6-methyl) heptanoic acid

FIGURE 3.20
Reaction sequence for the preparation of immobilized pepstatin A.

2. Place the beaker in a water bath that is maintained at 37°C, suspend the gel in 100 ml 50% ethanol, and stir with a paddle.
3. Add a pepstatin solution [100 mg pepstatin A (Sigma, St. Louis, Missouri) dissolved in 300 ml 50% ethanol)] to the gel suspension while stirring, followed by 3.0 gm EDC.
4. Maintain the pH of the reaction mixture at 5.5 for 1 hr by adding, drop by drop, 2 N HCl.
5. Stir the reaction mixture at 37°C overnight.
6. Finally, wash the gel successively with 1 L each of 50% ethanol, 25% ethanol, 20% acetone, water, 1 M NaCl, and 0.5 M NaCl containing 0.02% sodium azide. Immobilized pepstatin A can be stored in 0.5 M NaCl containing 0.02% sodium azide at 4°C.

3.1.4 Immobilized Dyes

The use of dye molecules as affinity ligands introduced a pseudo-affinity chromatography in which the structure of the dye molecule resembles a biospecific ligand. The dye is able to bind to the active site of a protein or enzyme by virtue of its similarity to a naturally occurring substrate. Several reactive dyes with application in the area of textile dyeing have been successfully reported for use in affinity chromatography applications.

Two major classes of dye molecules useful for affinity techniques belong to the cibacron and procion families. These dyes are characterized by reactive triazine rings with one or two replaceable chlorine atoms. This active triazine ring provides a convenient chemistry for immobilization to hydroxyl-containing supports, forming an ether bond between the dye and the matrix. The dye portion of the molecules usually consists of anthraquinone or naphthalene derivatives, often containing one or more sites of substitution with amines and/or sulfate groups. Many of the dyes also contain azo linkages between aromatic components of their structures.

It has been our experience that many dye molecules immobilized through triazine ether linkages slowly leach from the support. Prior to each use of these affinity matrices, extensively wash the gel with water and binding buffer until the washings are colorless.

3.1.4.1 Immobilized Cibacron Blue F3GA

Cibacron Blue F3GA (or the closely related Procion Blue MX-3G and MX-R) is the most popular dye used in affinity techniques. This dye contains an anthraquinone portion that successfully mimics the structure of certain enzyme substrates (Fig. 3.21). Cibacron Blue can bind enzymes that require adenylic cofactors, such as NAD^+, $NADP^+$, and even ATP. For this reason, this dye is an excellent general purpose ligand for the purification of many dehydrogenases and kinases. However, it also has utility in the isolation of many other types of enzymes that have no adenylic binding domains. Some of these are transferases, polymerases, nucleases, hydrolases, synthetases, and CoA-dependent enzymes. (For a book on reactive dyes in protein and enzyme technology, see Clonis *et al.,* 1987.)

Cibacron Blue has found great utility in the fractionation of blood proteins. Probably the most well-known application is its interaction with serum albumins but, with carefully designed elution conditions, it can also bind many minor components of plasma. Some of the proteins successfully purified using immobilized Cibacron Blue include various lipoproteins, IgG, α-fetoprotein, α-2-macroglobulin, the blood clotting factors II, IX, and X, a number of complement proteins, α1-proteinase inhibitor, antithrombin III, α1-antitrypsin, and α1-antichymotrypsin (Clonis *et al.,* 1987). Truly, Cibacron Blue, more so than any other immobilized ligand, has become almost a general purpose protein purification tool.

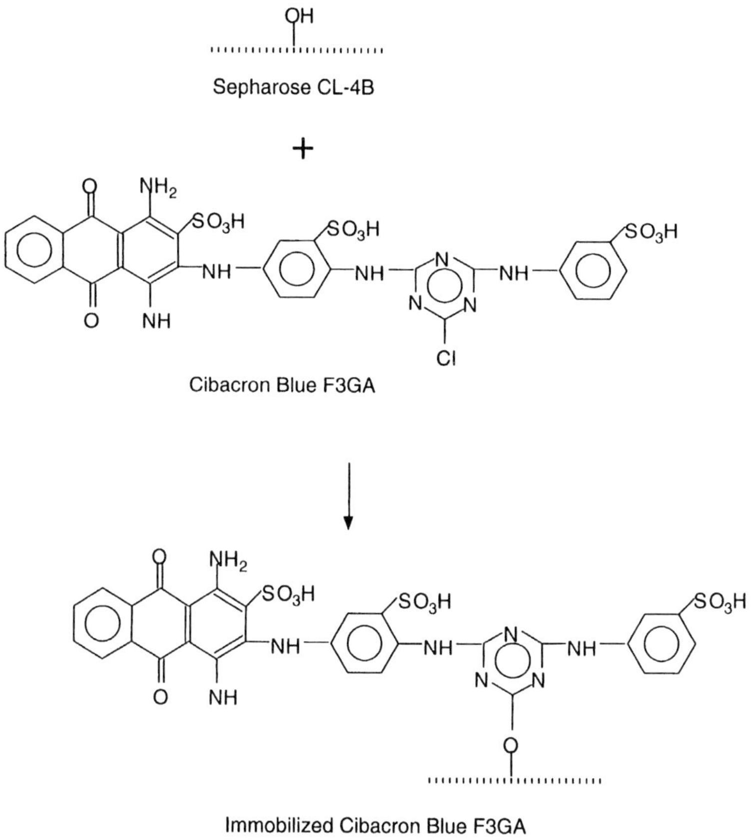

3.1 Small Ligands

Sepharose CL-4B (—OH)

+

Cibacron Blue F3GA (structure with anthraquinone-NH₂, SO₃H, NH linkages, triazine ring with Cl, and terminal benzene-SO₃H)

↓

Immobilized Cibacron Blue F3GA (same structure coupled via O to Sepharose)

FIGURE 3.21
Reaction sequence for the preparation of immobilized pepstatin A.

IMMOBILIZATION PROTOCOL

The following protocol is useful for coupling Cibacron Blue F3GA or Cibacron Brilliant Blue FBR-P to a hydroxyl-containing matrix. The brilliant blue variety is a purer dye because of sulfonation of the benzene ring only in the *meta* position.

Procion Brilliant Blue MX-R may also be immobilized using a similar procedure, except the dye is reacted at room temperature in sodium bicarbonate buffer, pH 8.6, for at least 24 hr. The dichlorotriazine ring of the

Procion dye reacts at a milder pH and temperature than the chlorotriazine of Cibacron Blue.

1. Wash Sepharose CL-6B (100 ml settled gel) with 2 L water, suction dry to a moist cake, and transfer to a 1-L three-necked round-bottom flask provided with a paddle stirrer.
2. Suspend the gel in 100 ml water and, while stirring, heat to 60°C using a heating mantle.
3. Slowly add Cibacron Blue F3GA solution (1 gm Cibacron Blue F3GA dissolved in 30 ml water) to the gel suspension and stir for 30 min 60°C.
4. Add 15 gm NaCl to the reaction mixture and stir for 1 hr at 60°C.
5. Increase the temperature of the reaction mixture to 80°C and add sodium carbonate (1.5 gm) to the gel suspension.
6. Continue the reaction for 2 hr at 80°C.
7. Cool the reaction mixture to room temperature and filter. Extensively wash the gel with water until the washings are colorless. Warm water often speeds this washing process. Finally, wash the gel with 2 L each of 1.0 M NaCl and water. Immobilized Cibacron Blue F3GA can be stored in 0.02% sodium azide at 4°C.

3.1.4.2 Preparation of Immobilized Procion Red HE-3B

Immobilized Procion Red HE-3B (Fig. 3.22) is a dye with group-specific ligand properties similar to those of Cibacron Blue, but with slightly different specificity. Like Cibacron Blue, it can be used as a general purpose purification tool for a wide range of enzymes and proteins. Procion Red, however, has higher binding affinity to $NADP^+$ dependent enzymes, whereas Cibacron Blue has greater affinity for NAD^+-dependent enzymes (Watson *et al.*, 1978; Clonis *et al.*, 1987). Thus, a combination of these dyes may be useful in a purification scheme.

Immobilized Procion Red HE3B is prepared essentially by the same method just described for Cibacron Blue F3GA.

3.1.4.3 Phenol Red and Thymol Blue

The procion and cibacron dyes just mentioned usually have reactive chlorotriazine rings that can be used for immobilization purposes. Phenol red and thymol blue, however, are examples of molecules that have no convenient "handles" for coupling to a matrix. Neither of these compounds has functionalities that can be reacted with the typical activation chemistries, other than the somewhat unreliable and messy diazonium method. In these situations, the Mannich reaction has primary advantages.

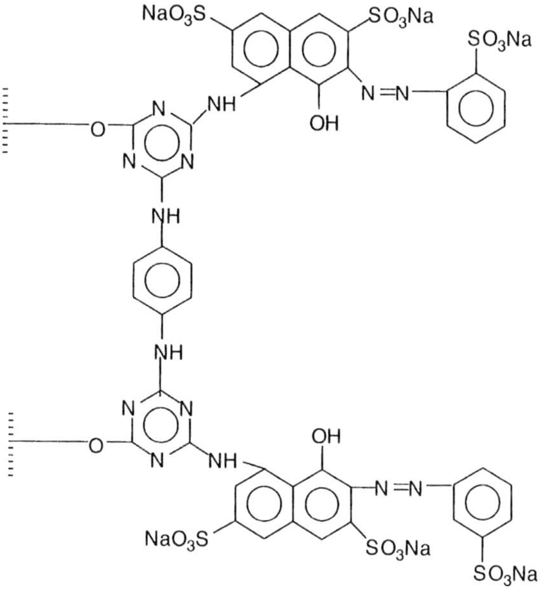

3.1 Small Ligands

Immobilized Procion Red HE3B

FIGURE 3.22
Partial structure of immobilized Procion Red HE3B.

The active hydrogens on the phenolic ring structures of these dyes may be condensed in the Mannich reaction (see Section 2.2.5.2, Chapter 2) with formaldehyde and an amine-containing matrix to yield the immobilized dye coupled through a secondary amine-linkage (Fig. 3.23). This reaction mechanism opens up the potential for investigating a fresh series of dye molecules for their utility in affinity chromatographic separations. Moreover, unlike the triazine linked dyes, those immobilized by the Mannich reaction show a markedly reduced tendency to leach off the support.

IMMOBILIZATION PROTOCOL

1. Wash 100 ml amine-containing support (DADPA–/ /–Sepharose CL-6B works well) with water and then with 0.1 M MES, 0.9 M NaCl, pH 4.7 (coupling buffer). The high salt content of the coupling buffer is to

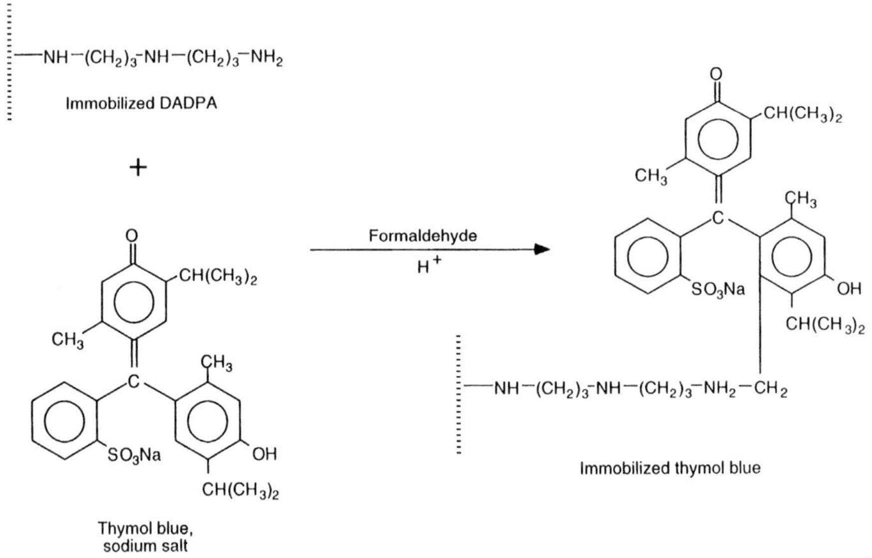

FIGURE 3.23

The dye thymol blue can be immobilized to an amine-containing support (DADPA) by condensation with formaldehyde in the Mannich reaction. The active hydrogens on the substituted phenolic ring react in high yield. The resulting affinity support has specificity for serum albumins.

---

prevent nonspecific binding, particularly of phenol red, to the amine-containing matrix. With some dyes, the salt strength may be reduced to 0.15 M or eliminated (especially in instances in which a 50% ethanolic solution is required to promote solubility; see subsequent text). Filter off excess buffer until the gel is left as a wet cake.

2. Prepare 100 ml dye solution by dissolving either phenol red or thymol blue at a concentration of 1–5 mg/ml in coupling buffer.
3. In a fume hood, add the gel to the dye solution in a round-bottom flask and mix with a paddle stirrer. Add 10 ml of a 37% formaldehyde solution. Heat with a heating mantle to 37–57°C (the higher the temperature, the greater the reaction rate).
4. React with constant stirring for at least 24 hr at elevated temperature.
5. Wash the gel extensively with water, 1 M NaCl, and again with water until no more noncovalently attached dye bleeds off the support.

Other dye molecules containing active hydrogens may be coupled using this method. For some dyes, a 50% ethanol solution in 0.1 M MES, pH 4.7, will aid solubility for coupling.

3.1.5 Metal Chelators

Metal chelate affinity chromatography is a technique by which proteins or other molecules can be separated based on their ability to form coordination complexes with immobilized metal ions (Porath *et al.*, 1975; Lonnerdal and Keen, 1982; Porath and Belew, 1983; Porath and Olin, 1983; Sulkowski, 1985; Kagedal, 1989). The metal ions are stabilized on a matrix through the use of chelation compounds that usually have multivalent points of interaction with the metal atoms. To form useful affinity supports, these metal ion complexes must have some free or weakly associated and exchangeable coordination sites. These exchangeable sites can then form complexes with coordination sites on proteins or other molecules (Fig. 3.24). Substances that are able to interact with the immobilized metals will bind and be retained on the column. Elution is typically accomplished by one, or a combination, of the following options: (1) lowering pH, (2) raising the salt strength, or (3) inclusion of chelation agents such as EDTA in the buffer. Often, through an appropriate combination of these three parameters, a gradient elution pattern can be developed to resolve many different binding components in a complex mixture. Thus, metal chelate affinity chromatography can provide a powerful general purpose separation technique to solve a number of purification problems, especially when biospecific affinity supports may not be available (Andersson, 1988; Arnold, 1991).

The interaction of proteins or peptides with metal chelate supports is primarily based on the formation of stable complexes with cysteine and histi-

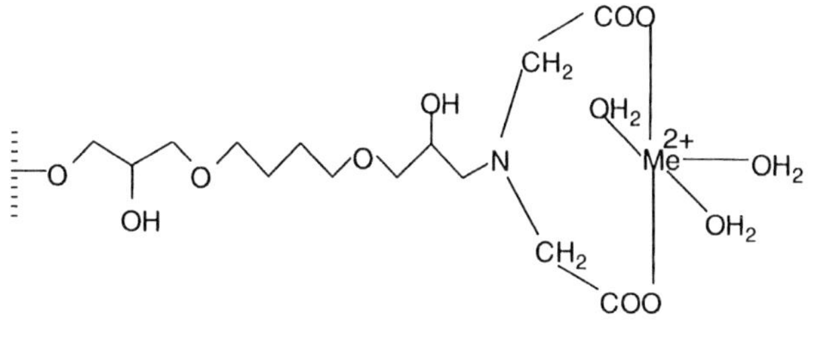

FIGURE 3.24
Schematic structure of an immobilized metal ion adsorbent (immobilized metal iminodiacetic acid).

dine residues. In neutral solutions, these amino acids will form strong coordination bonds with immobilized zinc (Zn^{2+}) or copper (Cu^{2+}) ions. Other metal ions that are useful for investigative purposes include cadmium, mercury, cobalt, nickel, and manganese, although each will have varying degrees of interaction with proteins. Also, group IIIA metal ions such as aluminum, gallium, indium, and titanium will have different effects and protein specificities. Cysteine and histidine residues will take part in the binding; phosphorylated proteins will specifically bind to chelated iron (Fe^{2+}) supports via their O-phosphate group (Andersson and Porath, 1986). Similar chelated iron supports also can interact with tyrosine-containing peptides. However, by far the most important metal chelate supports are built on complexes with copper, zinc, and nickel.

3.1.5.1 Preparation of Immobilized Iminodiacetic Acid

Immobilized iminodiacetic acid is perhaps the most well-known metal chelate affinity support. It is best coupled to a matrix using the epoxy activation procedure that reacts with the central secondary amine of the bis-acid (Fig. 3.25). The result gives two terminal carboxylate groups sticking out from the support that can form tight coordination complexes with metal ions.

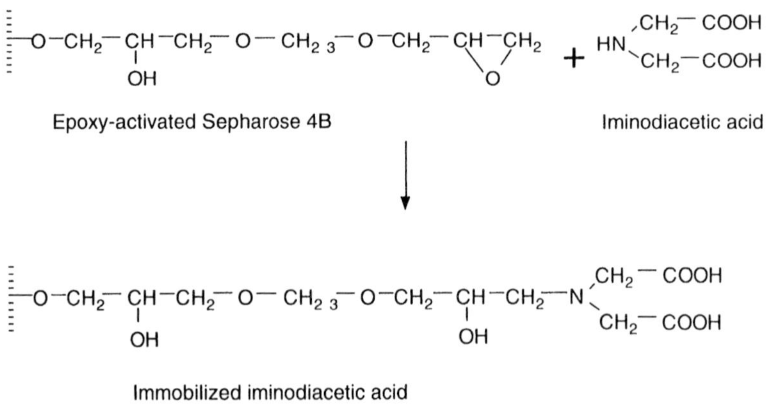

FIGURE 3.25

Reaction sequence for the preparation of immobilized iminodiacetic acid.

3.1 Small Ligands

IMMOBILIZATION PROTOCOL

1. Activate Sepharose 4B (100 ml settled gel) with 1,4-butanediol diglycidyl ether as described in Chapter 2 (Section 2.2.4.1).
2. Wash the epoxy-activated Sepharose 4B with 1 L water, suction dry to a moist cake, and add to an iminodiacetic acid solution in a round-bottom flask [dissolve 12.5 gm iminodiacetic acid, disodium salt (Aldrich, Milwaukee, Wisconsin) in 100 ml 2 M potassium carbonate].
3. Heat the reaction mixture using a heating mantle, to 60°C and stir with a paddle for 24 hr.
4. Filter and wash the reaction mixture with 2 L water. Immobilized iminodiacetic acid can be stored in 0.02% sodium azide at 4°C.

a. Preparation of the Metal Chelate

1. Pack a column with an appropriate amount of immobilized iminodiacetic acid gel and equilibrate with 10 mM HEPES, 50 mM NaCl, pH 7.5 (buffer A). Other buffers at neutral pH are also suitable. Nonionic detergents may be added as well.
2. Prepare a solution consisting of 1 mg/ml $ZnCl_2$ (or other suitable metal salt) in buffer A. Apply three column volumes of the zinc solution to the equilibrated gel. Other metal salts may be used in similar concentrations.
3. Wash with 4–5 column volumes of buffer A to remove excess nonchelated metal.

The column is now ready for metal chelate affinity chromatography.

3.1.5.2 Preparation of Immobilized Tris(carboxymethyl)ethylenediamine

An alternative chelating gel may be prepared having three carboxylate groups for forming coordination complexes. The ligand tris(carboxylmethyl)ethylenediamine (TED) is made by carboxymethylating immobilized ethylenediamine (Porath and Olin, 1983; Porath *et al.*, 1983). A hydroxylic support material is first activated with epichlorohydrin (see subsequent text) and coupled with ethylenediamine. This intermediate is then reacted with bromoacetic acid to give two acetate groups off the terminal primary amino group and one off the secondary amine (Fig. 3.26).

IMMOBILIZATION PROTOCOL

NOTE: All reactions described here should be carried out in a fume hood. Epichlorohydrin, ethylene diamine, and bromoacetic acid are toxic and corrosive.

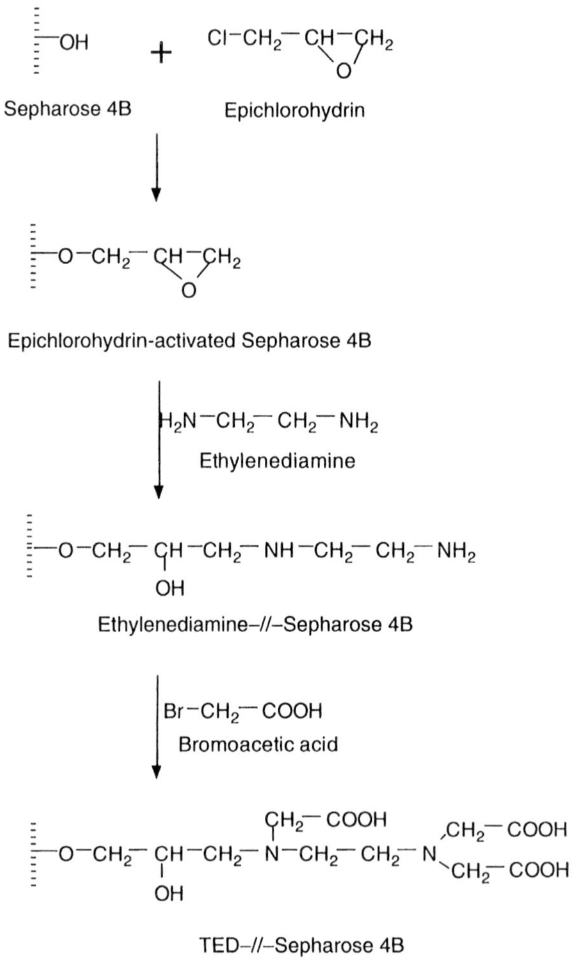

182 3. Immobilization of Ligands

```
╎─OH    +    Cl–CH₂–CH–CH₂
╎                     \ /
                       O
Sepharose 4B        Epichlorohydrin
                    │
                    ▼
╎─O–CH₂–CH–CH₂
╎            \ /
              O

Epichlorohydrin-activated Sepharose 4B

                    │  H₂N–CH₂–CH₂–NH₂
                    │  Ethylenediamine
                    ▼

╎─O–CH₂–CH–CH₂–NH–CH₂–CH₂–NH₂
╎        │
         OH

Ethylenediamine–//–Sepharose 4B

                    │  Br–CH₂–COOH
                    │  Bromoacetic acid
                    ▼

                              CH₂–COOH           CH₂–COOH
                              │                 /
╎─O–CH₂–CH–CH₂–N–CH₂–CH₂–N
╎        │                                      \
         OH                                      CH₂–COOH

TED–//–Sepharose 4B
```

FIGURE 3.26

Reaction sequence for the preparation of TED–//–Sepharose 4B.

---

1. Wash Sepharose 4B (100 ml settled gel) with 2 L water, suction dry to a moist cake, and transfer to a 2-L round-botton flask provided with a paddle stirrer.
2. Mix the gel with 37.5 ml epichlorohydrin, 345 ml 2M NaOH, and 1.275 gm sodium borohydride, and stir.
3. To the gel suspension, add 345 ml 2 M NaOH and 170 ml epichlorohydrin in small portions over a period of 1.5 hr.
4. Stir the reaction mixture overnight at room temperature.

5. Wash the gel successively with 2 L each of water, 0.1 M acetic acid, water, and 0.2 M sodium bicarbonate.
6. To the washed gel, add 225 ml 0.2 M sodium bicarbonate and 150 ml ethylenediamine, and stir the gel suspension for 24 hr at 50°C.
7. Filter and wash the gel with 2 L each of water, 0.1 M acetic acid, and water.
8. Suction dry the ethylenediamine intermediate gel to a moist cake and transfer to a 1-L beaker containing 16 gm bromoacetic acid, 50 ml 2 M NaOH, and 50 ml 1 M sodium bicarbonate, and stir.
9. Adjust the pH of the reaction mixture to 7.0 and stir the gel suspension overnight at room temperature.
10. Finally, wash the gel successively with 2 L each of water, 0.1 M acetic acid, and water. TED–/ /–Sepharose 4B can be stored in 0.02% sodium azide at 4°C.

3.1.6 Hydrophobic Ligands

Hydrophobic interaction chromatography takes advantage of the hydrophobic domains or binding sites on proteins and other molecules. Apolar groups coupled to a support material will bind to the hydrophobic regions of protein molecules. The association energy involved in this interaction is a function of the ordered structure of water molecules surrounding the apolar regions. In environments that stabilize the structure of water, the hydrophobic binding effect will be maximized. Conversely, in buffers that contain chaotropic salts or organic molecules that break down the orderly environment surrounding the apolar groups, the association energy will decrease. An additional factor in the strength of hydrophobic binding is the relative density of immobilized ligand. Low density hydrophobic supports will be mildly hydrophobic binders, whereas high density ligand coupling will promote strong hydrophobic interactions. Taking advantage of all this knowledge, chromatography supports and conditions may be designed to differentially separate a mixture of proteins based on their relative hydrophobic interaction potentials.

Elution conditions chosen from buffer salts of the Hofmeister series can provide selective binding and elution effects (see Section 5.2.2.3, Chapter 5). For instance, some proteins may bind well to a hydrophobic support in low chaotropic environment and be eluted as a sharp peak by a strongly chaotropic salt (e.g., NaSCN). Alternatively, a salt that increases the structure of water [such as high NaCl or $(NH_4)_2SO_4$ concentrations] may be used initially to promote binding of a weakly interacting protein. Removal of this salt will then often cause elution. Alternatively, low temperatures (i.e., 0–4°C) often promote stronger hydrophobic interactions, whereas elevated temperatures (i.e., 37°C) will weaken hydrophobic interactions. Clearly, the correct choice of chromatographic conditions strongly influences the success of the separation.

3.1.6.1 Preparation of Immobilized Octylamine

Immobilized octylamine is prepared from an amine-reactive support and results in an 8-carbon chain covalently attached to the matrix (Fig. 3.27). In this description, we have employed a CDI-activated support to take advantage of its ability to couple ligands in organic solvent environments. The acetone coupling medium provides excellent solubility for the octylamine ligand. The density of ligand coupled to the gel will be high for this protocol.

IMMOBILIZATION PROTOCOL

1. Activate Sepharose CL-6B (100 ml settled gel) with N,N'-carbonyl diimidazole as described in Chapter 2 (Section 2.2.1.3).
2. Wash CDI-activated Sepharose CL-6B with 1 L acetone, suction dry to a moist cake, and add to an octylamine solution (dissolve 13 gm octylamine in 100 ml acetone).
3. Stir the reaction mixture at room temperature for 24 hr in a well-ventilated hood.

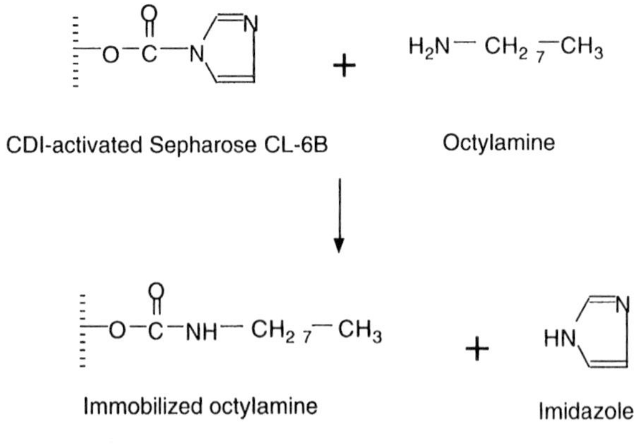

FIGURE 3.27
Reaction sequence for the preparation of immobilized octylamine.

4. Filter and wash the reaction mixture successively with 1 L each of acetone, 70% acetone, 35% acetone, water, 1.0 M NaCl, and water. Immobilized octylamine can be stored in 0.02% sodium azide at 4°C.

3.1.6.2 Preparation of Immobilized Benzylamine

Immobilized benzylamine is prepared from an amine-reactive support to yield a hydrophobic matrix with a dense aromatic surface character (Fig. 3.28). We use the CDI activation procedure (as for octylamine) to provide an organic coupling environment for the ligand.

IMMOBILIZATION PROTOCOL

Immobilized benzylamine is prepared essentially by the same method described for immobilized octylamine, except that 11 gm is used benzylamine.

3.1.7 Immobilization of Drugs and Receptor Ligands

Affinity supports prepared by the immobilization of drugs, their analogs or derivatives, and other ligands specific for certain receptor proteins are

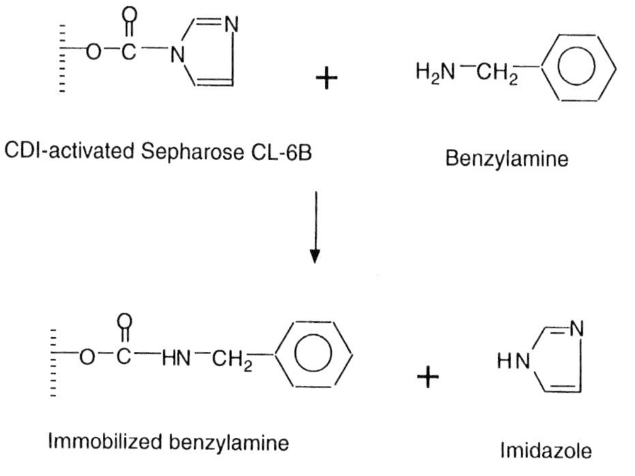

FIGURE 3.28
Reaction sequence for the preparation of immobilized benzylamine.

potentially powerful purification tools. Virtually every therapeutically active agent interacts with particular target molecules and systems in an organism. Immobilization of such components can provide routes to an understanding of the nature of the target molecule. Other immobilized systems not geared toward purification may nonetheless provide the key component of significant assay methods based on the specific interaction between the ligand and the receptor.

The immobilization problems associated with these ligands involve the coupling of complex organic molecules to an activated or spacer-arm-derivatized support. Sometimes appropriate functionalities are present on the ligand to allow easy immobilization. In other cases, however, the active hydrogen method of coupling using the Mannich reaction will be necessary. The following section provides some examples of how these ligands may be immobilized to various types of support materials.

3.1.7.1 Preparation of Immobilized Methotrexate

Methotrexate (amethopterin) is an antineoplastic agent that is an amino and methyl derivative of folic acid. In its immobilized form, the compound is coupled via its carboxylate groups to the spacer arm DADPA (Fig. 3.29). The affinity support is a useful tool in the preparation of dihydrofolate reductase (Whiteley *et al.,* 1977; Poe *et al.,* 1979; Then, 1979; Cayley *et al.,* 1981).

IMMOBILIZATION PROTOCOL

1. Wash immobilized DAH (100 ml settled gel), prepared as described earlier (Section 3.1.1.2), with 1 L water in a sintered glass funnel, and suction dry to a moist cake.
2. Prepare a methotrexate solution as follows. (**Caution:** Toxic and mutagenic.) Add 1 gm methotrexate to 100 ml 0.1 M NaCl and stir. Solubilization is achieved by adding a few drops of 1.0 M NaOH while continuing to stir.

NOTE: Methotrexate is not stable for a long period of time in alkaline solution. Use the solution immediately for coupling.

3. Add the gel to the methotrexate solution and stir with a paddle in a fume hood.
4. Adjust the pH to 6.5 by adding a few drops of 1.0 M HCl.
5. Add EDC (3 gm) to the gel suspension and maintain the pH of the reaction mixture at 6.5 for 1 hr by adding 1.0 M HCl dropwise. After 1 hr,

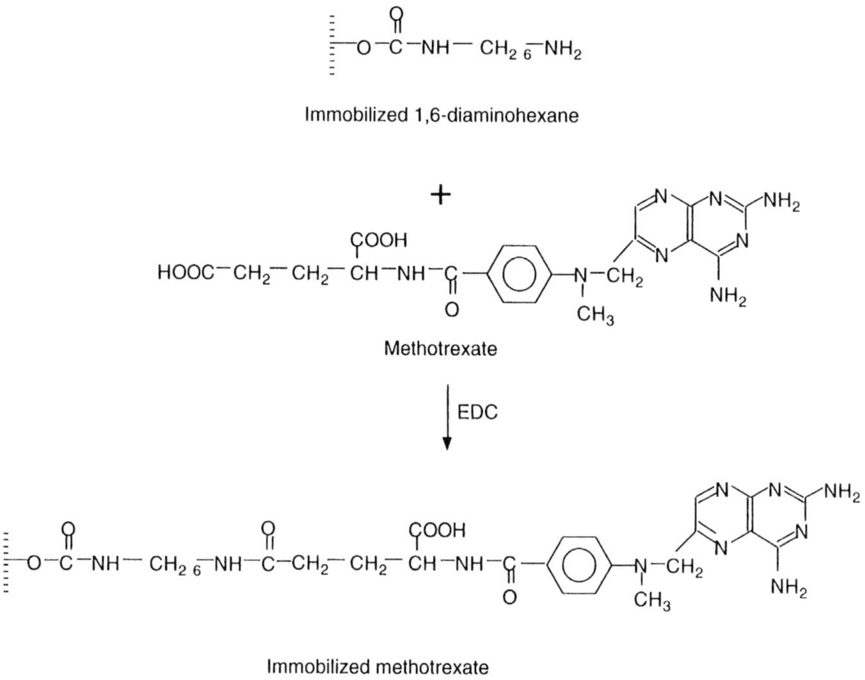

3.1 Small Ligands

FIGURE 3.29
Reaction sequence for the preparation of immobilized methotrexate

lower the pH to 6.0 and stir the reaction mixture at room temperature overnight.

6. Filter and wash the reaction mixture (on a sintered glass funnel) successively with 1 L each of water, 1 M NaCl, water, 1 M K_2HPO_4, water, 1 M K_2HPO_4, and water. Immobilized methotrexate has pale yellow color and can be suspended in 0.02% sodium azide and stored at 4° C.

3.1.7.2 Preparation of Immobilized Alprenolol

Catecholamine hormones exact their vasodilating effects through specific binding to β-adrenergic receptors, mainly located in bronchial passages and peripheral blood vessels. The purification of these receptors is necessary for critical structure and function studies.

The immobilization of the antagonist alprenolol to a chromatography matrix provides a powerful affinity support suitable for the purification of solubilized-adrenergic receptors in high yield. The strong interaction between the drug and receptor allows for extensive washes to be done prior to elution to eliminate nonspecifically bound proteins.

The affinity support is prepared from an epoxy-activated agarose gel by a modification of the procedures described in the literature (Caron et al., 1979). Briefly, dithiothreitol (DTT) is first reacted in excess with the oxirane end of an epoxy-activated support to provide terminal sulfhydryl groups. Alprenolol is then reacted with this intermediate gel under uv light to facilitate coupling through the propylene side chain of the drug by free radical addition across the double bond (Fig 3.30). This method provides a more direct route to the immobilized drug than those methods reported previously, and also results in a higher density of ligand on the affinity support.

IMMOBILIZATION PROTOCOL

1. Wash bisoxirane-activated Sepharose 4B (100 ml settled gel; prepared as described in Chapter 2, Section 2.2.4.1) with 500 ml 0.1 M sodium phosphate buffer, pH 7.0, containing 1 mM EDTA on a sintered glass funnel, and suction dry to a moist cake.
2. Add the washed gel to a DTT solution (dissolve 7.5 gm DTT in 100 ml 0.1 M sodium phosphate buffer, pH 7.0, containing 1 mM EDTA).
3. Stir the gel suspension at room temperature for 6 hr.
4. Filter the reaction mixture on a sintered glass funnel, wash the gel with 1 L water freshly bubbled with nitrogen. Suction dry to a moist cake.
5. Add the gel to an alprenolol solution (dissolve 1 gm alprenolol in 100 ml water freshly bubbled with nitrogen).
6. Expose the gel suspension to sunlight (or uv light) and pass air through the gel suspension for 6 hr.
7. Filter and wash the reaction mixture with 1 L each of 1 M NaCl and water.
8. Block the unreacted sulfhydryl groups on the gel as follows. Suspend the gel in 100 ml 0.2 M sodium bicarbonate containing 10 mM iodoacetamide and stir for 1 hr at room temperature. Finally, wash immobilized alprenolol with 2 L water and store in 0.02% sodium azide at 4°C.

3.1.7.3 Preparation of Immobilized 17β-Estradiol

The estrogen 17β-estradiol is a steroid hormone that acts on target cells to effect gene expression and protein synthesis rather than enzymatic or transport processes. Receptors for this hormone occur in vaginal epithelium, ovaries, uterus, fallopian tubes, and mammary tissue. For instance, the ini-

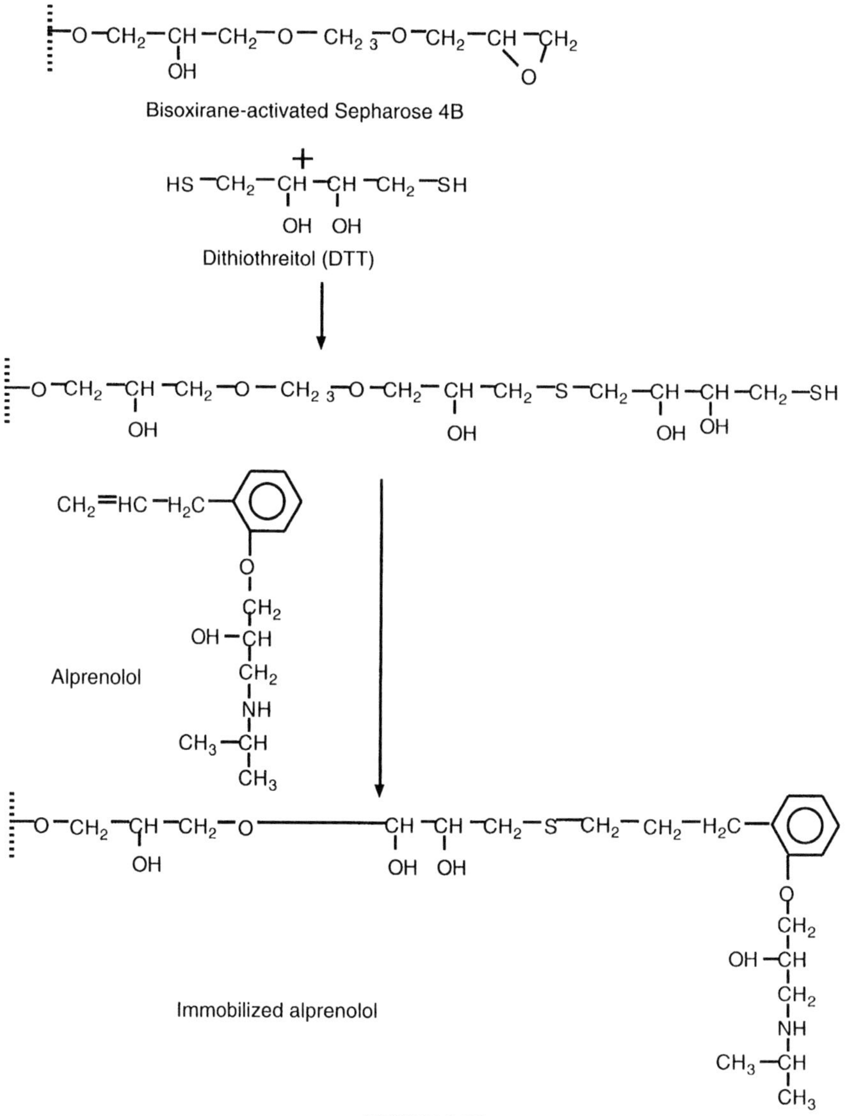

FIGURE 3.30
Reaction sequence for the preparation of immobilized alprenolol.

tial step in the stimulation of uterine cell growth by estradiol is the binding of this molecule to its specific receptor in the cytoplasm of the uterine cell. The steroid–receptor complex then effects gene expression. The density of estrogen receptors in mammary tissue are known to be increased in breast cancer.

The common structural characteristic of estrogen steroids is an aromatic A ring with a hydroxyl group attached to carbon 3. Thus, 17β-estradiol is a phenolic molecule, the properties of which may be utilized for immobilization (see subsequent text). Alternatively, the hemisuccinate derivative of estradiol may be used to couple the molecule in the opposite orientation (see subsequent text).

IMMOBILIZATION PROTOCOL

a. Coupling via the Hemisuccinate Derivative

The immobilization of β-estradiol via the hemisuccinate derivative will orient the molecule with its phenolic portion sticking out from the matrix (Parikh *et al.*, 1974) (Fig. 3.31).

1. Successively wash immobilized DADPA (100 ml settled gel; prepared as described in Section 3.1.1.1) on a sintered glass funnel with 1 L water, 1 L 30% dioxane, and 1 L 70% dioxane. Suction dry the gel to a moist cake.
2. Add the moist gel cake to a 100-ml solution of β-estradiol-17-β-hemisuccinate (Sigma, St. Louis, Missouri), dissolved at a concentration of 1 mg/ml in 70% dioxane, and stir.
3. Add EDC (200 mg) to the gel suspension, and adjust the pH of the reaction mixture to 4.5 by adding a few drops of 2.0 N HCl. After 30 min, again add EDC (200 mg) to the reaction mixture and continue the reaction for 24 hr at room temperature.
4. Finally, successively wash the gel with 1 L each of dioxane, 70% dioxane, 30% dioxane, water, 1.0 M NaCl, and water. Immobilized estradiol can be stored in 0.02% sodium azide at 4°C.

b. Coupling via the Mannich Reaction

The following procedure will immobilize β-estradiol (not the hemisuccinate form) via its number 4 active hydrogen on the phenolic ring (and, to a lesser extent, through its number 2 active hydrogen). This will result in the opposite end of the molecule pointing out from the support compared with the hemisuccinate derivative method just described (Fig. 3.32A). Other steroidal molecules that couple well in the Mannich procedure include androsterone and testosterone (Fig. 3.32B).

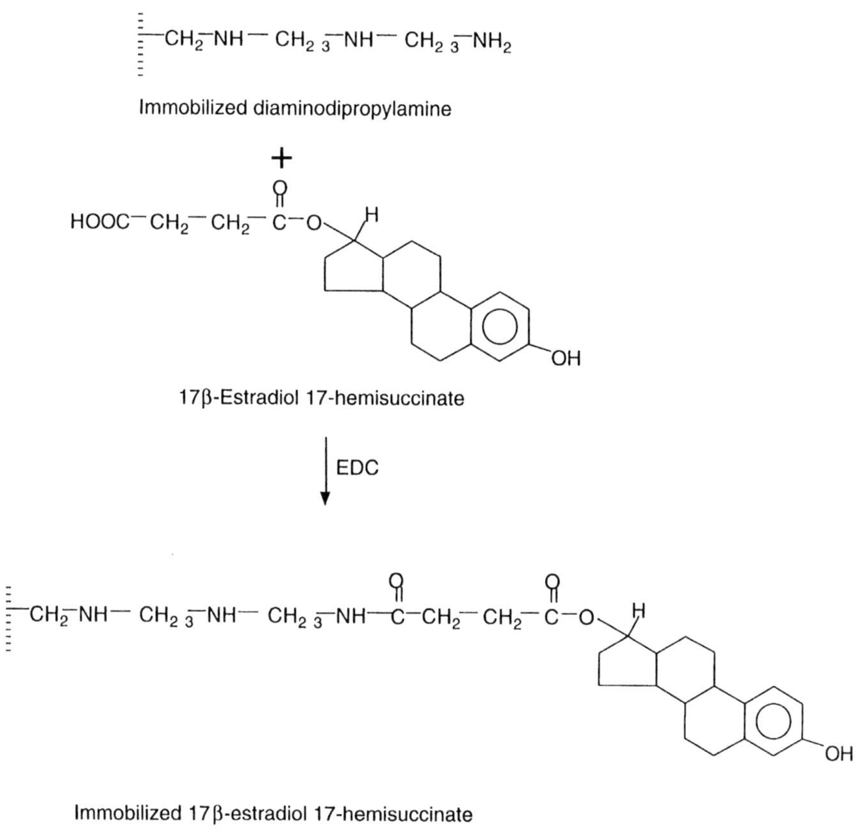

3.1 Small Ligands

```
┊—CH₂—NH—CH₂₃—NH—CH₂₃—NH₂
```

Immobilized diaminodipropylamine

+

HOOC—CH₂—CH₂—C(=O)—O—[17β-Estradiol structure]—OH

17β-Estradiol 17-hemisuccinate

↓ EDC

```
┊—CH₂—NH—CH₂₃—NH—CH₂₃—NH—C(=O)—CH₂—CH₂—C(=O)—O—[estradiol]—OH
```

Immobilized 17β-estradiol 17-hemisuccinate

FIGURE 3.31
Reaction sequence for the preparation of immobilized 17β-estradiol 17-hemisuccinate.

---

1. Wash immobilized DADPA (100 ml settled gel, prepared as described Section 3.1.1.1) on a sintered glass funnel with 1 L water, and then wash with 0.1 M MES, pH 4.7, containing 50% ethanol (buffer A). Suction the gel free of excess buffer to a moist cake.
2. Prepare a 100-ml solution of β-estradiol by dissolving it at a concentration of 2 mg/ml in buffer A.
3. Add the gel cake to the estradiol solution and place the slurry in a round-bottom flask in a fume hood. Add 10 ml 37% formaldehyde solution. Stir with a paddle and increase the temperature using a heating mantle to 37–57°C. The greater the temperature, the faster the rate of reaction.
4. React at elevated temperature for 24 hr.
5. Wash the gel extensively with buffer A and then with water.

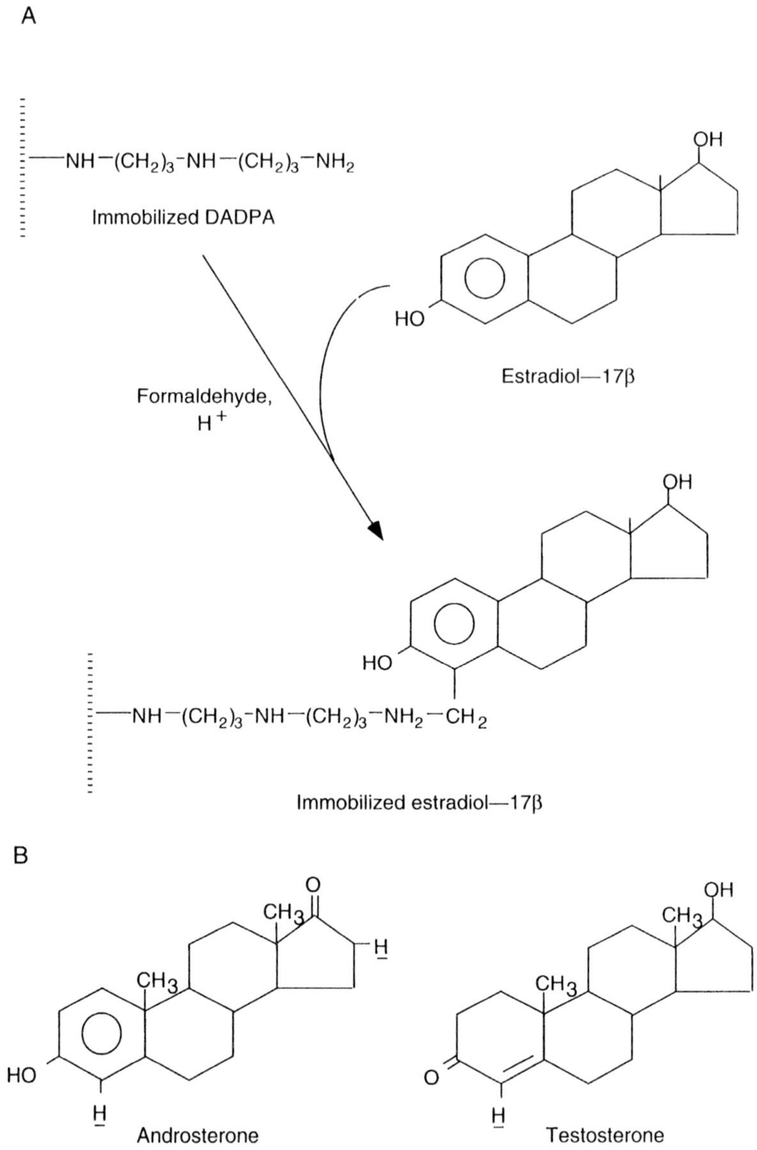

FIGURE 3.32

(A) Estradiol-17β can be immobilized to an amine-containing matrix by condensation with formaldehyde in the Mannich reaction. The coupling site is at the active hydrogens in the phenolic ring. (B) Other steroidal molecules containing active hydrogens can also be coupled to an amine-containing support using the Mannich reaction. Androsterone and testosterone are two steroids we have successfully immobilized in our labs.

3.1.7.4 Preparation of Immobilized Tetrahydrocannabinol

Tetrahydrocannabinols (THCs) are the active components in marijuana (hashish). The Δ^9 *trans* isomer is the major active form of THC. It has numerous documented physiological effects including euphoria, delirium, hallucinations, weakness, and drowsiness. THC also is used as a therapeutic agent in the treatment of some cancers and in hypertensive glaucomas.

The immobilization of THC may be useful for the purification of specific receptor proteins or for the affinity isolation of antibodies raised against the molecule. The 4' and 5' active hydrogens on the phenolic ring may be used in the Mannich reaction for coupling to an amine-containing support (Fig. 3.33).

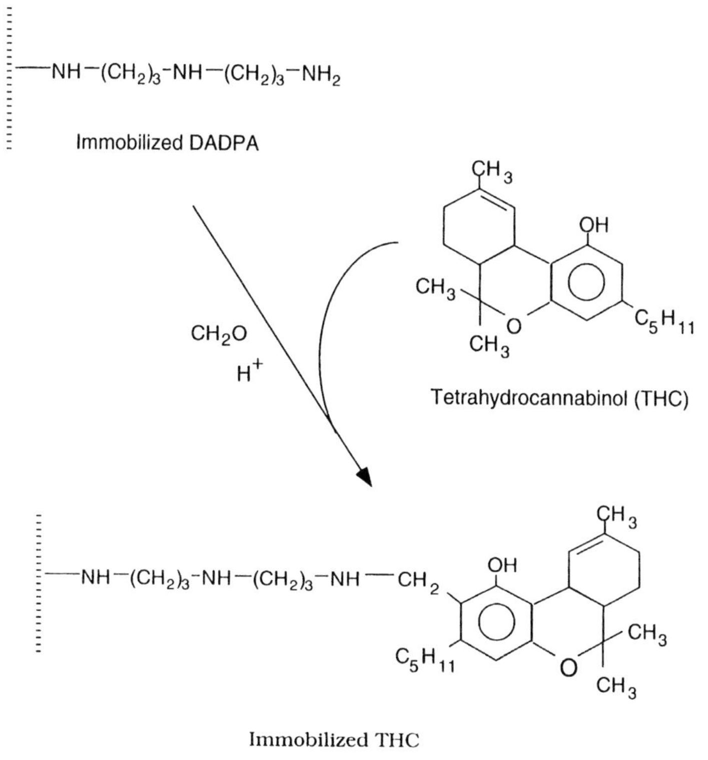

FIGURE 3.33
The drug THC can be immobilized by the Mannich reaction.

IMMOBILIZATION PROTOCOL

1. Wash immobilized DADPA (10 ml settled gel; prepared as described in Section 3.1.1.1) on a sintered glass funnel with 100 ml water, and then wash with 0.1 M MES, pH 4.7, containing 50% ethanol (buffer A). Suspend the gel in 5 ml buffer A.
2. Prepare a solution of THC by dissolving it at a concentration of 1 mg/ml in 10 ml absolute ethanol.
3. Add the gel slurry to the THC solution, mix well, and place the slurry in a 50-ml glass hypovial. Add 1 ml 37% formaldehyde solution, and incubate the reaction at 37–57°C. The greater the temperature, the faster the rate of immobilization.
4. React at elevated temperature for 24 hr with periodic mixing.
5. Wash the gel extensively with buffer A and then with water.

3.1.7.5 Immobilization of Ethynyl Steroids

Ethynyl steroid molecules contain a triple bonded ethynyl group on carbon 17. The terminal hydrogen atom on this group is sufficiently active to condense with formaldehyde and an immobilized amine in the Mannich reaction. The ethynyl group thus provides a convenient handle when phenolic active hydrogens or other constituents do not supply the functionalities necessary for coupling using other activation methods.

An example of this type of immobilization is given in the coupling of 17α-ethynylestradiol-3-methyl ether to DADPA–/ /–agarose (Fig. 3.34).

IMMOBILIZATION PROTOCOL

1. Wash 10 ml DADPA–/ /–Sepharose CL-6B (Section 3.1.1.1) with water and then with 50% methanol, 50% 0.5 M acetic acid/0.2 M sodium acetate solution (buffer A). Remove excess buffer by filtration and leave the gel as a wet cake.
2. Make a solution of 17α-ethynylestradiol-3-methyl ether by dissolving 20 mg in 10 ml buffer A.
3. Add the gel to the ethynylestradiol solution and transfer the slurry to a 30-ml hypovial. Then add 2 ml 37% formaldehyde.
4. Incubate the reaction mixture for 24 hr at 37–57°C. The higher the temperature, the faster the rate of coupling.
5. Wash extensively with buffer A to remove unreacted ethynylestradiol molecules and then into water.

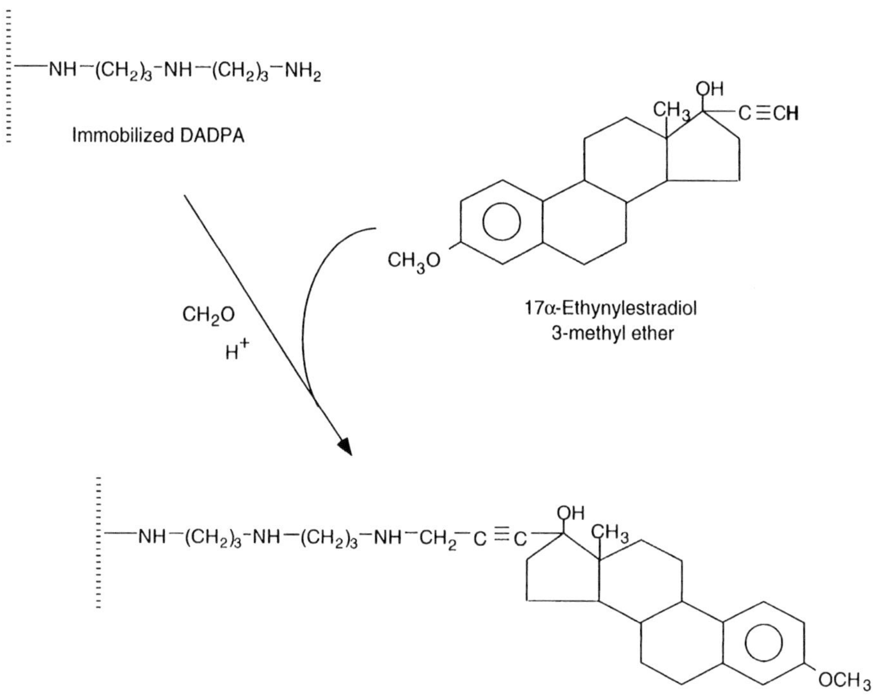

FIGURE 3.34
Ethynyl steroid molecules couple efficiently to amine-containing supports by the Mannich reaction. The active hydrogen at the triple bond is the principle site of immobilization.

3.2 GENERAL PROTEIN IMMOBILIZATION

This section describes a number of protocols that can be successfully used to immobilize proteins. Most of these procedures work very well on a wide variety of proteins with excellent coupling yields and good retention of activity. Specialized procedures dealing with enzymes and antibodies can be found in the sections that follow. (For a review of protein immobilization techniques, see Taylor, 1991.)

3.2.1 Preparation of Immobilized Serum Albumin

Serum albumins (especially BSA and HSA) are excellent proteins to use to investigate the suitability of an immobilization scheme. Albumins contain an abundance of amines, most are not easily denatured, and they can often be monitored to determine coupling yields by measuring the absorbance at 280 nm before and after the reaction. In most cases, albumins can be thought of as the "average" protein for immobilization studies.

3.2.1.1 Coupling of Albumin Using CNBr Activation

IMMOBILIZATION PROTOCOL

1. Wash CNBr-activated Sepharose 4B (10 ml settled gel; freshly prepared as described in Chapter 2, Section 2.2.1.1) with 100 ml ice-cold 0.1 M sodium bicarbonate, pH 8.5, in a sintered glass funnel, and suction dry to a wet cake.
2. Add the activated gel to an albumin solution (100–200 mg albumin dissolved in 10 ml ice-cold 0.1 M sodium bicarbonate, pH 8.5).
3. Mix the reaction gently in a shaker overnight at 4°C. We routinely use a 100-ml Erlenmeyer flask as a reaction vessel.
4. Filter and wash the reaction mixture successively with 200 ml each of 0.1 M sodium bicarbonate, pH 8.5, 1.0 M NaCl, and water.
5. Block the excess active groups on the gel as follows. Suspend the gel in 10 ml 1.0 M ethanolamine, pH 9.0, and mix gently in a shaker at room temperature for 1 hr. Filter and wash the gel suspension successively with 200 ml each of water, 1.0 M NaCl, and water. Immobilized albumin can be stored in 0.02% sodium azide at 4°C.

The coupling efficiency using this protocol is excellent. Within 24 hr, more than 95% albumin can be coupled to CNBr-activated Sepharose 4B. If desired, the coupling reaction can be carried out at room temperature. At room temperature, 95% albumin will be coupled to CNBr-activated gel within 6 hr.

3.2.1.2 Coupling of Albumin Using Periodate Activation

IMMOBILIZATION PROTOCOL

1. Wash periodate-activated Sepharose CL-6B (10 ml settled gel; prepared as described in Chapter 2, Section 2.2.1.4) with 100 ml 0.1 M sodium phosphate buffer, pH 7.0, in a sintered glass funnel and suction dry to a moist cake.
2. Add the gel to an albumin solution (100–200 mg albumin dissolved in 10 ml 0.1 M sodium phosphate buffer, pH 7.0).

3. Add NaCNBH$_3$ (60 mg) to the gel suspension and gently mix the reaction mixture in a shaker overnight at room temperature. We routinely use a 100-ml Erlenmeyer flask as a reaction vessel.

NOTE: NaCNBH$_3$ is highly toxic. The coupling reaction and the subsequent steps should be carried out in a well-ventilated hood.

4. Wash the coupled gel successively with 200 ml each of water, 1.0 M NaCl, and water.
5. Excess aldehyde groups on the gel may be blocked if desired. However, we have found that any remaining formyl groups do not contribute to nonspecific effects in the resulting affinity support. For this reason, we typically do not do a blocking step. To block, treat the support as follows. Suspend the gel in 10 ml 1.0 M Tris-HCl, pH 7.4, containing 60 mg NaCNBH$_3$ and mix gently in a fume hood at room temperature for 1 hr. Alternatively, the remaining aldehydes may be reduced to hydroxyls without adding Tris. This is done as follows. Adjust the pH of the reaction mixture to 9.0 by adding a few drops of 1.0 N NaOH. Add NaBH$_4$ (12.5 mg) to the reaction mixture and continue the reaction for 30 min at room temperature.

NOTE: Although robust proteins such as albumins are not harmed by the use of sodium borohydride, some proteins containing disulfide linkages (e.g., immunoglobulins) may be reduced and denatured by this treatment.

6. Finally, wash the gel successively with 200 ml each of water, 1.0 M NaCl, and water. Immobilized albumin can be stored in 0.02% sodium azide at 4°C.

The coupling efficiency using this protocol varies from 85% to 95%.

3.2.2 Preparation of Immobilized Avidin, Streptavidin, and Monomeric Avidin

One of the most useful interactions in immunochemistry involves the specific binding of biotin (vitamin H) to the egg white protein avidin or the similar bacterial protein streptavidin (Green and Toms, 1973). Avidin is a tetramer containing four identical subunits of molecular weight 15,000. Each subunit contains one high-affinity binding site for biotin with a dissociation constant of approximately 10^{-15} M. Streptavidin has an almost identical tetrameric structure and binding affinity for biotin. The binding is undisturbed by extremes of pH, buffer salts, or even chaotropic agents, such as guanidine hydrochloride (up to 3 M). The strength of the avidin–biotin interaction has provided researchers with a unique tool for immunoassays, receptor studies, immunocytochemical staining, and protein isolation. (For a book on avidin/biotin technology, see Wilchek and Bayer, 1990.)

Biotin (MW 244.31) normally exists bound to proteins (such as pyruvate carboxylase) through its valeric acid carboxylic group by an amide bond to lysine side-chain amines. *In vivo* it is a carrier of activated CO_2 through the N-1 nitrogen atom of the biotin ring. Proteins and other molecules may be artificially modified with biotin through selection of the appropriate activation chemistry off the valeric acid side chain. For instance, biotin that has its carboxylic group activated by an NHS ester will react with amino groups on proteins to give a stable biotin conjugate. A protein or other molecule modified in this manner is able to bind to avidin or streptavidin through its biotin "handle" (Fig. 3.35). A variety of activated biotins is now commercially available to construct almost any biotinylated molecule imaginable (Pierce; Molecular Probes; Sigma; Vector).

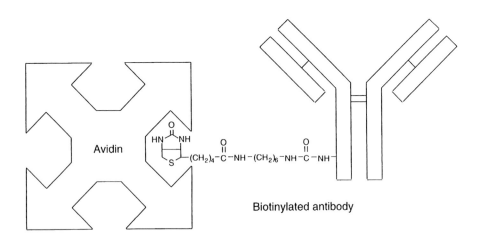

FIGURE 3.35

The affinity of avidin for biotin can be exploited to bind proteins or antibodies that have been modified by a biotin-containing reagent. The use of immobilized avidin can provide a powerful purification tool for isolating a specific antigen by the intermediate binding of a biotinylated antibody. The entire complex can be purified on immobilized avidin and the antigen–antibody interaction can be broken to isolate the target antigen.

Making use of immobilized avidin or streptavidin to bind these biotinylated molecules can provide a powerful purification tool to isolate a specific target protein. For instance, an antibody specific for some scarce protein in a physiological mixture can be biotinylated, incubated with the mixture, and the immune complex subsequently isolated by passing over the avidin or streptavidin column. The purification of receptor proteins also has been accomplished by biotinylating a ligand specific for the receptor.

The only disadvantage of this scheme for affinity isolations is the severe elution conditions necessary to release the bound biotinylated protein complex. Breaking this interaction will occur only under severe conditions, normally using 6 M guanidine hydrochloride, pH 1.5. In many cases, it may be possible to release the target molecule from the biotinylated component under milder conditions than required to break the biotin–avidin interaction but, in order to regenerate the column harsh conditions are still needed. For this reason, using a monomeric avidin column is often the best choice because of its moderated affinity constant for a biotinylated molecule (see subsequent text).

3.2.2.1 Coupling of Avidin or Streptavidin Using CNBr Activation

IMMOBILIZATION PROTOCOL

1. Wash CNBr-activated Sepharose 4B (10 ml settled gel; freshly prepared as described in Chapter 2, Section 2.2.1.1) with 100 ml ice-cold 0.1 M sodium bicarbonate, pH 8.5, in a sintered glass funnel, and suction dry to a wet cake.
2. Add the gel to an avidin or streptavidin solution (20–40 mg avidin or streptavidin dissolved in 10 ml 0.1 M sodium bicarbonate, pH 8.5).
3. Gently mix the reaction mixture on a shaker overnight at 4°C. We routinely use a 100-ml Erlenmeyer flask as the reaction vessel.
4. Filter and wash the reaction mixture successively with 200 ml each of 0.1 M sodium bicarbonate, pH 8.5, 1.0 M NaCl, and water.
5. The excess active groups on the gel are blocked as follows: Suspend the gel in 10 ml 1.0 M ethanolamine, pH 9.0, and mix gently on a shaker at room temperature for 1 hr. Wash the coupled gel successively with 200 ml each of water, 1.0 M NaCl, and water. The gel can be stored in 0.02% sodium azide at 4°C.

More than 95% of avidin or streptavidin can be coupled to CNBr-activated Sepharose using this protocol.

3.2.2.2 Coupling of Avidin or Streptavidin Using Periodate Activation

IMMOBILIZATION PROTOCOL

1. Wash periodate-activated Sepharose CL-6B (10 ml settled gel; prepared as described in Chapter 2, Section 2.2.1.4) with 100 ml 0.1 M sodium phosphate buffer, pH 7.0, in a sintered glass funnel, and suction dry to a moist cake.
2. Add the gel to an avidin or streptavidin solution (20–40 mg avidin or streptavidin dissolved in 10 ml 0.1 M sodium phosphate buffer, pH 7.0).
3. Add $NaCNBH_3$ (60 mg) to the gel suspension, and gently mix the reaction mixture on a shaker overnight at room temperature.

NOTE: $NaCNBH_3$ is highly toxic. The coupling reaction and the subsequent steps should be carried out in a well-ventilated hood. We routinely use a 100-ml Erlenmeyer flask as the reaction vessel.

4. Wash the coupled gel successively with 200 ml each of water, 1.0 M NaCl, and water.
5. Excess aldehyde groups on the gel may be blocked if desired. However, we have found that any remaining formyl groups do not contribute to nonspecific effects in the resulting affinity support. For this reason, we typically do not do a blocking step. To block, treat the support as follows. Suspend the gel in 10 ml 1.0 M Tris-HCl, pH 7.4, containing 60 mg $NaCNBH_3$, and mix gently in a shaker at room temperature for 1 hr. Alternatively, the remaining aldehydes may be reduced to hydroxyls without adding Tris. This is done as follows. Adjust the pH of the reaction mixture to 9.0 by adding a few drops of 1.0 N NaOH. Add $NaBH_4$ (12.5 mg) to the reaction mixture and continue the reaction for 30 min at room temperature.
6. Wash the gel successively with 200 ml each of water, 1.0 M NaCl, and water. The coupled gel can be stored in 0.02% sodium azide at 4°C.

More than 95% of avidin or streptavidin can be coupled to periodate activated Sepharose CL-6B using this protocol.

3.2.2.3 Preparation of Monomeric Avidin

Monomeric avidin is an extremely useful form of immobilized avidin, since its binding constant for biotin is significantly reduced from that of the native protein (Henrikson *et al.,* 1979; Gravel *et al.,* 1980; Beaty and Lane, 1982). The support is prepared from immobilized tetrameric avidin by dissociation of its subunits under denaturing conditions. The result leaves a significant amount of single-subunit avidin molecules on the gel, hence the name monomeric avidin. The monomers have a dissociation constant for biotinylated molecules that is more in the range of typical affinity interactions. Thus, it is possible to elute a bound biotinylated component using a counterligand buffer containing 2 mM biotin. Any tetrameric avidin molecules left on the gel that may have higher affinity binding potential are first blocked with biotin to eliminate their activity. Regenerating the gel with a low pH wash will free the lower affinity reversible sites for subsequent interaction with a biotinylated molecule, but keeps the high affinity sites blocked.

IMMOBILIZATION PROTOCOL

1. Wash freshly prepared CNBr-activated Sepharose 4B (Section 3.2.1.1) with cold water and then with cold 0.2 M sodium carbonate, pH 9.5. Drain the excess buffer from the support and leave the gel as a wet cake.
2. Dissolve 20–40 mg egg white avidin in 10 ml buffer A, and add the solution to the activated gel.
3. React overnight with stirring at 4°C.
4. Wash the gel with several volumes of water. Measuring the absorbance at 280 nm and determining the volume of these washes can give an indication of the amount of avidin coupled. Typically, we observe an 85–95% coupling yield.
5. Block remaining CNBr reactive sites on the gel by adding 10 ml 0.1 M ethanolamine, 0.1 M sodium phosphate, 0.15 M NaCl, pH 7.2.
6. React for 4 hr at room temperature.
7. Wash the gel thoroughly with water, 1 M NaCl, and again with water.
8. To create the monomeric form of immobilized avidin, wash the gel with several column volumes of 6 M guanidine hydrochloride. Then transfer the gel to a 10-ml solution of 6 M guanidine and stir for 2–3 hr at room temperature. Alternatively, the monomeric form of immobilized avidin may be created by washing the gel as a packed column. In this way, the absorbance at 280-nm may be monitored in the fractions coming off the column during the 6 M guanidine wash. As avidin subunits elute, the completeness of monomeric formation may be assessed by the peak of 280-nm absorbing protein coming off the column.
9. Wash the gel with several more gel volumes of 6 M guanidine-HCl and then with water, 1 M NaCl, and again with water.
10. Before use, the nonreversible or higher affinity biotin binding sites must be blocked. Wash the gel with at least four volumes of 2 mM biotin in 0.1 M sodium phosphate, 0.15 M NaCl, pH 7.2. This will block all biotin binding sites. To free the reversible or low affinity monomeric

sites, wash the gel with 4–6 volumes of 0.1 M glycine, 0.15 M NaCl, pH 2.8. Re-equilibrate the monomeric avidin support with PBS, pH 7.2.

The affinity gel is now ready for chromatography with a biotinylated molecule. After binding, the biotinylated complex may be eluted using 2 mM biotin in PBS, pH 7.2. The column can be regenerated many times by washing with 0.1 M glycine, 0.15 M NaCl, pH 2.8, and then with PBS, pH 7.2.

3.3 IMMOBILIZATION OF ENZYMES

Immobilized proteases and glycosidases are used extensively for the structural study of proteins and glycoproteins. They also provide a convenient mechanism for the preparation of many important protein fragments. In particular, immunoglobulin fragments prepared from enzymatic digestion have found major uses in histochemistry, tumor imaging, clinical diagnostics, preparation of immunoconjugates directed against malignant tumors, anti-antibody production, molecular studies, genetic engineering, and antigen binding studies. Enzymatic digestion even forms the basis for assaying the lot-to-lot purity of recombinant engineered proteins. Many of the current directions in biotechnology make use of enzymatic digestion as a critical component of the research. Perhaps of future significance will be the use of immobilized nucleases and restriction enzymes in unraveling and elucidating the structure and sequence of nucleic acids.

Immobilized enzymes have many advantages over their soluble counterparts. (1) They are more resistant to autolysis and therefore maintain activity over longer periods of time. (2) Their resistance to autolysis minimizes extraneous enzyme fragments in the protein digest. (3) They provide an easy way to regulate the digestion time by adjusting the flow rate through a column or simply removing the support material when the digest is done. (4) They are effective with even small volumes of proteins. (5) The digests made from immobilized enzymes are extremely reproducible. (6) The enzymes do not contaminate the protein mixture and therefore do not pose a problem in inactivation or removal.

The stability of these enzymes during the coupling reaction is the major factor that determines their successful immobilization. If at all possible, the active site of enzymes should be protected during the immobilization process. In addition, the pH and composition of the coupling buffer should be such that the enzyme does not readily inactivate.

This section describes the protocols that we have successfully used for the immobilization of some of the most important proteases and glycosidases commonly employed in biotechnology. However, this section is not meant to be a treatise on the manufacture and use of immobilized enzymes in general. Numerous applications of insolubilized enzymes can be found in the food and chemical industries, and already have many excellent texts written about them. In this book, we have purposely chosen examples of the use of immobilized enzymes in the biotechnology arena. The following examples describe their immobilization, whereas Chapter 5 describes protocols for their use in practical applications.

When immobilizing enzymes, we have found the Worthington Enzyme Manual to be an invaluable source of basic information regarding enzymes and their analysis. When purchasing enzymes for immobilization, we usually obtain the enzyme preparation with the highest specific activity and one that is free of salts that may interfere with coupling procedures (e.g., ammonium sulfate or Tris). If the commercially obtained enzyme preparation contains competing salts, dialyze against the appropriate coupling buffer prior to immobilization.

3.3.1 Preparation of Immobilized TPCK-Trypsin

Trypsin is a pancreatic serine protease that catalyzes the hydrolysis of protein, peptide, amide, or ester bonds at the carboxyl linkage of L-lysine or L-arginine residues. Trypsin hydrolysis of proteins produces peptides with lysine or arginine carboxy termini. (**NOTE:** These peptides may be isolated by using immobilized anhydrotrypsin, an inactivated form of trypsin; see Chapter 5.) The reported molecular mass of the enzyme is about 23,800 daltons, but it may exist in two forms: α-trypsin, which consists of two peptide chains joined by a disulfide bond, and β-trypsin, made up of a single polypeptide chain. Limited autolysis of the single chain enzyme is believed to lead to the production of the two-chain analog. Trypsin is a very basic protein (pI 10.5) with an optimum activity range of pH 7–9 (typical usage at pH 8). A 1 mg/ml solution will give an absorbance at 280 nm of 1.43 (1-cm path). During purification or handling, the protein is often diluted or dialyzed with cold dilute HCl (1 mM) to prevent degradation by autolysis.

Most highly purified preparations of native trypsin contain some chymotrypsin activity, perhaps as an endogenous property of the enzyme. To eliminate this activity, the enzyme is usually treated with L-(tosylamido-2-phenyl) ethyl chloromethyl ketone (TPCK), which irreversibly inhibits chymotryptic properties.

Activity assays usually use the substrates TAME (α-N-p-toluene sulfonyl-L-arginine methyl ester hydrochloride) or BAEE (α-N-benzoyl-L-arginine ethyl ester hydrochloride) in 10–80 mM Tris, pH 8.0, containing 20 mM CaCl$_2$. One unit of trypsin activity is defined as the hydrolysis of 1 μmol TAME/min (measured spectrophotometrically at 247 nm) at 25°C. Alternatively, using these substrates, the activity may be determined by pH stat.

Immobilized trypsin is useful in the preparation of immunoglobulin fragments, particularly from IgM or IgG antibodies (Chapter 5). It has also been used to prepare peptide digests of recombinant proteins to assess their lot-to-lot purity by analyzing the fragments produced with reverse-phase HPLC techniques.

IMMOBILIZATION PROTOCOL

1. Wash periodate-activated Sepharose CL-6B (20 ml settled gel; prepared as described in Chapter 2, Section 2.2.1.4) with 200 ml 0.1 M sodium phosphate buffer, pH 7.0, in a sintered glass funnel and suction dry to a wet cake.
2. Add the gel to a TPCK-trypsin solution containing 1 mM benzamidine·HCl (50 mg TPCK-trypsin and 3.2 mg benzamidine·HCl dissolved in 20 ml 0.1 M sodium phosphate buffer, pH 7.0).

NOTE: Benzamidine temporarily protects the active site of trypsin.

3. Add NaCNBH$_3$ (120 mg) to the gel suspension and gently mix the reaction mixture in a shaker overnight at room temperature. We routinely use a 100-ml Erlenmeyer flask as the reaction vessel.

NOTE: NaCNBH$_3$ is highly toxic. The coupling reaction and the subsequent steps should be carried out in a well-ventilated hood.

4. Excess aldehyde groups on the gel may be blocked if desired. However, we have found that any remaining formyl groups do not contribute to nonspecific effects in the resulting affinity support. For this reason, we typically do not do a blocking step. To block, treat the support as follows. Adjust the pH of the reaction mixture to 9.0 by adding a few drops of 1.0 N NaOH. Then add NaBH$_4$ (25 mg) to the reaction mixture and continue the reaction for 30 min at room temperature.
5. Finally, wash the gel successively with 200 ml each of water, 1.0 M NaCl, and water. This wash effectively removes the lightly bound benzamidine. Immobilized TPCK-trypsin can be stored in 50% glycerol containing 0.02% sodium azide at 4°C. We have stored immobilized TPCK-trypsin for up to 1 year without apparent loss of trypsin activity.

Immobilized TPCK-trypsin prepared by this protocol has an activity of 40–50 BAEE units per ml gel (40 BAEE units correspond to 1 mg active trypsin measured by pH stat). Apparently, 50% of trypsin activity is lost during immobilization, probably due to steric blockage of the active site. Losing part of the activity during the immobilization of many enzymes is typical.

3.3.2 Preparation of Immobilized Pepsin

Pepsin, isolated from porcine gastric mucosa, is an acid protease that has a broad substrate specificity, including an esterase activity. Although not completely nonspecific in its action, it is often difficult to predict the cleavage points within a protein. The enzyme preferentially cleaves on the carboxyl side of L-phenylalanine, L-leucine, or L-tyrosine, where the amino side residue is preferably, but not limited to, an amino acid containing a hydrophobic side chain. Despite its seeming lack of specificity, pepsin is very useful in the preparation of immunoglobulin fragments, particularly from IgG and IgM class antibodies. With most IgM molecules, it is possible to generate $F(ab')_2$, Fab, and Fv fragments, depending on conditions. For IgG antibodies, pepsin will yield $F(ab')_2$ fragments. In both cases, the Fc regions will undergo extensive degradation.

Pepsin is an extraordinarily acidic protein with a pI of only 1.0. As the name "acid protease" implies, its optimal pH for enzymatic activity is an equally low pH 1.0. The protein is, however, unstable above pH 6.0. The enzyme has a molecular mass of 35,000 daltons, and a 1 mg/ml solution will give an absorbance reading at 280 nm of 1.47 (1-cm path). Measurement of enzymatic activity may be done using denatured hemoglobin in 0.01 N HCl. One unit releases 0.001 A_{280} as TCA-soluble hydrolysis products per minute at 37°C (Worthington catalog).

IMMOBILIZATION PROTOCOL

1. Wash periodate-activated Sepharose CL-6B (20 ml settled gel; prepared as described in Chapter 2, Section 2.2.1.4) with 200 ml 0.1 M sodium phosphate, pH 4.5, in a sintered glass funnel and suction dry to a moist cake.
2. Add the gel to a pepsin solution (400 mg pepsin dissolved in 20 ml 0.1 M sodium phosphate, pH 4.5).

NOTE: Pepsin loses its activity above pH 6.0, so the coupling reaction should be carried out below this point.

3. Add NaCNBH$_3$ (120 mg) to the gel suspension and gently mix the reaction mixture in a shaker overnight at room temperature. We routinely use a 100-ml Erlenmeyer flask as the reaction vessel.

NOTE: NaCNBH$_3$ is highly toxic. The coupling reaction and the subsequent steps should be carried out in a well-ventilated hood.

4. Excess aldehyde groups on the gel may be blocked if desired. However, we have found that any remaining formyl groups do not contribute to nonspecific effects in the resultant affinity support. For this reason, we typically do not do a blocking step. To block, treat the support as follows. Add 1.0 M ethanolamine·HCl, pH 4.4 (10 ml), to the reaction mixture and mix for 2 hr at room temperature. There is no need to add additional sodium cyanoborohydride.
5. Wash the gel successively with 200 ml each of 0.1 M sodium phosphate, pH 4.5, 0.1 M sodium phosphate, pH 4.5, containing 1.0 M NaCl, and 0.1 M sodium acetate buffer, pH 4.5. Immobilized pepsin can be stored in 50% glycerol containing 0.1 M sodium acetate, pH 4.5, at 4°C. The immobilized enzyme can be stored for up to 1 year without any loss of pepsin activity.

3.3.3 Preparation of Immobilized Papain

Papain is a sulfhydryl protease from *Carica papaya* latex. Its native form is unreactive until activated by a mild disulfide reducing agent such as cysteine. The enzyme catalyzes the hydrolysis of numerous peptide, amide, and ester linkages. In general, at the hydrolysis point, the carboxyl side amino acid residue should be arginine, lysine, glutamine, histidine, glycine, or tyrosine. The amino side amino acid next to this residue should preferentially have a nonpolar side chain. With such a broad specificity, most proteins should be extensively degraded. Papain has important application, however, in the production of Fab fragments from IgG class antibodies (Chapter 5).

Papain is a single polypeptide with a molecular mass of 23,000 daltons. A 1 mg/ml solution of the enzyme will give an absorbance at 280 nm of 2.4 (1-cm path). It has an isoelectric point of 9.6 and an optimum range of enzymatic activity of pH 6–7. Enzyme activity may be measured by pH stat using BAEE (*N*-α-benzoyl-L-arginine ethyl ester; Sigma).

IMMOBILIZATION PROTOCOL

1. Wash periodate-activated Sepharose CL-6B (20 ml settled gel; prepared as described in Chapter 2, Section 2.2.1.4) with 200 ml 0.1 M sodium phosphate buffer, pH 7.0, using a sintered glass filter funnel and suction dry to a moist cake.
2. Add the activated gel to a papain solution (60 mg papain dissolved in 20 ml 0.1 M sodium phosphate buffer, pH 7.0).
3. Add $NaCNBH_3$ (120 mg) to the gel suspension and gently mix the reaction mixture on a shaker overnight at room temperature.

NOTE: $NaCNBH_3$ is highly toxic. The coupling reaction and the subsequent steps should be carried out in a well-ventilated hood. We routinely use a 100-ml Erlenmeyer flask as the reaction vessel.

4. Excess aldehyde groups on the gel may be blocked if desired. However, we have found that any remaining formyl groups do not contribute to nonspecific effects in the resulting affinity support. For this reason, we typically do not do a blocking step. To block, treat the support as follows. Add 1.0 M ethanolamine·HCl, pH 4.4 (10 ml), to the reaction mixture and gently mix for 2 hr at room temperature. There is no need to add additional sodium cyanoborohydride.
5. Finally, wash the gel successively with 200 ml each of 0.1 M sodium phosphate, pH 4.4, 0.1 M sodium phosphate, pH 4.5, containing 1.0 M NaCl, and 0.1 M sodium acetate buffer, pH 4.5. Immobilized papain can be stored in 50% glycerol containing 0.1 M sodium acetate, pH 4.5, at 4°C. Immobilized papain should be activated with cysteine before use (see Chapter 5).

3.3.4 Preparation of Immobilized *Staphylococcus aureus* V8 Protease

Staphylococcus aureus protease (SAP) is a useful enzyme in protein investigations and is considered to have a specificity opposite to that of trypsin. SAP specifically cleaves peptide bonds on the carboxyl side of aspartic or glutamic acid residues. Under certain conditions, the enzyme can be limited to cleavage at glutamoyl bonds (Worthington catalog). SAP has been used extensively in protein chemistry and peptide mapping. Used in conjunction with immobilized trypsin, SAP can form the basis of powerful assay methods to document the purity and lot-to-lot reproducibility of a protein preparation.

IMMOBILIZATION PROTOCOL

1. Wash periodate-activated Sepharose CL-6B (10 ml settled gel; prepared as described in Chapter 2, Section 2.2.1.4) on a fritted glass filter funnel with 100 ml water and 100 ml 0.1 M sodium phosphate buffer, pH 7.0 (coupling buffer). Suction dry the gel to a wet cake.
2. Add the wet cake to a 10-ml solution of SAP (Pierce; Worthington; Boehringer Mannheim) at a concentration of 1 mg/ml in coupling buffer.
3. Add $NaCNBH_3$ (60 mg) to the gel suspension and gently mix the reaction mixture on a shaker at 4°C for 24 hr.

NOTE: $NaCNBH_3$ is highly toxic. The coupling reaction and the subsequent steps should be carried out in a well-ventilated hood. We routinely use a 100-ml Erlenmeyer flask as the reaction vessel.

4. Excess aldehyde groups on the gel may be blocked if desired. However, we have found that any remaining formyl groups do not contribute to nonspecific effects in the resulting affinity support. For this reason, we typically do not do a blocking step. To block, treat the support as follows. First, wash the gel with 200 ml coupling buffer. Then block the unreacted aldehyde groups of the gel by incubation with 10 ml 1.0 M Tris-HCl and 63 mg $NaCNBH_3$ at room temperature for 1 hr.
5. Finally, successively wash the gel with 50 ml coupling buffer, 100 ml 1.0 M NaCl, 100 ml water, and 100 ml 0.02% sodium azide. Immobilized SAP is stored in 0.02% sodium azide at 4°C. Preparations stored for up to 4 months did not show any detectable decrease in specific activity.

The coupling efficiency of this procedure can be determined to be nearly 85% by quantitative amino acid analysis of the gel. The overall efficiency of immobilized SAP with respect to enzymatic activity (compared with the soluble enzyme) has been calculated from kinetic analysis of the hydrolysis of performic acid-oxidized RNase A. The specific activity of immobilized SAP is found to be 27% that of the soluble enzyme (Sahni *et al.*, 1991).

3.3.5 Preparation of Immobilized β-D-Galactosidase

β-Galactosidases occur in many types of microorganisms, plants, and animals. The enzyme catalyses the hydrolysis of β-D-galactoside into galactose and alcohol. It consists of four identical subunits, each with its own active site. Each subunit has a molecular mass of 135,000 daltons; therefore, the intact β-galactosidase molecule is a rather large 540,000 daltons. A 1 mg/ml solution of the enzyme will give an absorbance at 280 nm of 2.09 (1-cm path).

β-Galactosidase has been used extensively as an analytical tool in the determination of lactose, as an enzyme label in immunoassays, and as a processing agent in the food industry. Most industrial applications require that the enzyme be in an immobilized form.

Enzyme activity may be measured spectrophotometrically at 405 nm using the substrate ONPG (*o*-nitrophenyl β-D-galactopyranoside). Optimal enzyme activity is observed in the pH range 6–8 (Worthington catalog).

IMMOBILIZATION PROTOCOL

1. Wash periodate-activated Sepharose CL-6B (10 ml settled gel; prepared as described in Chapter 2, Section 2.2.1.4) on a sintered glass funnel with 100 ml water and 100 ml 0.1 M sodium phosphate buffer, pH 7.0, containing 1 mM $MgCl_2$ (coupling buffer). Suction dry the gel to a wet cake.
2. Add the wet cake to a 10-ml solution of β-D-galactosidase at a concentration of 2 mg/ml in coupling buffer.
3. Add sodium cyanoborohydride (60 mg) to the gel suspension and gently mix the reaction mixture on a shaker at 4°C for 24 hr.

NOTE: sodium cyanoborohydride is highly toxic. Therefore, the coupling reaction and the subsequent steps should be carried out in a well-ventilated hood. We routinely use a 100-ml Erlenmeyer flask as the reaction vessel.

4. Excess aldehyde groups on the gel may be blocked if desired. However, we have found that any remaining formyl groups do not contribute to nonspecific effects in the resulting affinity support. For this reason, we typically do not do a blocking step. To block, treat the support as follows. Wash the gel with 200 ml coupling buffer and block the unreacted aldehyde groups of the gel by incubation with 10 ml 1.0 M Tris-HCl, pH 7.4, and 60 mg sodium cyanoborohydride at room temperature for 30 min.
5. Finally, successively wash the gel with 50 ml coupling buffer, 100 ml 1.0 M NaCl, and 100 ml water. Immobilized β-D-galactosidase can be stored in 50% glycerol containing 0.02% sodium azide and 1 mM $MgCl_2$ at 4°C. Preparations of immobilized β-D-galactosidase stored for up to 6 months did not show any detectable decrease in specific activity.

Immobilized β-D-galactosidase prepared by this protocol has an activity of approximately 350 ONPG units per ml gel (350 ONPG units correspond to 1 mg β-D-galactosidase). Apparently, 50% of the enzyme activity is lost during immobilization.

3.4 PEPTIDE ANTIGENS, ANTIBODIES, AND IMMUNOGLOBULIN BINDING PROTEINS

Immunochemical preparations often require the purification of specific antibody or antigen molecules (Harlow and Lane, 1988). Affinity chromatography plays an important role in almost all antibody purification schemes (Manil et al., 1986), whether they use polyclonal antisera, monoclonal ascites, cell culture supernatant as the source. For instance, immobilized antigens may be used to isolate specific antibody molecules easily from a mixture of many different immunoglobulins in serum (Vunakis and Langone, 1980). Immobilized antibody binding proteins can be used to purify the Ig fraction from serum or to isolate monoclonals from cell culture broths or ascites fluid pools. Certain immobilized immunoglobulin binding proteins can interact with specific classes of antibody molecules, pulling out only an IgM type from all others. Immobilized antibodies themselves can be made into powerful immunoaffinity supports that have the capability of binding and purifying almost any biological molecule imaginable.

In this section, the preparation of some of the most representative immunoaffinity supports is described. The immobilization methods used here will result in stable affinity matrices with extremely low nonspecific binding. We have routinely used all these methods in solving many of our own purification problems.

3.4.1 Immobilization of Peptide Antigens

The immobilization of antigen molecules for the purification of specific antibody is a broad subject that could encompass all the immobilization methods discussed in this book. At least in theory, any antigen can be coupled to a support by one or more of the activation and coupling methods that we have outlined throughout the earlier chapters. However, since peptide antigens are perhaps the most dominant type of antigen employed, a special section on how to immobilize these molecules was deemed necessary.

Although peptides may be accurately thought of as small protein fragments, their properties are often markedly different from those of larger protein molecules. Since the peptides used for antibody production are often the critical short sequences found on the immunogenic portions of larger proteins, these segments may have solubility properties that vary widely from those of the parent polypeptide. Peptide sequences that have a preponderance of hydrophobic residues may be especially difficult to solubilize in an aqueous environment. For immobilization purposes, methods frequently must be chosen to provide an organic phase in the reaction medium to aid solubility.

The immobilization chemistries useful in peptide coupling include those reactive to amines, carboxylic groups, sulfhydryls, and active hydrogens on aromatic side-chain residues. The proper choice of coupling methodology can result in a peptide that is better able to interact with specific antibody molecules since its orientation on the support is enhanced. Careful site-directed immobilization techniques are typically considered to be better than methods that result in random coupling. Today, synthetic peptide molecules may be constructed to have an appropriate coupling group at one end of the molecule to insure a certain orientation for antibody presentation. For instance, adding a cysteine residue to one end allows a peptide to be coupled only through the side-chain sulfhydryl group. This causes the rest of the molecule to protrude from the gel, free to interact with specific antibody. Reduction of disulfide residues in a peptide molecule also can provide known reactive sites for immobilization. The presence of just one tyrosine residue can give a definite point of attachment if a reactive hydrogen coupling chemistry is chosen for immobilization. In general, however, due to the natural abundance of amines and carboxyls in peptides, coupling via these groups usually results in many orientations of the peptide molecule on the matrix. For more defined orientation, immobilization ideally should be done through groups that are present only once in a peptide, regardless of the particular functionality involved in coupling.

3.4.1.1 Coupling Peptides via Amine or Carboxylate Groups

As discussed earlier, peptide immobilization done through available amine or carboxylate groups often results in many different orientations of the molecule on the matrix. This is especially true if many of these groups are present on the molecule. Although the exact structure of the coupled peptide may not be as well defined as it is when using a site-directed strategy, antibody interaction with such a ligand usually will not suffer. Therefore, coupling through the C- and N-termini as well as through side-chain carboxylate and amine groups (aspartate, glutamate, and lysine amino acids) is often the easiest route for immobilization (Fig. 3.36).

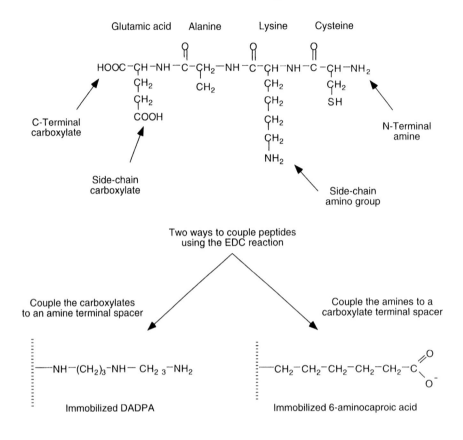

FIGURE 3.36

Small peptide molecules easily can be immobilized to either amine-containing matrices or carboxylate-containing matrices using the carbodiimide reaction with EDC. The abundance of amines and carboxylic acids on peptides makes this procedure very efficient. The only disadvantage of using EDC is the potential for polymerization of peptides during coupling. If the isolation of a specific antibody to the peptide is the goal, however, such polymerization is often not detrimental to antibody recognition.

With small peptides, spacer arms should be used to allow adequate antibody access and interaction. The spacer serves to extend the ligand from the surface of the support (see Section 3.1.1) and allows the much larger immunoglobulin to orient for binding. When coupling through amine groups on the peptide molecule, choose spacers that terminate in carboxylate functionalities; use the opposite logic for coupling through carboxylate groups on peptides. For larger peptides, direct attachment to an activated matrix will give satisfactory results.

IMMOBILIZATION PROTOCOL

a. Coupling via Amine or Carboxylate Groups Using EDC

The following protocol involves the use of the water-soluble carbodiimide EDC to form an amide bond between the primary amine spacer DADPA and the peptide ligand. To couple to amine groups on the peptide, the succinylated form of immobilized DADPA may be used with no further modifications in the overall methodology. One word of caution should be given when coupling peptides with both amines and carboxylate functions using this protocol: EDC will effect not only the covalent attachment of the peptide to the support, but will also cause peptide polymerization, both in solution and extending from the matrix. For antibody purification, this result is not detrimental. In fact, polymerized peptides often are more immunogenic than the small peptide monomer from which they are made. However, for the purification of some other receptor protein that may recognize only the native peptide structure, such polymerization may cause nonspecific interactions or complete lack of recognition. In these instances, it is best to use another method of immobilization, for example, direct coupling (see next section).

1. Wash 10 ml immobilized DADPA (prepared according to Section 3.1.1.1) into 0.1 M MES, pH 4.7 (coupling buffer). If the peptide ligand is insoluble in a completely aqueous solution, include up to 50% methanol or 50% ethanol in this buffer to aid solubility. Filter off excess buffer until the gel remains as a wet cake.
2. Dissolve the peptide to be coupled in 10 ml coupling buffer. Again, up to 50% of either methanol or ethanol may be added to effect solubilization. For particularly insoluble peptides, dissolving first in methanol or ethanol and then diluting in coupling buffer may give better results. In general, the more concentrated the peptide solution, the more ligand will be coupled. Since many peptides are of relatively low molecular weight, using a concentration of 1–5 mg/ml in the coupling buffer is usually sufficient.
3. Mix the peptide solution with the gel and add 300 mg EDC.
4. React for at least 4 hr. Stir with a paddle stirrer or by gentle mixing on a rocker.

5. Wash the gel with coupling buffer (containing the organic phase, if added), water, 1 M NaCl, and water.

b. Coupling via Amine Groups Directly to a Matrix

Peptide coupling via amine groups directly to an activated matrix may be done by any of the amine coupling chemistries discussed previously (see Chapter 2). If the peptide molecule is water soluble, perhaps the best method of direct immobilization involves the periodate oxidation/reductive amination procedure on agarose. For organic soluble peptides, choose one of the activation methods that can be done in organic solvents (e.g., CDI, FMP, or Tresyl). React 1–5 mg peptide per ml support, and follow the generalized protocols given under the particular activation chemistry section.

3.4.1.2 Coupling Peptides via Sulfhydryl Groups

The presence of a sulfhydryl group in a peptide molecule provides an advantageous point of attachment that eliminates the problems of polymerization or uncertain orientation of the immobilized ligand. Many peptides constructed synthetically for immunization purposes have a cysteine residue purposely attached at one end that is then used for conjugation to a carrier protein. This same sulfhydryl functionality can be used for immobilization to an affinity support to subsequently purify the antibodies so produced. The use of an activation chemistry such as iodoacetyl, bromoacetyl, DVS, epoxy, or maleimide (see Sections 2.2.2.1, 2.2.2.2, 2.2.2.4, and 2.2.2.5, Chapter 2) will provide the site-directed qualities necessary for coupling sulfhydryls, producing a covalent thioether bond. The use of activation chemistries such as pyridyl disulfide (Section 2.2.2.3, Chapter 2) or TNB-thiol (Section 2.2.2.6, Chapter 2) will also immobilize sulfhydryl-containing peptides, but with reversible disulfide linkages. Regardless of which of these methods is chosen, the coupling chemistry will provide an endogenous spacer to push the peptide away from the matrix in an orientation better suited to interacting with specific immunoglobulins. Figure 3.37 shows the chemistry of attachment via sulfhydryl active supports.

The second scheme for immobilizing a peptide through sulfhydryl groups involves reducing intra- or intermolecular disulfide bonds. When peptides contain a cystine residue or when an existing solitary sulfhydryl is in the oxidized form (a common occurrence with peptides stored for any length of time), these may be reduced by using an immobilized sulfhydryl reductant (Section 3.8). The use of an immobilized reductant is easier than using a soluble one such as DTT because the problems associated with the

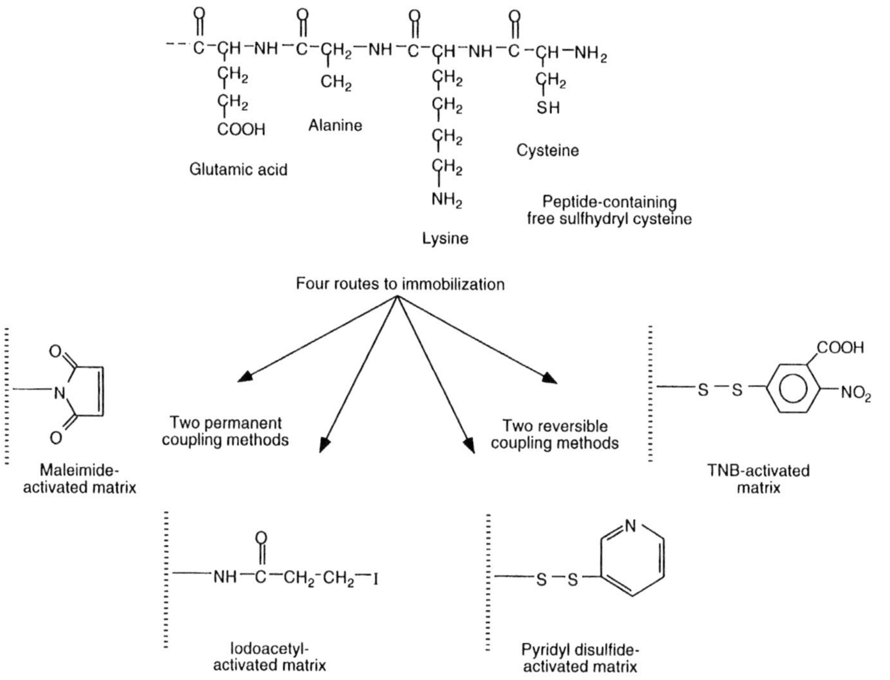

FIGURE 3.37
Coupling of peptide molecules through available sulfhydryl residues (cysteine) can take four potential routes. Maleimide or iodoacetyl activation methods will result in permanent covalent attachment through thioether linkages. In contrast, pyridyl disulfide- or TNB-thiol-activated matrices will provide reversible attachment of sulfhydryl ligands.

subsequent removal of the reducing agent are avoided. The peptide bearing a free sulfhydryl can then be coupled to the appropriate activated matrix.

IMMOBILIZATION PROTOCOL

1. Dissolve 1–5 mg sulfhydryl-containing peptide (or at least 50–100 µmol) in 10 ml cold coupling buffer (0.1 M sodium phosphate, 0.15 M NaCl, pH 7.2). Ellman's reagent (Section 4.1.9, Chapter 4) may be used to verify that the cysteine sulfhydryls are indeed in the reduced state prior to coupling. If a disulfide group needs to be reduced for immobilization, follow the protocol described for the immobilized reductant homocysteine (Section 3.8).

2. Wash 10 ml maleimide-activated gel (prepared as described in Section 2.2.2.2, Chapter 2) with cold coupling buffer and filter to a wet cake.
3. Immediately mix the peptide solution with the wet gel cake.
4. Mix the reaction by rocking and allow it to react at 4°C for at least 2 hr.
5. Wash the affinity support with coupling buffer, 1 M NaCl, and water. Store the gel as an aqueous slurry containing a preservative at 4°C.

3.4.1.3 Coupling Peptides via Active Hydrogens

Tyrosine residues represent another good target for site-directed immobilization of peptide molecules, since these amino acids are usually present in low molar amounts. Directing the coupling reaction toward the phenolic side chain of tyrosine will insure that the peptide will be presented to an antibody or other binding molecule in a known orientation.

Potentially, two choices are available for active hydrogen coupling: the diazonium reaction or the Mannich condensation reaction with amine plus formaldehyde (see Chapter 2). Unfortunately, neither of these reactions is 100% reliable with peptide molecules. Some peptides seem to couple readily and in high yields, whereas others have poor reactivity. This characteristic of the diazonium reaction may be due to its difficult-to-control reactivity, but low yields in the Mannich reaction are typically because of competing amines on the peptide molecule. Since N-terminal and lysine side-chain amines potentially participate in the Mannich reaction, this interference renders the method almost impractical. A possible solution to avoid participation of peptide amines in the Mannich reaction is to protect the amines with an easily removable group (e.g., trifluoroacetyl). We have not attempted this approach in our laboratories. Thus, the diazonium reaction is probably the best choice for specific coupling of tyrosine residues in peptides. Figure 3.38 illustrates the chemistry of the diazonium coupling reaction for peptides.

IMMOBILIZATION PROTOCOL

1. Convert 10 ml of a precursor gel containing aminophenyl groups to the diazonium form using sodium nitrite and HCl at 4°C, according to the protocol given in Chapter 2 (Section 2.2.5.1). Immediately wash the gel with cold coupling buffer (0.1 M sodium borate, pH 9.0) and then drain to a wet cake.
2. Prepare 10 ml of a peptide solution (3–5 mg/ml) in cold coupling buffer and quickly add to the wet gel. For peptides that are only sparingly soluble in aqueous environments, the borate coupling buffer may be made 50% ethanolic to aid solubility. The peptide itself may be dissolved first

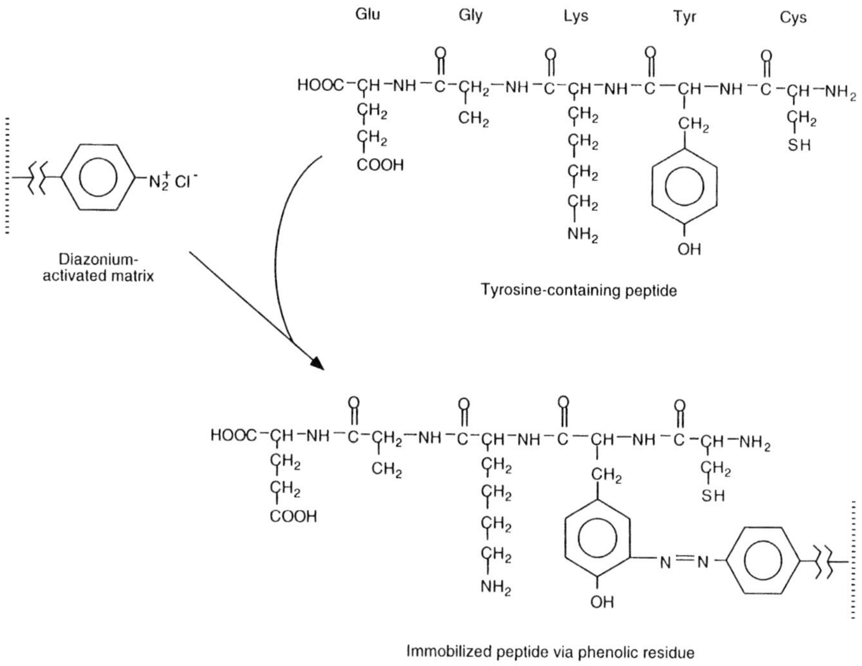

3.4 Peptide Antigens, Antibodies

FIGURE 3.38

Tyrosine-containing peptides can be immobilized through the phenolic side chain of the amino acid using diazomium chemistry.

---

in ethanol and then diluted in coupling buffer to further aid its dissolution.
3. React overnight at 4°C using a rocker or gentle shaker.
4. Wash the gel extensively with coupling buffer to remove unreacted ligand. If ethanol was included to aid solubility, use the ethanolic buffer. Some decrease in flow rate through the gel may be noticed, especially if severe cross-linking occurred in the matrix during activation and coupling.

3.4.2 Immobilization of Antibody Molecules

The immobilization of polyclonal antibodies usually proceeds with excellent yields and good retention of antigen binding activity. To obtain the best specificity in the resulting immunoaffinity matrix, only affinity purified antibodies should be used. These are antibodies that have been isolated from antisera by affinity chromatography using the corresponding immobilized antigen. Thus, these preparations contain only that population of antibody molecules in serum that has the desired antigen specificity. The immobilization of whole immunoglobulin fractions should be avoided, since many different antibody specificities are represented in the mixture; together they may cause nonspecific binding properties in the affinity support.

Monoclonal antibodies should also be affinity purified from cell culture supernatant or ascites prior to immobilization. Monoclonals that can be successfully affinity purified are usually stable enough to undergo coupling to an activated matrix. Occasionally, however, a particular monoclonal will be partially or totally inactivated through the coupling reaction. Sometimes the activity loss is caused by the blocking of antigen binding sites or by conformational changes in complementarity-determining regions. If the antigen binding site is merely blocked during immobilization, this problem may be overcome by choosing an appropriate site-directed activation and coupling procedure for antibodies (discussed subsequently). On the other hand, some monoclonals are too labile to undergo immobilization, regardless of the coupling method. When working with monoclonals, trial and error is often necessary to determine whether immobilization is feasible or not.

The structure of an antibody molecule can provide insight into the best methods to use for immobilization (Roitt, 1977; Goding, 1986). A detailed illustration of the antibody structure is shown in Figure 3.39A; the simplified stick version is shown in Figure 3.39B. The most basic immunoglobulin (Ig) molecule is composed of two light and two heavy chains held together by disulfide bonds. The light chains are disulfide bonded to the heavy chains in the C_L and C_H^1 regions, respectively, whereas the heavy chains are disulfide bonded to one another in the hinge region. There are two identical antigen binding sites on each molecule, each formed by the variable regions of one light and one heavy chain. Enzymatic digestion of IgG with papain (Fig. 3.40A) produces two fragments, each containing an antigen binding site (called Fab fragments), and one large fragment containing only heavy chains (called Fc, for "fragment crystallizable") (Coulter and Harris, 1983). When the antibody molecule is cleaved enzymatically with pepsin, one large and several smaller fragments are produced (Rousseaux et al., 1983). The antigen binding fragment, termed $F(ab')_2$, in this case retains both binding sites and is held together by the disulfide bonds in the hinge region. The Fc fragment, however, is degraded into many pieces by this enzyme (Fig. 3.40B).

The two heavy chains in the immunoglobulin molecule are identical; each has a molecular mass of about 50,000–75,000 daltons (depending on the class of antibody). Likewise, the two light chains of an antibody mole-

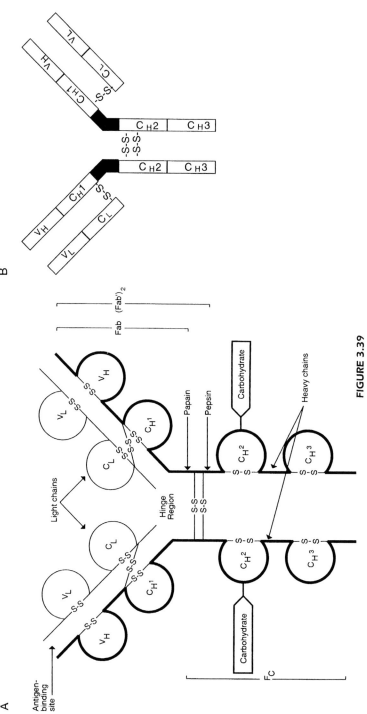

FIGURE 3.39

(A) Schematic drawing of a typical IgG molecule. The circular intrachain structures are the major domains, the H designation corresponding to the heavy chains and the L designation to the light chains. The lower portion of the heavy chain pairs is called the Fc region, which contains most of the effector functions of the immunoglobulin (for example, complement activation). The upper heavy/light chain region is called Fab or F(ab')$_2$ is bivalent due to the retention of the interchain disulfides in the hinge region. Carbohydrate may be present at various locations, but primarily off the C_H2 domains. (B) The typical IgG molecule, shown detailed in Fig. 3.39A, is often represented as a stick structure. The simplest versions in this volume have no domain labels and use lines to represent the disulfide bridges shown here.

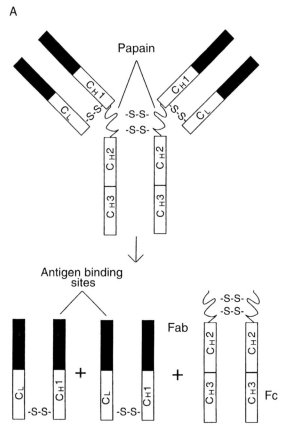

FIGURE 3.40A

Papain cleavage of IgG molecules occurs above the hinge region disulfides, liberating two Fab fragments (each containing one antigen binding site) and one intact Fc piece.

---

cule are identical; each has a molecular mass of about 25,000 daltons. An intact IgG molecule thus will have a molecular mass in the range of 150,000–160,000 daltons.

There are actually two forms of light chain that may be incorporated into immunoglobulins (designated κ and λ) and five different heavy chain varieties (designated γ, μ, α, ε, and δ). The particular heavy chain identifies the immunoglobulin by class, either IgG, IgM, IgA, IgE, or IgD. Most antibodies prepared for immunoaffinity or assay work will be of the IgG or IgM class. Any antibody molecule belonging to a particular class of immunoglobulin may have either light chain variety (λ or κ), but only one heavy chain

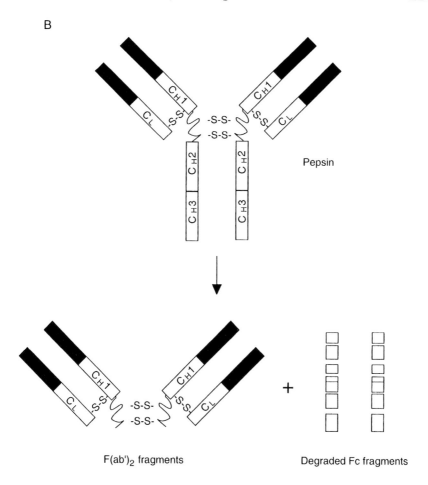

FIGURE 3.40B

Pepsin cleavage of IgG antibodies occurs below the hinge region disulfides, producing one F(ab')$_2$ fragment containing two antigen binding sites. The Fc region is typically extensively degraded.

---

type. IgG, IgE, and IgD class antibodies consist of the basic Ig monomeric structure, whereas IgM molecules are pentameric constructs of the Ig monomer (Fig. 3.41). IgA can be either a singlet, doublet, or triplet of Ig-type molecules. Both IgM and IgA contain (in their polymeric forms) a single J chain subunit, a very acidic polypeptide of molecular mass 15,000 daltons that is rich in carbohydrate. The heavy chains of immunoglobulins are also glycosylated, usually in the C_H2 domain of the Fc fragment.

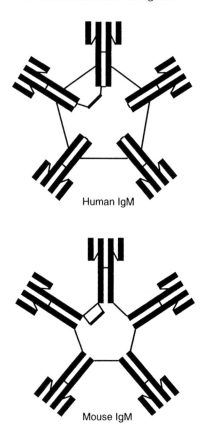

FIGURE 3.41
The pentameric IgM molecule contains five Ig units bonded together by disulfide bridges and the J chain. The structural differences between mouse and human IgM are reflected in the different fragments produced by enzymatic cleavage.

Antibody molecules thus have a number of different functionalities available for immobilization. They may be coupled via lysine and N-terminal amines or through C-terminal, aspartic, and glutamic carboxylic residues as in general protein immobilization procedures. Alternatively, the specific fragmentation of antibody molecules through enzymatic cleavage or disulfide reduction can provide discrete coupling sites away from the antigen binding region. Mild carbohydrate oxidation can also provide formyl groups that may be exploited for immobilization. The main factor in choosing which activation and coupling method to use typically involves the most important criterion, the retention of antigen binding activity. In the following sections, protocols for each of these methods are presented.

3.4.2.1 Preparation of Immobilized Anti-Human Serum Albumin

Anti-human serum albumin (anti-HSA) is an IgG antibody with all the idiosyncrasies encountered when immobilizing intact polyclonal immunoglobulins. If treated with some care, most immunoglobulins prepared from antisera will behave in a robust manner and survive immobilization with antigen binding activity preserved. The procedures outlined in this section can be used with the great majority of polyclonal antibodies with little or no difficulty. Both of these methods couple antibodies through their available primary amine groups and do not result in a specific immobilized orientation.

Either of the protocols described next can be used to purify human serum albumin according to the method in Section 5.1.4, Chapter 5.

IMMOBILIZATION PROTOCOL

a. Using CNBr Activation

1. Wash 10 ml freshly prepared CNBr-activated agarose (Section 2.2.1.1, Chapter 2) with cold water and then with cold coupling buffer (0.1 M sodium phosphate, 0.15 M NaCl, pH 7.5). Drain off excess buffer until the gel is left as a wet cake. Use immediately.
2. Dissolve anti-HSA in 10 ml cold coupling buffer at a concentration of 10–20 mg/ml. Stir the solution gently, being careful to limit the degree of frothing. If any insolubles are evident, centrifuge or filter prior to adding the solution to the gel.
3. Mix the activated gel with the antibody solution and stir overnight at 4°C.
4. Wash the gel with coupling buffer, water, 1 M NaCl, and again with water.

b. Using Periodate Oxidation/Reductive Amination

1. Wash 10 ml periodate-oxidized agarose (Section 2.2.1.4, Chapter 2) with cold water and then with cold coupling buffer (0.1 M sodium phosphate, 0.15 M NaCl, pH 7.0). Drain off excess buffer until the gel is left as a wet cake.
2. Dissolve anti-HSA in 10 ml cold coupling buffer at a concentration of 10–20 mg/ml. Stir the solution gently, being careful to limit the degree

of frothing. If any insolubles are evident, centrifuge or filter prior to adding the solution to the gel.
3. Mix the activated gel with the antibody solution and stir overnight at 4°C.
4. Wash the gel with coupling buffer, water, 1 M NaCl, and again with water.

3.4.2.2 Preparation of Immobilized Anti-α_2-Macroglobulin Using Immobilized Protein A Cross-Linked with DMP

The main concern with traditional methods of antibody immobilization is the retention of antigen binding activity. Even stable antibody preparations can lose effective activity if their antigen binding sites are blocked by the coupling process. To overcome the potential to obstruct these regions, immobilized protein A (Section 3.4.4.1) can be used to orient the immunoglobulin with its antigen binding sites pointing away from the matrix. The Fc binding properties of protein A leave the antigen specific sites of the immunoglobulin molecule free to interact with antigen (Fig. 3.42). Cross-linking this complex with a bifunctional cross-linker such as dimethyl pimelimidate (DMP) creates a stable immunoaffinity support with excellent retention of antigen binding activity (Schneider *et al.*, 1982).

Unfortunately, using protein A to form this immobilized antibody complex gives considerable nonspecific binding properties to the resulting affinity support. In the example described next, DMP was used to cross-link a polyclonal antibody specific for human α_2-macroglobulin to an immobilized protein A support. This immunoaffinity support then was used in our laboratories to attempt the purification of α_2-macroglobulin from human serum.

Normal serum will contain about 2–3 mg/ml α_2-macroglobulin, but up to 20 mg/ml various immunoglobulins. With this level of interfering IgG molecules, very little α_2-macroglobulin was observed to bind (data not shown). To check the degree of IgG binding to the protein A component of the support, purified human IgG was applied in a separate experiment and found to bind at a level of 3–5 mg/ml gel. Thus, enough protein A immunoglobulin binding sites were available to obliterate the specific interaction between the coupled antibody and α_2-macroglobulin.

Given these results, we have found that this method of antibody immobilization works well only if there are no immunoglobulins in the solution from which the antigen is to be purified. In instances in which a serum antigen is to be isolated, it is probably best to use another method of antibody coupling to avoid the nonspecificity that protein A contributes to the support. Of all the methods of antibody immobilization, the protein A approach probably results in the greatest nonspecific binding potential.

IMMOBILIZATION PROTOCOL

1. Wash 10 ml immobilized protein A (prepared as described in Section 3.4.4.1) with water and then with antibody binding buffer (50 mM sodium borate, pH 8.2). Drain the gel of excess buffer until it is left as a wet cake.
2. Dissolve the antibody to be coupled (up to 12–15 mg/ml gel) in 10 ml antibody binding buffer.

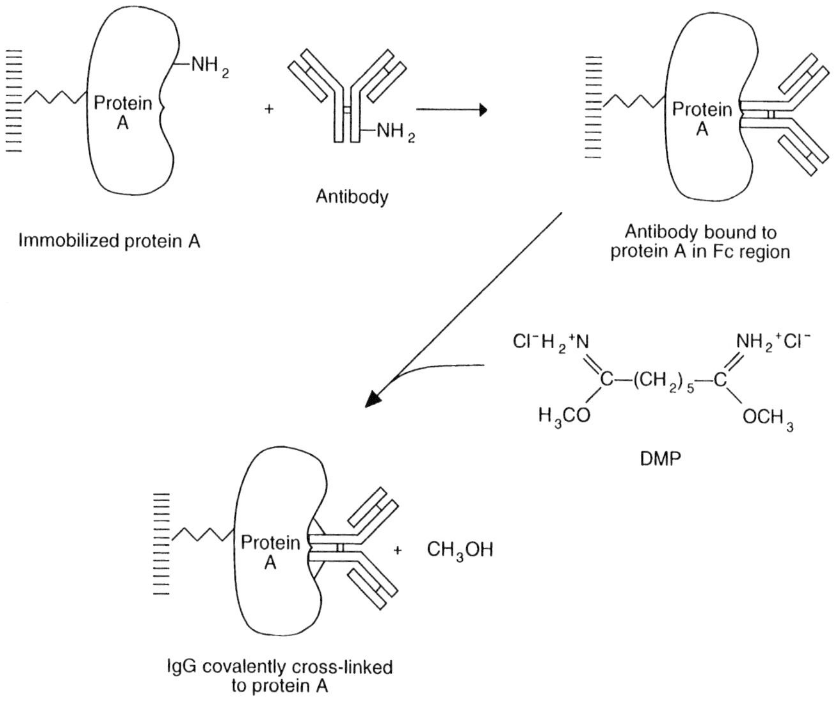

FIGURE 3.42

Immobilized protein A can be used to immobilize an antibody molecule by taking advantage of the natural affinity of protein A for immunoglobulins. Incubation of a specific antibody with a protein A matrix will bind the antibody in the Fc region, away from the antigen binding sites. Subsequent cross-linking of this complex with DMP (dimethyl pimelimidate) yields a covalently attached antibody with the antigen binding sites facing outward and free to interact with antigen.

3. Add the antibody solution to the gel cake, and mix by gentle rocking for 30 min.
4. Wash the gel with 5 gel volumes of antibody binding buffer and 1 volume of 0.2 M triethanolamine, pH 8.2 (cross-linking buffer). Drain to a wet cake.
5. Dissolve 66 mg DMP (Pierce, Rockford, Illinois) into 10 ml cross-linking buffer. Immediately add this solution to the gel cake and mix by gentle rocking.
6. React for 1 hr at room temperature.
7. Wash the gel with 5 bed volumes of water and drain to a wet cake.
8. To block any remaining cross-linker active sites, add 10 ml 0.1 M ethanolamine, pH 8.2, to the gel cake. React for 10 min at room temperature.
9. Wash the gel extensively with water, 1 M NaCl, and then with 0.1 M glycine, pH 2.8, to remove noncovalently bound antibody from the protein A support. Finally, wash with water.

3.4.2.3 Site-Directed Immobilization of Antibody

The term "site-directed immobilization" of an antibody merely means to couple to known positions within the three-dimensional structure of the immunoglobulin (Domen et al., 1990). By proper selection of the immobilization chemistry and a knowledge of antibody structure (discussed earlier), the molecule can be oriented on the support so its bivalent binding potential for antigen can be realized fully.

Two immobilization methods are especially useful in this regard. The disulfides in the hinge region that hold the heavy chains together can be reduced selectively with 2-mercaptoethylamine (2-MEA) to form two half-immunoglobulin molecules, each containing an antigen binding site (Palmer and Nissonoff, 1963). Alternatively, the same disulfides can be reduced with 2-MEA from the F(ab')$_2$ fragments formed from pepsin digestion. Following the recommended protocol, the treatment will cause only limited reduction of the disulfides between the heavy and light chains, thus retaining antigen binding activity. Immobilization of the resulting free sulfhydryls using any of the activation chemistries designed to couple these groups (Section 2.2.2, Chapter 2) will provide covalent attachment away from the antigen binding site. Figure 3.43A shows the reduction of intact antibody and F(ab')$_2$ fragments yielding fragments containing one antigen binding site per molecule. These reduced fragments can be coupled to a sulfhydryl-reactive support such as iodoacetyl-activated agarose (Fig. 3.43B). The advantage of using reduced F(ab')$_2$ fragments over using reduced whole antibody is based on the fact that most of the Fc fragment is not present to give potentially non-antigen-specific interactions. The fol-

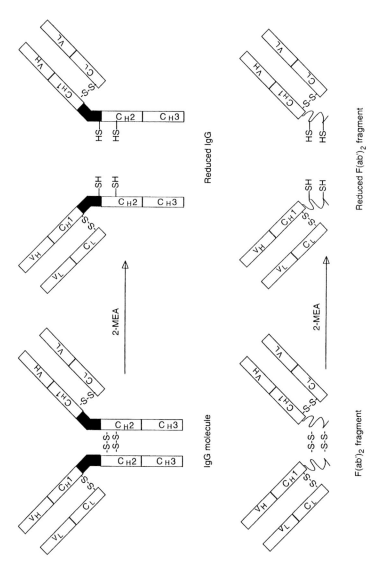

FIGURE 3.43A

2-Mercaptoethylamine (2-MEA) is a gentle immunoglobulin-reducing agent that will primarily cleave disulfides in the hinge region of IgG. With intact antibody, 2-MEA will yield two half molecules, each containing an intact heavy and light chain. Using F(ab')₂ fragments, 2-MEA reduction will produce two fragments, each containing one antigen binding site. Reduction of either species will generate free sulfhydryl groups suitable for immobilization to sulfhydryl reactive supports.

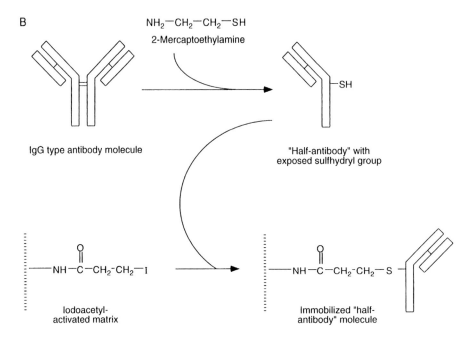

FIGURE 3.43B

Site-directed immobilization of antibody can be accomplished through sulfhydryl residues if the immunoglobulin is first reduced with 2-MEA. Iodoacetyl-activated matrices will couple the reduced antibody in the hinge region away from the antigen binding area.

lowing protocol describes the immobilization of these reduced species to iodoacetyl-activated support materials.

The second method of site-directed immobilization takes advantage of the carbohydrate chains coming off the heavy chains in the $C_H{}^2$ domain. Mild oxidation of the sugar residues using sodium periodate will generate formyl groups. Hydrazide activation chemistry (Section 2.2.3.1, Chapter 2) then can be used to immobilize the antibody specifically through these modified carbohydrate residues. This method results in the coupling of intact antibody molecules and usually gives the highest yield of antigen binding site activity (Fig. 3.44).

Table 3.1 shows the results of a comparison of three immobilization methods for coupling rabbit anti-HSA molecules. As can be seen from the molar ratio of antibody coupled to the resulting amount of antigen (HSA) bound (column capacity determined by affinity overload analysis), the site-directed methods yield supports with higher activity than the random cou-

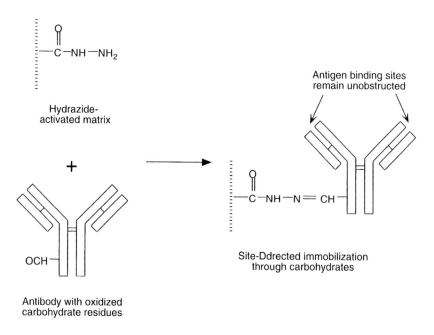

FIGURE 3.44
Mild oxidation of antibody carbohydrate chains with sodium *meta*-periodate results in formyl groups that can be specifically coupled to a hydrazide-activated support. This method of site-directed immobilization results in covalent attachment away from the antigen binding sites.

pling approaches of immobilizing through amine groups (by reductive amination). In fact, only the site-directed method of coupling to the oxidized carbohydrate residues using hydrazide chemistry resulted in the theoretical maximum (for a bivalent immunoglobulin) of two antigen molecules bound for every IgG molecule immobilized (Domen *et al.*, 1990).

TABLE 3.1
Comparison of Three Activation Methods for Coupling Immunoglobulins to Solid Supports

Coupling chemistry	Coupling efficiency (%)	Ratio of immobilized antibody to bound antigen
Reductive amination	79	1:1
Iodoacetate	74	1:1.2
Hydrazide	90	1:2

COUPLING ANTIBODIES THROUGH SULFHYDRYL RESIDUES

1. Dissolve 1–10 mg intact antibody (IgG) or F(ab')$_2$ fragments in 2 ml 0.1 M sodium phosphate, 0.15 M NaCl, 5 mM EDTA, pH 6.0.
2. Add 12 mg 2-MEA to the antibody solution and incubate at 37°C for 1.5 hr.
3. Desalt the reduced antibody solution using a 15-ml Sephadex G-25 column equilibrated with 50 mM Tris, 0.15 M NaCl, 5 mM EDTA, pH 8.5. Monitor the protein elution peak by measuring the absorbance at 280 nm. Pool the fractions containing protein (the first peak off the column).
4. Equilibrate 2 ml iodoacetyl-activated gel (Section 2.2.2.1, Chapter 2) with 10 ml 50 mM Tris, 0.15 M NaCl, 5 mM EDTA, pH 8.5 (coupling buffer).
5. Add the reduced antibody fractions from Step 3 to the gel cake from Step 4.
6. React with gentle mixing for 1 hr at room temperature.
7. Wash the gel with at least 5 volumes of coupling buffer.
8. Block unreacted iodoacetyl sites by mixing the gel with 2 ml 0.05 M cysteine in 50 mM Tris, 0.15 M NaCl, 5 mM EDTA, pH 8.5. React for 30 min at room temperature with mixing.
9. Wash the gel thoroughly with water, 1 M NaCl, and again with water.

COUPLING ANTIBODIES THROUGH OXIDIZED CARBOHYDRATE RESIDUES

1. Dissolve 1–10 mg antibody (IgG) per ml in 2 ml 0.1 M sodium phosphate, pH 7.0.
2. Add 10 mg sodium *meta*-periodate to the antibody solution and gently mix for 30 min at room temperature.
3. Purify the oxidized antibody by desalting on a 15-ml Sephadex G-25 equilibrated with 0.1 M sodium phosphate, pH 7.0. Pool the first peak off the column (monitor A_{280}), which contains the oxidized antibody.
4. Wash 2 ml hydrazide-activated gel (Section 2.2.3.1, Chapter 2) with 10 ml 0.1 M sodium phosphate, pH 7.0 (coupling buffer).
5. Add the pooled oxidized antibody fractions to the gel cake and react with gentle mixing for 6 hr at room temperature.
6. Wash the gel thoroughly with coupling buffer, water, 1 M NaCl, and again with water.

3.4.2.4 Immunoglobulin Coupling to Polystyrene

Many forms of polystyrene are available for the immobilization of antibody molecules for use in immunoassay procedures. The most popular of these options are the 96-well microtiter plates and $1/8$ in. and $1/4$ in. polystyrene balls. Antibodies (or antigens) attached to the surfaces of these devices can function in heterogeneous immunoassays designed to separate and measure almost an infinite variety of analytes. The process of antibody immobilization on these nonporous surfaces usually involves either a noncovalent passive adsorption phenomenon or the use of a specific modification and coupling chemistry that can covalently bind the molecules. The following sections discuss the most popular of these immobilization methods. Reference should also be made to Chapter 1, which discusses the various forms of polystyrene available.

Passive Adsorption

Passive adsorption techniques for immobilizing antibody molecules on polystyrene generally work well and are used more frequently than covalent attachment procedures (Tijssen, 1985). Noncovalent coating of antibody on plates or balls is a poorly understood process that is believed to involve primarily hydrophobic interactions. The adsorption proceeds independent of the molecular weight, pI, or charge of the antibody or protein being coated. Mild denaturation of protein molecules usually results in greater adsorption efficiency, presumably due to the revealing of inner hydrophobic domains that can bind more easily to the polystyrene surface.

Antibody molecules can be noncovalently immobilized on polystyrene up to a density of about 1.5 ng/mm^2. This density represents approximately one-third of the surface area available as coated with protein molecules. Additional attachment usually involves protein-to-protein interactions, not plastic-to-protein adsorption. Layering caused by protein–protein interaction should be avoided, because such secondary interactions are inherently unstable compared with the strength of adsorption directly onto the plastic surface. Presumably, steric hindrance inhibits any increase in direct surface interaction beyond the reported one-third coverage limit.

The concentration of antibody solution used to passively coat polystyrene ultimately determines the density of surface coverage and the degree of secondary protein-to-protein layering. The best results seem to occur when antibody solutions of 1–10 µg/ml are used to treat the plastic. Concentrated protein solutions should be avoided, since severe layering will occur and protein may continually leach off the surface during assays.

Noncovalent adsorption of antibody molecules proceeds with little difficulty in a wide range of buffer conditions. The most common procedures recommend 50 mM sodium carbonate at a pH between 9.2 and 9.6. However, a PBS buffer (10 mM sodium phosphate, 0.15 M NaCl), pH 7.2, or a TBS buffer (10 mM Tris, 0.15 M NaCl), pH 8.5, often works just as well. The only major precaution in buffer composition is to avoid components that have the potential to compete for the hydrophobic adsorption sites on the plastic surface. Do not include detergents such as Triton X-100, Tween 20, or NP-40 in the coating buffer, since they will bind to the surface better than the antibody molecules will.

Time and temperature also affect the efficiency of noncovalent adsorption processes. In general, coating at 4°C should be done for at least 18 hr to obtain optimal densities of coverage. At 37°C, however, the coating process may be complete within 90 min.

IMMOBILIZATION PROTOCOL

a. Adsorption of Antibody

1. Prepare an antibody solution from a purified specific antibody (monoclonal or polyclonal) at a concentration of 1–10 µg/ml in 50 mM sodium carbonate, pH 9.5. For a 96-well plate, a 20-ml solution will be sufficient to dispense 150 µl per well with enough extra solution to properly pipette with a multichannel pipetter. For coating polystyrene beads or balls, make enough coating solution to fully immerse the balls in the antibody.
2. Add 150 µl antibody solution to each well of a microtiter plate. Alternatively, submerge polystyrene balls in the antibody solution, making sure they are completely covered.
3. Incubate at 4°C for at least 18 hr, or at 37°C for at least 90 min.
4. Wash the plates or balls at least five times with PBS, pH 7.2, containing 0.05% Tween 20. Add aliquots of 250 µl of each wash solution to the wells of the microplate. For beads or balls, immerse them in wash solution and let them incubate for several minutes. Then remove the solution from the wells or balls and repeat.

b. Blocking Remaining Binding Sites

Coating polystyrene with antibody or protein molecules does not eliminate passive adsorption sites on the plastic surface completely. These excess potential sites of nonspecific binding must be blocked before specific antigen–antibody interactions can be measured in an assay. A blocking step usually involves coating with a nonrelevant molecule that does not interfere with the desired immunoassay. The most popular coating substances include proteins such as BSA, gelatin, casein, or a 5% solution of Carnation® instant milk (see Chapter 4, section 4.2.2.2, as well as the following references: Towbin et al., 1979; Johnson et al., 1984; Hauri and Bucher, 1986; Vogt et al., 1987; Douglas et al., 1988; Harlow and Lane, 1988; Zimmerman and Van Regenmortel, 1989). In some cases, instant milk may interfere with antibody–antigen binding. Sometimes a PBS buffer containing 0.05% Tween 20 will block the nonspecific sites as well as and faster than a protein blocker. For most assay systems built on polystyrene immobilized antibodies, trial and error will reveal the best blocking reagent to use to obtain the lowest nonspecific binding potential. Commercially available blocking agents also can be used to prevent nonspecific binding (Pierce, Rockford, Illinois). See also the discussion in Chapter 4 on nonspecific binding for further details on blocking buffers.

1. Add to each well at least 150 µl 0.25–1% BSA solution in PBS, pH 7.2, containing 0.05% Tween 20. Alternatively, 0.25% gelatin or 5% Carnation instant milk may be used in the same buffer. For balls or beads, submerge them completely in the blocking buffer.
2. Incubate for at least 30 min at room temperature. For some systems, complete blocking may occur in as little as 5 min, but a 30-min incubation will insure good coverage.
3. Wash the plate or balls with the PBS-Tween buffer as described in Step 4.

Covalent Attachment to Polystyrene Beads or Balls

Chemical modification procedures for polystyrene make it possible to covalently couple antigens, antibodies, or other proteins to the ball or bead surfaces. The noncovalent adsorption method just described works for most applications and is still the most commonly used technique for immobilization. However, passive adsorption is not a leakproof technique. Moreover, antibodies passively adsorbed onto polystyrene may lose their activity on storage, presumably because of denaturation on the hydrophobic surface. Some reports have indicated that up to 70% of the immobilized ligand may be lost during an immunoassay procedure if it is only passively adsorbed on the surface of the plastic (Engvall et al., 1971; Lehtonen and Viljanen, 1980). Particularly, when a small ligand (e.g., peptide) is employed, leakage rates may be severe or passive adsorption may not occur at all. This would be true for microplates, balls, or any other polystyrene device. It seems logical that covalent attachment would yield a much more stable immobilized ligand, although it may not always show higher detectability in immunoassays. Covalent attachment, however, is often the only way to immobilize small molecules.

Chapters 1 and 2 describe the major chemistries that can be used to modify the polystyrene surface of solid beads and provide the appropriate functionalities to couple immunoglobulins and other molecules. The following sections describe the corresponding ligand coupling protocols.

IMMOBILIZATION PROTOCOLS

a. Coupling to Alkylamine Beads Using Sulfo-SMCC

Alkylamine beads contain hexanediamine spacers that create primary amino groups on the surface for the immobilization of ligands. After reaction of these beads with the heterobifunctional cross-linker Sulfo-SMCC [sulfosuccinimidyl-4-(N-maleimidomethyl)cyclohexane-1-carboxylate], the NHS ester end covalently couples to the amines but leaves a terminal maleimide group to react with sulfhydryl-containing ligands. Figure 3.45 shows the reactions involved in this coupling procedure.

1. 10 polystyrene beads derivatized with alkylamine functionalities giving terminal primary amino groups ($^1/_4$ in.; Pierce, Rockford, Illinois) are washed with water on a small Buchner funnel (without a filter pad) and then with ice cold activation buffer. The activation buffer of choice to promote efficient NHS ester coupling with Sulfo-SMCC is 0.1 M sodium phosphate, 0.15 M NaCl, pH 7.2. Other buffer salts may be used provided they contain no amines that will compete with the conjugation reaction. Immerse the beads in a minimum volume of this cold buffer and mix by slow gentle rocking at 4°C. The low temperature will reduce the hydrolysis rate of the maleimide groups.
2. To the bead mixture, add 43.6 mg Sulfo-SMCC (Pierce) and gently mix to dissolve. This level of cross-linker is about 10 μmol/bead, which in turn is a 3-fold excess over the 3 μmol/bead level of alkylamine functionality. The beads should be handled with great care since bead-on-bead grinding during the mixing processes may cause surface degradation and loss of chemical functionality.
3. Continue mixing the reaction for 1 hr at 4°C.
4. Quickly wash the beads with cold coupling buffer, 1 M NaCl, and again with buffer to remove uncoupled cross-linker and the reaction by-products. The beads now will contain terminal maleimide groups and should

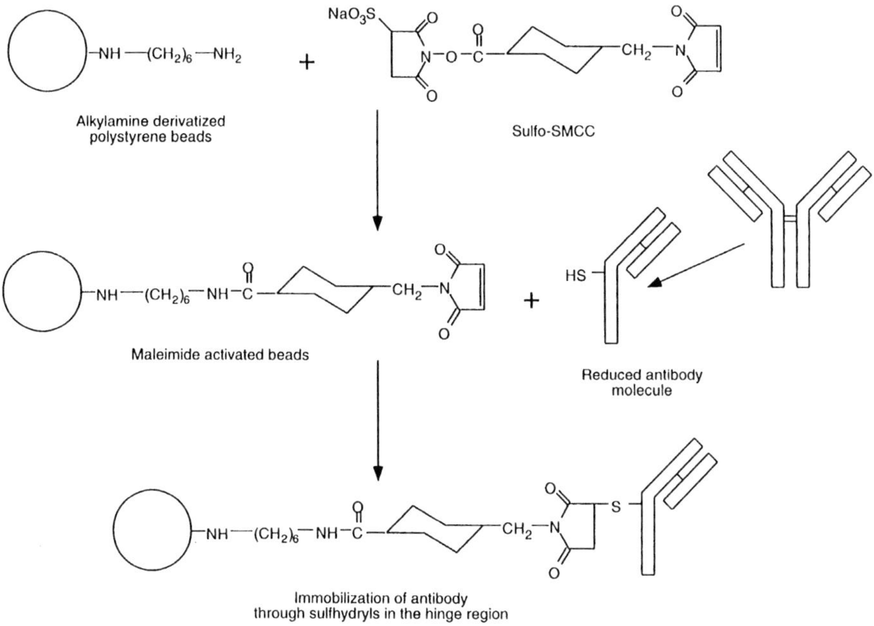

3.4 Peptide Antigens, Antibodies

FIGURE 3.45
Alkylamine-derivatized polystyrene beads containing the spacer 1,6-diaminohexane can be activated with Sulfo-SMCC [sulfosuccinimidyl-4-(N-maleimidomethyl)cyclohexane-1-carboxylate] to yield a sulfhydryl reactive matrix. These activated beads can then be coupled to reduced antibody molecules.

---

be used immediately for coupling to an antibody through generated sulfhydryl groups.

5. Immerse the maleimide-activated beads just prepared in a minimum volume of cold coupling buffer (0.1 M sodium phosphate, 0.15 M NaCl, pH 7.2) containing the reduced antibody to be immobilized. The antibody reduction protocol is described in Section 3.4.2.3 and will result in either half-antibody molecules (cleaved at the disulfide bonds between the heavy chains in the hinge region) or reduced F(ab')$_2$ fragments, depending on what is desired.
6. Mix the reaction by gentle rocking and allow it to continue at 4°C for at least 2 hr.
7. Wash the beads with coupling buffer, 1 M NaCl, and water. Store them as an aqueous slurry containing a preservative at 4°C.

Before using the antibody-coupled beads in an assay procedure, block any passive adsorption sites on the bead surfaces using the same basic procedure described for coating polystyrene microtiter plates earlier (Section 3.4.2.4).

Other conjugation reagents also may be used in conjunction with alkylamine beads to form a covalent bond with a ligand. Any bifunctional cross-linking reagents that will react with amines on one end and another functionality on an antibody or protein on the other can be used in a protocol similar to the one just described. In addition, the water-soluble carbodiimide EDC can be used to form amide linkages with carboxylate groups on proteins. Reference to the appropriate sections in Chapter 2 will provide additional options for immobilization to alkylamine beads.

b. Coupling to Hydrazide Beads Using Glutaraldehyde

Polystyrene beads containing terminal hydrazide functionalities (Pierce, Rockford, Illinois) can be reacted with glutaraldehyde to give formyl coupling sites for reaction with amine-containing ligands. Antibody molecules may be immobilized to these aldehyde groups in the presence of sodium cyanoborohydride, which promotes reduction of Schiff bases to form secondary amine bonds. Figure 3.46 shows the reactions involved in this coupling procedure.

1. Add 5 ml 12.5% glutaraldehyde solution (v/v in 0.1 M sodium phosphate, pH 7.0) to 25 hydrazide beads and mix gently for 2 hr.
2. Wash the beads on a Buchner funnel without a filter pad with 100 ml water followed by 20 ml 0.1 M sodium phosphate, pH 7.0 (coupling buffer).
3. Add the beads to a 0.5 mg/ml solution of IgG in coupling buffer. Make sufficient antibody solution to fully immerse the beads.
4. Add 1 mg sodium cyanoborohydride and mix gently overnight at room temperature or 4°C.
5. Wash the beads with 100 ml coupling buffer followed by 50 ml 0.1 M sodium bicarbonate, pH 9.5.
6. Add the beads to 5 ml 0.1 M sodium bicarbonate, pH 9.5, containing 1 mg sodium borohydride to reduce the uncoupled aldehydes. React for at least 15 min.
7. Wash the beads with 100 ml 0.1 M sodium carbonate, pH 9.5, followed by 100 ml water. Store the IgG beads at 4°C, either dry or as an aqueous slurry containing a preservative.

Before using the antibody-coupled beads in an assay procedure, block any passive adsorption sites on the bead surfaces using the same basic pro-

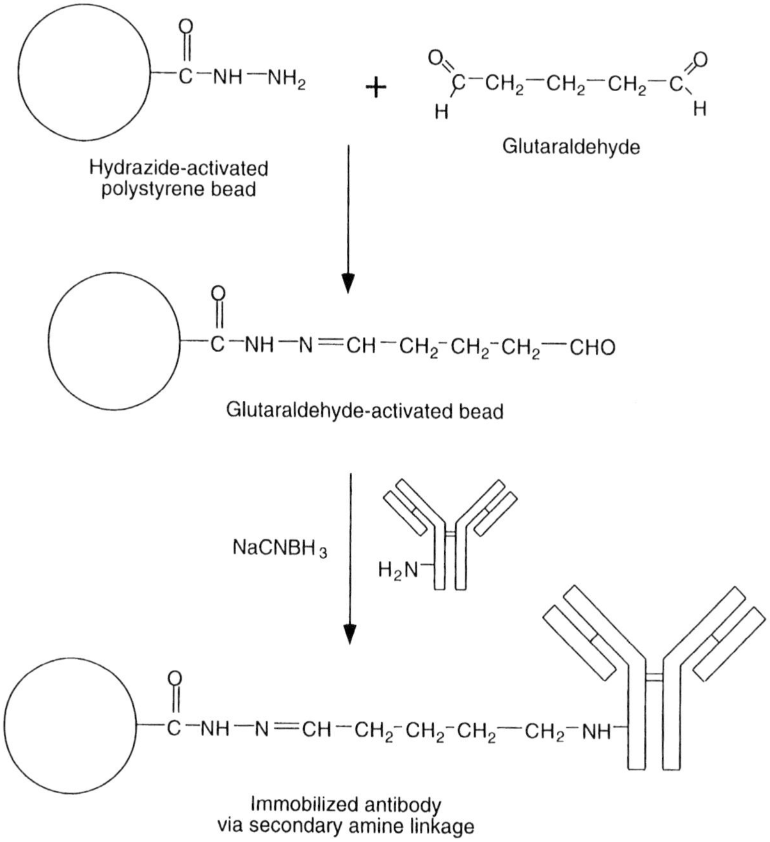

3.4 Peptide Antigens, Antibodies

FIGURE 3.46

Hydrazide-activated polystyrene beads can be further derivatized with glutaraldehyde to yield formyl functionalities suitable for coupling ligands by reductive amination. Sodium cyanoborohydride will specifically reduce the Schiff bases formed between an amine-containing ligand and the aldehyde bead.

cedure described for coating polystyrene microtiter plates earlier (Section 3.4.2.4).

c. Coupling to Hydrazide Beads through Oxidized Carbohydrates

Another method of immobilizing antibodies to hydrazide beads involves mild oxidation of the Fc region carbohydrate residues followed by formation of a hydrazone linkage with the bead. This procedure is similar to the site-directed coupling to a hydrazide chromatography support discussed in Section 3.4.2.3 earlier. The chemistry of this coupling reaction is shown in Figure 3.47.

1. Dissolve 1–10 mg antibody (IgG) per ml in 2 ml 0.1 M sodium phosphate, pH 7.0.
2. Add 10 mg sodium *meta*-periodate to the antibody solution and gently mix for 30 min at room temperature.
3. Purify the oxidized antibody by desalting on a 15-ml Sephadex G-25 column (Pharmacia) equilibrated with 0.1 M sodium phosphate, pH 7.0. Pool the first peak off the column (monitor A_{280}), which contains the antibody.
4. Wash 10 hydrazide activated beads (Pierce) with 10 ml 0.1 M sodium phosphate, pH 7.0 (coupling buffer). Drain.

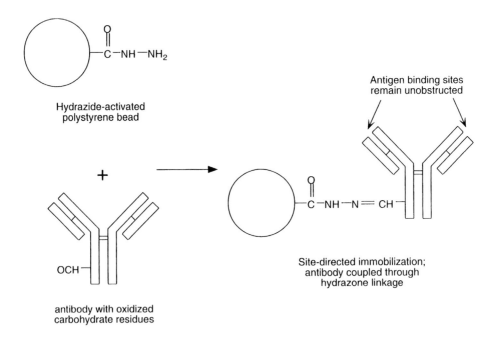

FIGURE 3.47
Hydrazide-activated polystyrene beads can be used to immobilize antibody molecules in a site-directed manner by coupling to carbohydrate residues that have been oxidized by periodate treatment. The result is covalent attachment away from the antigen binding sites, producing the highest possible antibody activity.

5. Add the pooled oxidized antibody fractions to the beads and react with gentle mixing for at least 6 hr at room temperature.
6. Wash the beads thoroughly with coupling buffer, water, 1 M NaCl, and again with water. Store the antibody-coupled beads at 4°C, either in a dry state or in an aqueous solution containing a preservative.

Before using the antibody-coupled beads in an assay procedure, block any passive adsorption sites on the bead surface using the same basic procedure described for coating polystyrene microtiter plates earlier (Section 3.4.2.4).

Covalent Attachment to Polystyrene Microtiter Plates

Several methods are available for the covalent coupling of antibodies or antigens to the polystyrene wells of a microplate. The most important criterion for choosing an activation and coupling chemistry is maintenance of the optical characteristics of the well bottom. The strategy must not coat the surface with a compound that adds absorptivity in the spectral region in which the enzyme immunoassay is to be detected or measured. Similarly, the activation chemistry or solvent requirements must not dissolve or cloud the polystyrene surface of the well.

Current covalent coupling strategies may be divided into two main categories: (1) the intermediate noncovalent coating of a polymer on the well surface that contains functional or reactive groups that can be coupled to a ligand or (2) the direct covalent modification of the surface with a compound that also has such functionalities. The first method involving passive coating of a functional polymer on the surface again poses all the problems associated with noncovalent linkages discussed previously. Although coatings of silicone and other polymers can be used to form a layer of functionality on the wells, the polymer–ligand complex can still leach off the surface. For this reason, the discussion that follows is concerned only with those methods that form a true covalent linkage between the plate and the ligand to be immobilized.

a. Glutaraldehyde Modification of Polystyrene Plates

Cross-linking reagents useful for immobilizing proteins may be attached to the plate surface if they contain a suitable functionality on one end. A simple example of this approach is to use glutaraldehyde-treated plates to immobilize protein via their amino groups (Barrett, 1977). Plates can be precoated with a covalently coupled layer of glutaraldehyde at low pH and subsequently coupled to protein at pH 8–9.5. The reaction is done without addition of sodium borohydride or sodium cyanoborohydride to reduce

Schiff bases. Presumably the procedure creates a polymeric coating of glutaraldehyde on the well surfaces through a process of free radical vinyl polymerization. Protein or antibody molecules may be coupled to this layer in 100 mM sodium phosphate buffer, pH 8.0, by incubating for 3 hr at 37°C. Therefore, this procedure, although it uses a traditional cross-linking agent, probably also involves a graft copolymerization on the polystyrene surface. A brief protocol follows.

1. Prepare a 2% solution of glutaraldehyde in 0.1 M sodium phosphate, pH 5.0, and add 200 μl to each well of a microplate.
2. Incubate for at least 4 hr at room temperature.
3. Wash the plate thoroughly with the phosphate buffer.
4. Add 200 μl of a 2–10 μg/ml solution of antibody in 0.1 M sodium phosphate, pH 8.0.
5. Incubate for at least 3 hr at 37°C.
6. Wash with 0.1 M sodium phosphate, pH 8.0, 0.15 M NaCl.
7. Block excess reactive sites by adding 200 μl per well 0.1 M lysine in 0.1 M sodium phosphate, pH 8.0.
8. Wash with PBS buffer, pH 7.2, containing 0.05% Tween 20.

b. Glutaraldehyde Modification of Aminophenyl Groups

Another method of using this cross-linking reagent involves first derivatizing a microplate surface to create aminophenyl residues, then modifying these groups with glutaraldehyde to form reactive formyl functionalities. Neurath and Strick (1981) developed this protocol for the covalent immobilization of antibody molecules.

1. Add methanesulfonic acid to the wells of a microplate and incubate overnight at room temperature in a fume hood.
2. Wash thoroughly with water and dry.
3. In a fume hood, place the plate in a heating block and add 200 μl of a 1:1 mixture of glacial acetic acid and fuming HNO_3 to each well.
4. Incubate at 40°C for 4 hr.
5. Wash thoroughly with water.
6. Return the plate to the heating block at 50°C and add to each well 200 μl 0.5% $Na_2S_2O_4$ (sodium hydrosulfite, also called sodium dithionite, Aldrich, Milwaukee, Wisconsin) in 0.5 N NaOH.
7. Incubate at 50°C for 1 hr.
8. Wash the plate extensively with water and 50 mM sodium phosphate, pH 8.5 (coupling buffer).
9. Add to each well 200 μl of a 1% glutaraldehyde solution in coupling buffer and incubate for 2 hr at room temperature and overnight at 4°C.

10. Wash the plates extensively with coupling buffer.
11. To each well, add 200 µl of a 100 mg/ml solution of IgG in coupling buffer, and incubate overnight at 4°C.
12. Wash as before. Block excess formyl groups with 1 M glycine in coupling buffer for 4 hours at room temperature.
13. Wash with 10 mM sodium phosphate, 0.15 M NaCl, pH 7.2. The plate may be stored in the same buffer if a preservative is added.

c. Use of Photoactivatable Cross-Linking Reagents

A cross-linking reagent containing a photochemical coupling group on one end may be used to modify the surface of polystyrene microplates. If the cross-linker contains another suitable reactive group on its other end, for example, an amine- or sulfhydryl-reactive functionality, then antibodies can be immobilized covalently. BioMetric Systems (BSI; Eden Prairie, Minnesota) uses this approach to immobilize ligands on plastic surfaces. The protein or antibody molecules are first modified with the photoactivatable cross-linker using an amine-reactive end such as an NHS ester (Fig. 3.48). The modified antibody is then photoreacted with the polystyrene plate surface by exposure to strong light in the range of 320–350 nm. The result is a covalently immobilized immunoglobulin that is extremely stable and will not leak off during the washing and binding steps of an assay. An example of this approach is described in the following procedure using the cross-linking reagent sulfosuccinimidyl-6-(4′-azido-2′-nitrophenyl-amino)-hexanoate (Sulfo-SANPAH, Pierce).

1. 1–10 mg specific antibody is reacted with a 10-fold molar excess of Sulfo-SANPAH dissolved in 2 ml 0.1 M sodium phosphate, pH 7.2. Sulfo-SANPAH is best measured by taking an appropriated aliquot of a 1 mg/ml stock solution prepared fresh in reaction buffer. The reaction is done at 4°C overnight in the dark.
2. Dialyze or desalt the protein mixture using 0.1 M sodium phosphate, 0.15 M NaCl, pH 7.2, to remove excess reactants. Protect the modified antibody from light.
3. Add the modified antibody solution at a concentration of 100 µg/ml to the wells of a microplate. Incubate at 37°C for 30 min.
4. Irradiate with a uv light source with a primary wavelength range of 320–350 nm. Usually a strong light source such as a camera flood light is required. Illuminate the plate for at least several minutes.
5. Wash the plate extensively with PBS, pH 7.2, containing 0.05% Tween 20, and store in the same buffer if a preservative has been added.

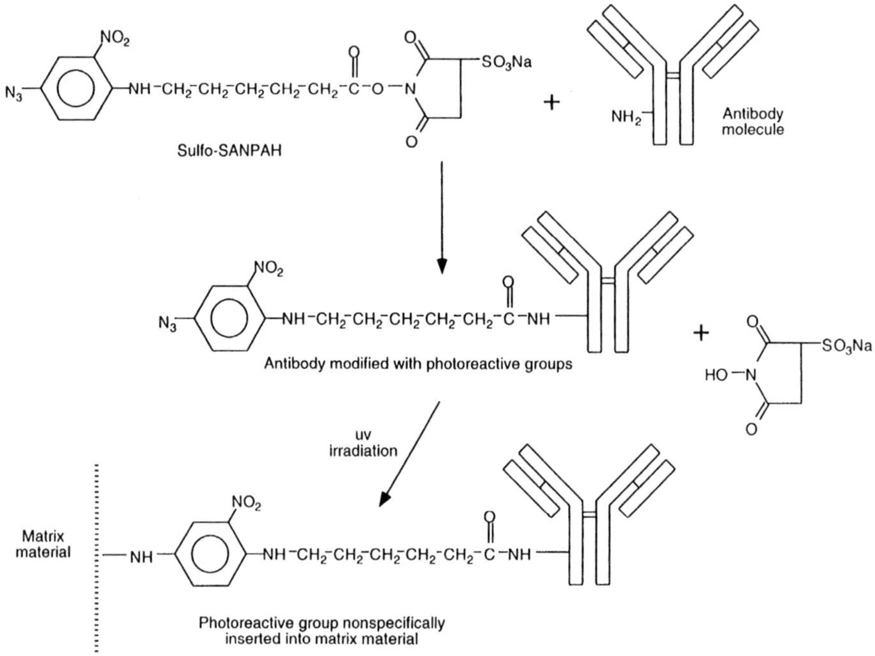

FIGURE 3.48
The use of a photoreactive cross-linking agent to immobilize ligands. The NHS ester end of Sulfo-SANPAH [sulfosuccinimidyl-6-(4′-azido-2-nitrophenylamino)hexanoate] is first reacted with an antibody molecule, coupling to the available amine groups on the proteins. Next, the modified immunoglobulin is mixed with the matrix and photolyzed with uv light for several minutes. The photoreactive group will nonselectively insert into the matrix structure.

---

d. Use of Functional Vinyl Monomers to Couple to the Plate Surface

Vinyl monomers typically used in polymerization reactions can often be made to graft copolymerize in a uv-catalyzed free radical addition reaction to the surface of a microplate. The careful choice of a vinyl monomer with another functionality that can be used to couple to protein or antibody molecules can provide a useful covalent linking reagent for immobilization purposes. The best monomers are those that are either water soluble or soluble in a solvent that does not affect the polystyrene microplate. Two strategies are possible for the use of these compounds. The vinyl monomer can be conjugated first to the antibody; then the complex is photoreacted with the plate surface in a manner similar to the protocol described for phenylazide cross-linkers. Alternatively, the monomers can be used to modify the

3.4 Peptide Antigens, Antibodies 243

plate surface by graft copolymerization; the secondary functionalities subsequently are used to couple to the antibody.

A commercial example of this methodology is found in CovaLink plates (Nunc). Through an invention by the Danish company GlueTech, Nunc makes use of a psoralen derivative that is covalently linked to the well surface. The psoralen molecules are grafted onto the polystyrene by a uv-catalyzed free radical reaction that leaves the optical clarity of the plastic unaffected. The particular derivative used by Nunc contains an aliphatic spacer arm terminating in a secondary amine (Fig. 3.49). The secondary amine may be reacted with cross-linking reagents containing an NHS ester to provide additional functionalities suitable for immobilizing ligands such

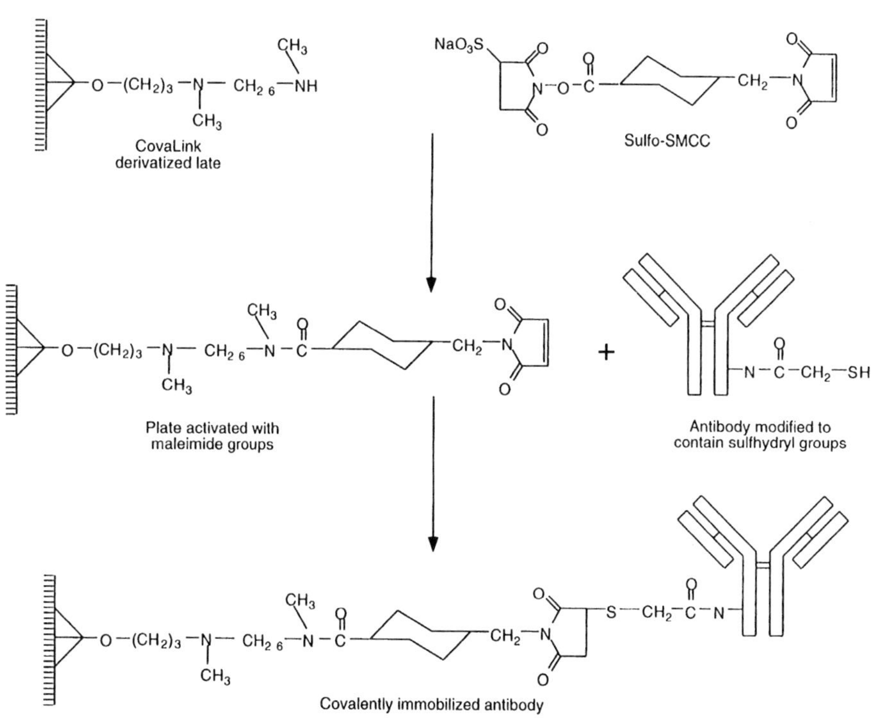

FIGURE 3.49

CovaLink polystyrene microplates (Nunc) contain a secondary amine function which can be activated with Sulfo-SMCC. The resultant maleimide-activated surface can be used to immobilize sulfhydryl-containing ligands. In this case, an antibody molecule that has been modified to contain free –SH groups is coupled, resulting in a stable thioether linkage.

as antibodies. A generalized protocol involving the use of such a cross-linking reagent to immobilize immunoglobulin molecules is given here.

1. Dissolve the cross-linking reagent Sulfo-SMCC (Pierce) at a concentration of 100 µg/ml in 0.1 M sodium phosphate, 0.15 M NaCl, pH 7.2 (coupling buffer). Prepare 20 ml per microplate.
2. Immediately transfer 200 µl of the Sulfo-SMCC solution to each well of the plate.
3. Incubate overnight at room temperature.
4. Wash the plate at least four times with coupling buffer containing 0.05% Tween 20.
5. Reduce a purified immunoglobulin with 2-mercaptoethylamine as described in Section 3.4.2.3. Prepare at least 2 mg reduced IgG per plate.
6. Add the reduced IgG to the activated plate at a concentration of 100 µg/ml in coupling buffer containing 0.05% Tween 20.
7. React overnight at room temperature.
8. Remove the antibody solution from the wells, and block the unreacted maleimides by adding 100 µg/ml cysteine in coupling buffer containing 0.05% Tween 20. Incubate for 2 hr at room temperature.
9. Remove the cysteine solution, and block nonspecific binding sites remaining on the plastic surface by following the protocol described earlier in Section 3.4.2.4.
10. Wash the plate at least four times with the PBS-Tween buffer.

3.4.3 Immunoglobulin Binding Proteins

Immunoglobulin binding proteins are unique proteins that are able to specifically interact with immunoglobulin molecules at locations other than the antigen binding sites. In other words, these proteins can be used to purify whole classes of immunoglobulins, not just a population with a distinct antigenic specificity. When immobilized immunoglobulin binding proteins are used in conjunction with immobilized antigens in immunoaffinity chromatography, powerful separation methods can be developed that can aid in the isolation of virtually any antibody molecule. The following sections describe the most useful of these proteins, including a new one specific for IgM class molecules that has not been reported previously (Mallia *et al.*, 1992).

3.4.3.1 Preparation of Immobilized Protein A

Protein A is a cell wall component produced by several strains of *Staphylococcus aureus*. It consists of a single polypeptide chain of molecular mass

42,000 daltons and contains little or no carbohydrate. Protein A is able to bind specifically to the Fc region of immunoglobulin molecules, especially to those of the IgG class (Ey et al., 1978; Lindmark et al., 1983). The protein contains four high-affinity binding sites (approx. $K_a = 10^8$ M^{-1}) capable of interacting with the Fc region in antibodies of several species. The molecule is relatively heat stable and retains its native conformation even after exposure to denaturing reagents such as 4 M urea, 4 M thiocyanate, and 6 M guanidine hydrochloride.

The recombinant form of the protein (Pierce, Repligen) is a genetically truncated version with a molecular mass of approximately 32,000 daltons. Nonessential regions are removed from the carboxy terminus to give a protein containing 301 amino acids, 28 of which are lysines. This recombinant protein A exhibits the same affinity as the native protein for IgG molecules, but has considerably lower nonspecific binding properties in immunoglobulin applications. Purified recombinant protein A has a maximal absorbance at 275 nm; a 1 mg/ml solution will yield an absorbance of about 0.2 at 280 nm (1-cm path). The extinction coefficient of nonrecombinant forms of the protein may be somewhat smaller than this value.

Immobilized protein A has been used extensively as an affinity support for the purification of a wide variety of IgG molecules from many different species of mammals. However, the interaction between protein A and IgG is not equivalent for all species. Even within a species, protein A interacts very well with some subgroups of IgG and not as well with others. For instance, human IgG1, IgG2, and IgG4 bind strongly, whereas IgG3 does not bind at all (or very slightly). In mice, IgG1 does not bind well to protein A, but IgG2a, IgG2b, and IgG3 bind tightly.

Protein A is typically immobilized through its amine groups. Several immobilization options are available for coupling these proteins to solid supports, including periodate oxidation/reductive amination, CNBr, tresyl, and FMP. These methods produce high capacity affinity supports with very low nonspecificity because of the coupling chemistry. A well-documented observation is that, by controlling the amount of protein A added to the coupling reaction, the capacity of the support can be controlled. In general, reacting 2–4 mg of protein A per ml gel will produce an affinity support of "average" capacity (with respect to other as commercial sources). Such a matrix will be able to bind 12–15 mg human IgG per ml gel. Increasing the amount of protein A in the coupling media to 6–8 mg/ml gel will result in a high capacity support able to bind 35–40 mg human IgG per ml gel. Supports with much higher capacities than these can be created, but their practical value is dubious because of the high cost of protein A.

Perhaps the easiest way to obtain immunoglobulin binding affinity supports is through a variety of commercial sources, including Pierce, Pharmacia, BioRad, Sigma, Sterogene, Repligen, and many others.

IMMOBILIZATION PROTOCOL

1. Wash 10 ml periodate-oxidized glycidol-treated Toyopearl HW-65f (Section 2.2.1.4, Chapter 2) with several bed volumes of 0.1 M sodium phosphate, pH 7.0 (coupling buffer). Drain the gel to a wet cake.
2. Dissolve 30–60 mg protein A (either recombinant or native protein) in 10 ml coupling buffer and add to the wet gel cake.
3. In a fume hood, add 60 mg sodium cyanoborohydride (toxic) to the gel slurry and gently mix at room temperature for at least 4 hr. Alternatively, the reaction may be mixed overnight at 4°C.
4. Wash the gel thoroughly with coupling buffer, water, 1 M NaCl, and again with water.

The reaction yield should be greater than 85% coupling of protein. The support may be stored indefinitely as a 50% slurry in an aqueous solution containing a preservative (e.g., 0.02% sodium azide).

3.4.3.2 Preparation of Immobilized Protein G

Protein G is a bacterial cell wall protein isolated from group G streptococci. Like protein A from *Staphylococcus aureus,* protein G binds to most mammalian IgGs, primarily through their Fc regions (Bjorck and Kronvall, 1984; Akerstrom and Bjorck, 1986). There is some evidence as well of an interaction between protein G and Fab fragments, since immobilized protein G (unlike protein A) can not be used to separate Fabs from Fc fragments after enzymatic digestion of IgG. DNA sequencing of nonrecombinant protein G identifies two locations at which immunoglobulins are bound to the protein and also identifies sites for albumin and cell-surface binding. These albumin and cell-surface binding sites have been eliminated from the recombinant form of protein G to reduce the nonspecific binding potential when the protein is used in an affinity support. With the albumin binding site removed, immobilized protein G can be used to isolate IgG molecules from serum and other sources with high purity.

The unique immunoglobulin binding characteristics of protein G may be used for the purification and identification of those mammalian monoclonal and polyclonal IgGs that do not bind well to protein A. Protein G is reported to bind with greater affinity than protein A to most mammalian IgG molecules (Akerstrom and Bjorck, 1986). There are, however, several species, such as dog and guinea pig, for which protein A has greater affinity. Protein G does bind with significantly greater affinity to several IgG subclasses of human (IgG3) and rat (IgG2a) immunoglobulins. Another difference between the two immunoglobulin binding proteins is that, unlike protein A, protein G does not bind human IgM, IgD, and IgA. The differences in binding characteristics between protein A and protein G may be explained by different compositions in their antibody binding sites. It is interesting that, although their three-dimensional structures are somewhat similar, their amino acid compositions are quite different.

The immobilization of protein G can be accomplished in a manner similar to that of protein A using a variety of potential amine-reactive activation chemistries (see Section 3.4.3.1). In this example, another polymeric support material is used that is suitable for large-scale purification of immunoglobulins.

IMMOBILIZATION PROTOCOL

1. Wash 10 ml periodate-oxidized glycidol-treated Trisacryl GF2000 LS (Section 2.2.1.4, Chapter 2) with several bed volumes of 0.1 M sodium phosphate, pH 7.0 (coupling buffer). Drain the gel to a wet cake.
2. Dissolve 30–60 mg protein G (either recombinant or native protein; Pharmacia) in 10 ml coupling buffer and add to the wet gel cake.
3. In a fume hood, add 60 mg sodium cyanoborohydride (toxic) to the gel slurry and gently mix at room temperature for at least 4 hr. Alternatively, the reaction may be mixed overnight at 4°C.
4. Wash the gel thoroughly with coupling buffer, water, 1 M NaCl, and again with water.

The matrix used here and the one described for the immobilization of protein A will both provide rigid supports with high flow rates. However, the Trisacryl matrix used for protein G immobilization has a larger bead size and will, thus, provide higher gravity flow rates than the Toyopearl support used for protein A. Both supports are suitable for medium pressure work.

3.4.3.3 Preparation of Immobilized Protein A/G

As described earlier, protein A originates from *Staphylococcus aureus* and protein G from *Streptococcus* sp. Both proteins have the ability to specifically bind to the Fc portion of the constant region of immunoglobulins originating from a broad array of species. Although there is some overlap in the specificity ranges of these two proteins for immunoglobulin classes, subclasses, and species type, they do have considerable specificity differences. Gene fusion of the Fc binding domains of protein A and protein G has led to the production of a structurally and functionally chimeric protein with a more extended binding specificity than either immunoglobulin binding protein alone (Eliasson *et al.*, 1988). During fusion, the protein G gene sequence encoding the albumin binding site has been eliminated, thus insuring a low nonspecific binding character. This Fc binding fusion protein, known as protein A/G (Pierce), is expressed in *Bacillus* sp. and is secreted into the surrounding medium during fermentation.

Protein A/G is a 45,000 to 47,000-dalton protein containing 442 amino acids, 43 of which are lysines. The protein has an isoelectric point of 4.2 as determined by chromatofocusing. Protein A/G binds immunoglobulins in a pH-independent manner from pH 5 to pH 8. The effect of pH on protein A binding to IgGs is overcome through the fusion of the protein G component (Fig. 3.50).

Protein A/G thus combines the advantages, affinities, and specificities of both protein A and protein G into a unique composite molecule. The result is perhaps the ideal IgG binding protein with a broader range of interactions and applications than either protein A or protein G alone. The immobilized form of this protein provides an affinity support capable of purifying IgGs from all the common species used in immunochemical studies.

Figure 3.51 shows a comparison of the relative affinities of protein A, protein G, and protein A/G for the immunoglobulins from 10 different spe-

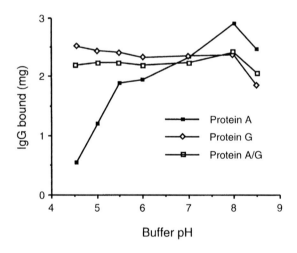

FIGURE 3.50

The effect of pH on the binding of immunoglobulins to immobilized protein A, immobilized protein G, and immobilized protein A/G. The relative pH-independent binding character of protein G is inherited by the chimeric recombinant protein A/G.

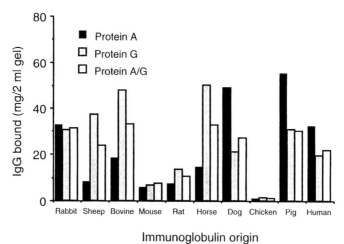

FIGURE 3.51
Comparison of the ability of three immobilized immunoglobulin-binding proteins to bind the IgG from the serum of 10 different species. In most cases, the chimeric protein A/G performs somewhere between the capacity of protein A and that of protein G.

cies. It is interesting to note that, in some instances, protein A outperforms protein G whereas in other cases the reverse is true. In almost every case, however, protein A/G averages these differences to produce an affinity matrix that performs at a capacity between those of protein A and protein G individually. In Figure 3.52, these three immunoglobulin binding supports are further compared with respect to their relative binding efficiencies of mouse IgG subclasses.

IMMOBILIZATION PROTOCOL

1. Wash 10 ml periodate-oxidized Sepharose CL-4B (Section 2.2.1.4 in Chapter 2) with several bed volumes of 0.1 M sodium phosphate, pH 7.0 (coupling buffer). Drain the gel to a wet cake.
2. Dissolve 30–60 mg protein A/G in 10 ml coupling buffer and add to the wet gel cake.
3. In a fume hood, add 60 mg sodium cyanoborohydride (toxic) to the gel slurry and gently mix at room temperature for at least 4 hr. Alternatively, the reaction may be mixed overnight at 4°C.
4. Wash the gel thoroughly with coupling buffer, water, 1 M NaCl, and again with water.

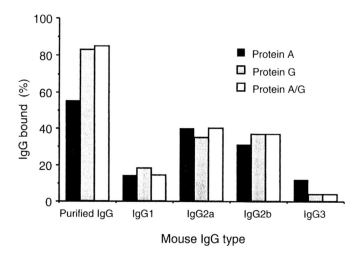

FIGURE 3.52
The relative efficiency of immobilized protein A, immobilized protein G, immobilized protein A/G to bind the four mouse IgG subclasses and purified whole mouse IgG is compared. The separations were done using a Waters protein purification unit using 1-ml columns, a linear flow of 0.5 ml/min, and a sample size of 1 mg/100 μl. The percentage of the applied IgG that was bound is proportional to the relative strength of the affinity interaction between the IgG subclass and the immobilized protein.

3.4.3.4 Preparation of Immobilized Mannan Binding Protein

Various mammalian sera contain a binding protein that binds mannose and N-acetylglucosamine residues (Kawasaki *et al.*, 1989). This binding protein, known as mannan-binding protein (MBP), has been shown to activate the complement system through the classical pathway. The gross structure of MBP is remarkably similar to that of C1q and it is believed that MBP functions like C1q. Very recently it has been shown (Nevens *et al.*, 1992) that immobilized MBP specifically binds mouse monoclonal IgM from mouse ascites fluid (patent pending, Pierce Chemical). The binding of IgM to MBP is calcium dependent, and IgM bound to immobilized MBP can be eluted with EDTA.

MBP may be isolated according to the published protocol described in Section 5.1.13.2, Chapter 5. The final step in the purification process results in an MBP solution that is used directly in Step 3. The described immobilization protocol uses the standard CNBr activation scheme, which results in good IgM binding activity. Other methods of protein immobilization are also suitable for coupling this protein.

IMMOBILIZATION PROTOCOL

1. Activate Sepharose 4B (110 ml settled gel) with 22 gm CNBr (see Section 2.2.1.1, Chapter 2).
2. After CNBr activation, wash the activated gel with 2 L ice-cold water and 1 L ice-cold 0.1 M sodium bicarbonate, pH 8.5.
3. Suction dry the CNBr-activated Sepharose 4B to a moist cake and add to the MBP solution (obtained from the final stage of the MBP purification process; Section 5.1.3.2, Chapter 5).
4. Stir the gel suspension at 4°C for 20 hr.
5. Filter and wash the reaction mixture with 200 ml 1.0 M NaCl and 200 ml water. The filtrate may be saved to determine the unreacted MBP by measuring the absorbance at 280 nm.
6. Block the excess reactive groups on MBP–/ /–Sepharose 4B as follows. Suction dry the gel to a moist cake and add to 100 ml 1.0 M ethanolamine, pH 9.0.
7. Stir the gel suspension at room temperature for 1 hr.
8. Finally, wash the MBP–/ /–Sepharose 4B successively with 2 L each of water, 1.0 M NaCl, water, 10 mM Tris-HCl buffer, pH 7.4, containing 1.25 M NaCl and 2 mM EDTA, and water.
9. MBP–/ /–Sepharose 4B can be stored in 0.02% sodium azide at 4°C.

The coupling efficiency is found to be 90% and 3.2 mg MBP are coupled per ml Sepharose 4B.

3.4.3.5 Preparation of Immobilized Protein B

Protein B is another immunoglobulin binding protein of bacterial origin, with specificity for the Fc region of human IgA (Faulmann *et al.*, 1991). The IgA binding protein is isolated as a subset of the β antigen protein expressed on the surface of group A streptococci. Protein B can interact with both the human IgA$_1$ and IgA$_2$ subclasses, as well as with secretory IgA, but not with the IgA fraction from mouse or other mammalian species.

Protein B can be immobilized to chromatography supports and other solid surfaces for immunoassay use without loss of activity. Passive adsorption or covalent coupling to polystyrene microplates is also possible. An affinity support using this protein can be prepared using conventional amine-reactive chemistries, the best of which are probably the CNBr or reductive amination techniques.

IMMOBILIZATION PROTOCOL

1. Wash 10 ml periodate-oxidized Sepharose CL-4B (Section 2.2.1.4, Chapter 2) with several bed volumes of 0.1 M sodium phosphate, pH 7.0 (coupling buffer). Drain the gel to a wet cake.
2. Dissolve 30–60 mg protein B in 10 ml coupling buffer and add to the wet gel cake.
3. In a fume hood, add 60 mg sodium cyanoborohydride (toxic) to the gel slurry and gently mix at room temperature for at least 4 hr. Alternatively, the reaction may be mixed overnight at 4°C.
4. Wash the gel thoroughly with coupling buffer, water, 1 M NaCl, and again with water.

3.5 IMMOBILIZATION OF LECTINS

Immobilized lectins have been used extensively for the isolation of glycoconjugates and glycoproteins with specific carbohydrate structures. The separation of glycosylated molecules by lectin affinity chromatography depends on the specific interaction of the glycosylated portion with the appropriate immobilized lectin that recognizes certain structural characteristics within the carbohydrate tree. In lectin affinity chromatography, the glycosylated protein of interest is allowed to bind to the appropriate immobilized lectin column and the bound material is subsequently eluted with counterligands of low molecular weight, usually the coresponding mono- or disaccharides. Lectin affinity chromatography is an excellent tool for separating glycoconjugates in complex mixtures (Merkle and Cummings, 1987). The technique can supplement conventional methods of purification of glycoconjugates, such as ion-exchange chromatography, size-exclusion chromatography, or HPLC.

Table 3.2 lists some of the most commonly used immobilized lectins for affinity chromatography and their corresponding carbohydrate binding specificities.

In this section, we describe the procedures that we have successfully used for the immobilization of lectins, with special reference to concanavalin A and jacalin. One important point to consider when immobilizing lectins is that the carbohydrate binding sites should be protected during immobilization by incorporating their specific binding sugar to a final concentration of 0.1 M in the coupling buffer solution. Usually a pH of the

3.5 Immobilization of Lectins

TABLE 3.2
Immobilized Lectins and Their Carbohydrate-Binding Specificity

Immobilized lectin	Carbohydrate-binding specificity [a]
Concanavalin A	α-D-Man, α-D-Glc
Jacalin	α-D-Gal
Lentil	α-D-Man, α-D-Glc
Pea	α-D-Man, α-D-glc
Peanut	D-Gal and its derivatives
Phytohemagglutinin E_4	GalNAc
Phytohemagglutinin L_4	GalNAc
Pokeweed	GlcNAc-β(1→4)-GlcNAc
Ricinus communis lectin 60	β-D-Gal, D-GalNAc
Ricinus communis lectin 120	β-D-Gal
Soybean	D-Gal, D-GalNAc
Ulex europaeus I	α-L-Fuc-(→2)Gal
Wheat germ	D-GlcNAc

[a] Man, Mannose; Glc, Glucose: Gal, galactose: GalNac, N-acetyl galactosamine; GlcNAc, N-acetyl glucosamine: Fuc, fucose.

coupling buffer between 7.0 and 8.0 is preferred for optimal activity after immobilization.

3.5.1 PREPARATION OF IMMOBILIZED CONCANAVALIN A

Concanavalin A (Con A) is a hemagglutinin from the common jack bean *Canavalia ensiformis*. Immobilized Con A shows specific affinity for molecules containing α-D-mannopyranosyl, α-D-glucopyranosyl, and sterically related residues. Con A extensively used in the isolation, fractionation, and structural characterization of glycoproteins and glycopeptides (Merkle and Cummings, 1987). Immobilized Con A is also used for the separation and purification of polysaccharides, glycolipids, hormones, hormone receptors, and enzyme–antibody conjugates (Renthal *et al.*, 1976; Dulaney, 1979; Lis and Sharon, 1984).

IMMOBILIZATION PROTOCOL

1. Wash periodate-activated Sepharose CL-6B (20 ml settled gel; prepared as described in Section 2.2.1.4, Chapter 2) with 200 ml water and 200 ml

0.1 *M* sodium phosphate buffer, pH 7.0 (coupling buffer), in a sintered glass funnel, and suction dry to a moist cake.
2. Add the gel to a Con A solution (400 mg Con A dissolved in 20 ml coupling buffer containing 0.1 *M* methyl α-D-mannopyranoside).
3. Add 120 mg sodium cyanoborohydride to the gel suspension and gently mix the reaction mixture on a shaker overnight at room temperature.

NOTE: Sodium cyanoborohydride is highly toxic. The coupling reaction and the subsequent steps should be carried out in a well-ventilated hood. We routinely use a 100-ml Erlenmeyer flask as the reaction vessel.

4. Excess aldehyde groups on the gel may be blocked if desired. However, we have found that any remaining formyl groups do not contribute to nonspecific effects in the resulting affinity support. For this reason, we typically do not do a blocking step. To block, treat the support as follows. Wash the coupled gel with 200 ml coupling buffer and then suspend in 20 ml 1.0 *M* Tris-HCl, pH 7.4, containing 120 mg sodium cyanoborohydride. Gently mix the reaction at room temperature for 1 hr.
5. Finally, wash the gel successively with 200 ml each of water, 1.0 *M* NaCl, and water. Immobilized Con A can be stored in 0.02% sodium azide at 4°C.

3.5.2 Preparation of Immobilized Jacalin

The seeds of the jack fruit (*Artocarpus integrifolia*) contain a hemagglutinating lectin called jacalin that preferentially binds to nonreducing α-D-galactosyl groups. The binding and precipitating specificities of jacalin for heavy chains of human immunoglobulins facilitate its use as a diagnostic (IgA subclass typing) and preparative (purification of IgA and IgD; removal of IgA from biological samples and preparations) tool. For a review on jacalin and its applications in immunochemistry and cellular immunology, see Aucouturier *et al.* (1989). Immobilized jacalin has been used successfully for the isolation of human IgA.

IMMOBILIZATION PROTOCOL

1. Wash periodate-activated Sepharose CL-6B (10 ml settled gel; prepared as described in Section 2.2.1.4, Chapter 2) successively with 100 ml each of water and 0.1 *M* sodium phosphate buffer, pH 7.0 (coupling buffer), and suction dry to a moist cake.

2. Add the gel to a jacalin solution (50 mg jacalin dissolved in 10 ml coupling buffer containing 0.1 M melibiose).
3. Add 60 mg sodium cyanoborohydride to the gel suspension and gently mix the reaction mixture on a shaker overnight at room temperature.

NOTE: Sodium cyanoborohydride is extremely toxic. The coupling reaction and all subsequent operations should be carried out in a well-ventilated hood. We routinely use a 100-ml Erlenmeyer flask as the reaction vessel.

4. Excess aldehyde groups on the gel may be blocked if desired. However, we have found that any remaining formyl groups do not contribute to nonspecific effects in the resulting affinity support. For this reason, we typically do not do a blocking step. To block, treat the support as follows. Wash the coupled gel with 200 ml coupling buffer and then suspend in 20 ml 1.0 M Tris-HCl, pH 7.4, containing 60 mg sodium cyanoborohydride. Gently mix the reaction on a shaker at room temperature for 1 hr.
5. Finally, wash the gel successively with 100 ml water, 1.0 M NaCl, and water. Immobilized jacalin can be stored in 0.02% sodium azide at 4°C.

3.6 IMMOBILIZED NUCLEIC ACIDS

Nucleic acids and polynucleotides immobilized on solid supports are very useful tools to investigate the enzymes and proteins that interact with them, as well as to isolate their complementary strands. Affinity chromatography using immobilized nucleic acids is a versatile technique that has found various applications in molecular biology, genetics, and biochemistry (Potuzak and Dean, 1978). Table 3.3 lists the most commonly used methods for immobilization of nucleic acids.

3.6.1 Adsorption onto Cellulose

Adsorption onto cellulose is a very simple, mild, and nondestructive procedure for attachment of single- and double-stranded DNA to cellulose. To prepare DNA–/ /–cellulose, a DNA solution is mixed with dry cellulose. The DNA–cellulose paste is air-dried and subsequently lyophilized. Although this is a convenient method to prepare immobilized DNA, the DNA is bound to the support reversibly and slowly leaks from the matrix in aqueous solution.

TABLE 3.3
Immobilization Methods for Nucleic Acids and Polynucleotides

Adsorption onto cellulose
Ultraviolet irradiation
Entrapment within cellulose acetate, agar, or polyacrylamide
CNBr activation
Azo coupling
Carbodiimide activation
Cyanuric chloride activation
Coupling with carboxymethyl-cellulose
Epoxy activation
Periodate oxidation of RNA
Reversible complex formation with immobilized boronic acid

DNA–/ /–cellulose prepared by the adsorption method has been used successfully for the purification of DNA-binding proteins. However, for applications involving nucleic acid hybridization, this preparation is not recommended because of its limited stability when exposed to elevated temperatures, formamide, and low ionic strength buffers.

IMMOBILIZATION PROTOCOL (ALBERTS ET AL., 1968)

1. Purify the cellulose used for the preparation DNA–cellulose as follows. Mix 30 gm cellulose with 600 ml 1.0 N HCl for 10 min. Filter the support on Whatman No. 1 filter paper in a Buchner funnel and wash extensively with water. Mix the cellulose with 600 ml 1.0 N HCl for 10 min, then wash extensively with water until washings are neutral. Finally, air dry the support.
2. Mix 15 mg DNA in 20 ml 10 mM Tris-HCl, pH 7.4, containing 1 mM EDTA, with 5 gm purified cellulose until a thick paste results. Stir the paste to uniform consistency and spread on a Petri dish to a thin layer to facilitate drying.
3. Dry the paste and break the mixture into small pieces. Subsequently, grind the dry powder with a mortar, and lyophilize for 48 hr.
4. Suspend the dry powder in 100 ml 10 mM Tris-HCl, pH 7.4, containing 0.15 M NaCl and 1 mM EDTA, and store at 4°C for 24 hr.
5. Filter the DNA–/ /–cellulose in a Buchner funnel and wash with 200 ml Tris-NaCl-EDTA buffer. Finally, suspend the matrix in 20 ml Tris-NaCl-EDTA buffer and store frozen.

Prior to use, DNA–cellulose should be washed with the buffer to be used in the chromatography.

3.6.2 Ultraviolet Irradiation Technique

Ultraviolet (uv) irradiation is the first method reported to covalently link DNA to cellulose (Britten, 1963; Litman, 1968). In this procedure, DNA is physically trapped in a cellulose matrix and subsequently cross-linked by uv irradiation.

The uv irradiation method is efficient and, unlike the cellulose adsorption method just described, the resulting DNA–cellulose linkage is very stable when exposed to elevated temperatures and strong desorbing conditions.

IMMOBILIZATION PROTOCOL

1. Wash cellulose with 1 N HCl and water as described in Section 3.6.1.
2. Mix DNA (16 mg in 8 ml 10 mM NaCl) with 1 gm cellulose until a thick paste results. Stir the paste to uniform consistency and spread on a Petri dish in a thin layer to facilitate drying.
3. Dry the paste and break it into small pieces. Next, grind the mixture with a mortar, and suspend in 25 ml absolute ethanol.
4. Irradiate the alcohol suspension by placing a low pressure mercury lamp at a distance of about 10 cm from the surface of the mixture. Irradiate (about 100,000 ergs/mm^2) for 15 min with gentle stirring.
5. Filter the gel suspension on Whatman No. 2 filter paper in a Buchner funnel, and wash the DNA–/ /–cellulose five times by suspending it in 50 ml 10 mM NaCl for 10 min followed by filtration. Finally, spread the immobilized DNA on a filter paper to dry in air. DNA–/ /–cellulose can be stored at room temperature for several months.

Ultraviolet irradiation methods have been used successfully to immobilize RNA to cellulose and to immobilize RNA, DNA and polynucleotides to vinyl, nylon, and fiberglass supports.

3.6.3 Entrapment in Cellulose Acetate, Agar, or Polyacrylamide

High-molecular-weight single-stranded DNA can be immobilized by trapping it in matrices of cellulose acetate, agar, or polyacrylamide during polymerization or gel formation (Bolton and McCarthy, 1962). The method involves the entrapment of DNA in the pores of the matrices as they form, resulting in physically bound nucleotide chains in the gel structure. Polyacrylamide is often used as a matrix to immobilize DNA by this method. DNA–/ /–polyacrylamide is made by dissolving DNA in the acrylamide solution and subsequently initiating cross-link polymerization of the mixture. DNA–polyacrylamide can be used for column chromatography (if beads are formed during the polymerization process) and is stable when exposed to elevated temperatures during hybridization conditions.

3.6.4 CNBr Activation

CNBr-activated Sepharose 4B has been used successfully to immobilize a wide spectrum of nucleic acids and synthetic polynucleotides (Lindberg and Persson, 1971; Poonian *et al.*, 1971). Table 3.4 lists the various nucleic acids and polynucleotides that have been immobilized using CNBr-activated Sepharose 4B.

3.6.4.1 Preparation of Immobilized HeLa DNA Using CNBr Activation

IMMOBILIZATION PROTOCOL

1. Activate Sepharose 4B (30 ml settled gel) with CNBr as described in Chapter 2, Section 2.2.1.1.
2. Wash the fresh CNBr-activated Sepharose 4B with 300 ml 50 mM potassium phosphate buffer, pH 8.0 (coupling buffer), in a sintered glass funnel, and suction dry to a moist cake.
3. Immediately add the gel to a HeLa DNA solution (denatured) (30 A_{260} units dissolved in 30 ml coupling buffer).

TABLE 3.4
Nucleic Acids and Polynucleotides Immobilized Using CNBr-Activated Sepharose 4B

Polynucleotides	Nucleic acids
Poly(A)	Single-stranded RNA
Poly(C)	Single-stranded DNA
Poly(I)	tRNA
Poly(U)	
Poly(I:C)	
Poly(dA–T)	

4. Gently mix the gel suspension on a shaker at 4°C for 48 hr.
5. Pour the reaction mixture into a suitable column and wash the gel extensively with coupling buffer until nucleic acid no longer appears in the wash (as determined by absorbance at 260 nm).
6. Finally, wash the gel with 300 ml 0.5 M potassium phosphate, pH 8.0. DNA–Sepharose 4B can be stored in 10 mM Tris-HCl, pH 7.4, containing 0.1 M NaCl, 1 mM EDTA, and 0.02% sodium azide at 4°C.

3.6.4.2 Preparation of Immobilized Poly(A) Using CNBr Activation

IMMOBILIZATION PROTOCOL

1. Activate Sepharose 4B (10 ml settled gel) with CNBr as described in Chapter 2, Section 2.2.1.1
2. Wash fresh CNBr-activated gel with 100 ml 0.2 M 4-morpholinoethanesulfonic acid buffer (MES), pH 6.0 (coupling buffer), in a sintered glass funnel, and suction dry to a moist cake.
3. Add the gel to a poly(A) solution [60 mg poly(A) dissolved in 30 ml coupling buffer].
4. Gently mix the gel suspension in a shaker at 4°C for 24 hr.
5. Pour the reaction mixture into a suitable column and extensively wash the gel with 10 mM sodium phosphate buffer, pH 7.0, containing 0.15 M NaCl until poly(A) no longer appears in the wash (as determined by absorbance at 260 nm). Poly(A)–Sepharose 4B can be stored in 10 mM Tris-HCl, pH 7.4, containing 0.1 M NaCl, 1 mM EDTA, and 0.02% sodium azide at 4°C.

3.6.4.3 Preparation of Immobilized Single-Stranded RNA Using CNBr Activation

The method of preparation is the same as that described earlier for the preparation of poly(A)–Sepharose 4B except that single-stranded RNA is used instead of poly(A).

3.6.5 Diazonium Coupling

Single-stranded DNA or RNA can be immobilized to a hydroxylic matrix through a diazotized arylamine that couples primarily with guanine and uracil residues (Fig. 3.53). Azo coupling is highly efficient and the resulting nucleic acid–Sepharose has high capacity and good stability.

260   3. Immobilization of Ligands

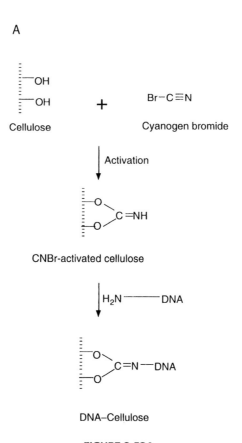

FIGURE 3.53A
Reaction sequence for the preparation of DNA–cellulose using CNBr.

3.6.5.1 Preparation of Immobilized Single-Stranded DNA Using Diazonium Activation

IMMOBILIZATION PROTOCOL

NOTE: This protocol involves the preparation of a diazonium group on Sepharose and the subsequent coupling of DNA. An alternative method of developing a diazonium activated matrix is given in Chapter 2. Either approach will work well for this application.

1. Activate Sepharose CL-6B (100 ml settled gel) with sodium periodate as described in Chapter 2, Section 2.2.1.4.

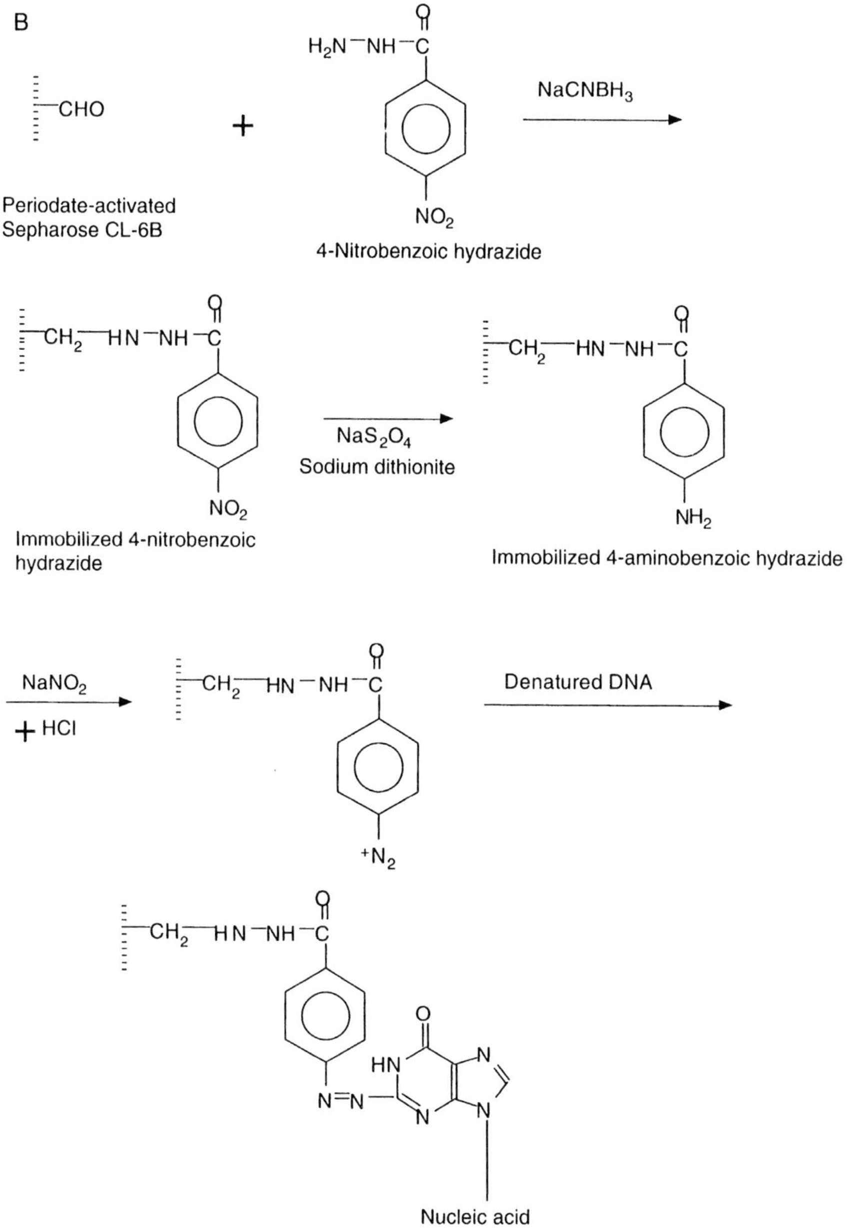

3.6 Immobilized Nucleic Acids

FIGURE 3.53B

Immobilization of single-stranded DNA or RNA onto an hydroxylic matrix through a diazotized arylamine.

2. Wash the periodate-activated Sepharose with 1 L water in a sintered glass funnel, suction dry to a moist cake, and add to a 4-nitrobenzoic acid hydrazide solution (Aldrich) (2.2 gm dissolved in 350 ml 50% ethanol).
3. Stir the gel suspension and add sodium cyanoborohydride (2.2 gm). Adjust the pH of the reaction mixture to 4.0 by adding a few drops of 6.0 N HCl.
4. Continue the reaction for 5 hr at room temperature.
5. Block the excess active aldehyde groups on the coupled gel as follows. Increase the pH of the reaction mixture to 8.5 by adding a few drops of 2.0 N NaOH. Add sodium borohydride (100 mg) and stir for 30 min at room temperature.
6. Finally, filter and wash the gel suspension successively with 2 L each of ethanol, 50% ethanol, and water. Immobilized 4-nitrobenzoic acid hydrazide has a pale yellow color.
7. Reduce the nitro groups of immobilized 4-nitrobenzoic acid hydrazide with sodium dithionite (sodium hydrosulfite) as follows. Wash the gel with 500 ml 0.5 M sodium bicarbonate, suction dry to a moist cake, and add to a sodium dithionite solution (12 gm sodium dithionite dissolved in 200 ml 0.5 M sodium bicarbonate).
8. Stir the gel suspension at room temperature for 1 hr. The gel will become colorless, indicating the reduction of nitro groups.
9. Finally, filter and wash the reaction mixture successively with 2 L each of 0.5 M sodium bicarbonate and water. The intermediate gel, immobilized 4-aminobenzoic hydrazide, can be stored indefinitely in 0.02% sodium azide at 4°C.
10. Diazotize the aryl amine groups of the gel and couple to single stranded DNA as follows. Wash immobilized 4-aminobenzoic hydrazide (5 ml settled gel) (note that only a portion of the gel is used here) with 50 ml ice-cold 0.3 N HCl and suspend in 5 ml ice-cold 0.3 N HCl.
11. Add ice-cold sodium nitrite solution (1.0 ml made by dissolving 50 mg sodium nitrite per ml water) to the gel suspension and mix gently in a shaker at 4°C for 15 min.
12. Quickly wash the diazotized gel with 50 ml ice-cold 0.3 N HCl and 50 ml ice-cold 50 mM sodium phosphate buffer, pH 8.0 (coupling buffer).
13. After washing, suction dry the diazotized gel to a moist cake and immediately add to the denatured single-stranded DNA solution (1 mg/ml in coupling buffer).
14. Mix the gel suspension in a shaker at 4°C for 24 hr.
15. Wash the coupled gel with 100 ml coupling buffer.
16. Block unreacted diazo groups on immobilized DNA as follows. Suspend the gel in 10 ml 0.5 M glycine, pH 8.0, and gently mix in a shaker

for 2 hr at room temperature. Finally, wash the immobilized DNA with 100 ml coupling buffer and store in 10 mM Tris-HCl, pH 7.4, containing 0.15 M NaCl and 0.02% sodium azide at 4°C.

3.6.6 Carbodiimide Coupling

Carbodiimides are widely used as condensing agents in the synthesis and immobilization of nucleic acids. The carbodiimide coupling method for the preparation of DNA–Sephadex G-200 and oligo(dT)-cellulose is described in this section. In principle, in the presence of carbodiimide, the 5′-phosphate end of a nucleotide condenses with the hydroxyl groups of solid supports (Fig. 3.54).

3.6.6.1 Preparation of DNA–/ /–Sephadex G-200 Using Carbodiimide Coupling

IMMOBILIZATION PROTOCOL (WEISSBACH AND POONIAN, 1974)

1. Swell Sephadex G-200 (5 gm) overnight in 250 ml deionized water. Wash the gel on Whatman No. 1 filter paper in a Buchner funnel successively with 1 L 1.0 M NaCl, 1 L 10 mM Tris-HCl, pH 7.4, 2 L deionized water, and 500 ml ethanol. Dry the washed Sephadex in an oven at 60°C and grind it with a mortar and pestle to a fine powder.
2. Dissolve calf thymus DNA (100 mg) in 15 ml 50 mM sodium 2-(N-morpholinoethane)sulfonate (MES) buffer, pH 6.0, by gentle mixing and periodic sonication.
3. Add 1 gm CMC [1-cyclohexyl-3-(2-morpholinethyl) carbodiimide metho-p-toluene-sulfonate] to the DNA solution and mix to dissolve.
4. Spread the reaction mixture on a Petri dish to a thin layer and sprinkle 1 gm dry Sephadex G-200 evenly over the bottom.
5. Place the Petri dish containing the DNA–Sephadex mixture in an oven at 42°C for 8 hr and subsequently transfer to a water-saturated atmosphere at 20°C for 24 hr.
6. Swell the DNA–Sephadex in 10 mM Tris, 15 mM sodium citrate, 0.15 M NaCl, pH 7.4, for 4 hr.
7. Filter and successively wash the gel in a Buchner funnel with 1 L each of water, 1.0 M NaCl, and 10 mM Tris-HCl buffer, pH 7.4, containing 0.15 M NaCl. DNA–Sephadex G-200 can be stored in 10 mM Tris-HCl buffer, pH 7.4, containing 0.15 M NaCl, 1 mM EDTA, and 0.02% sodium azide at 4°C.

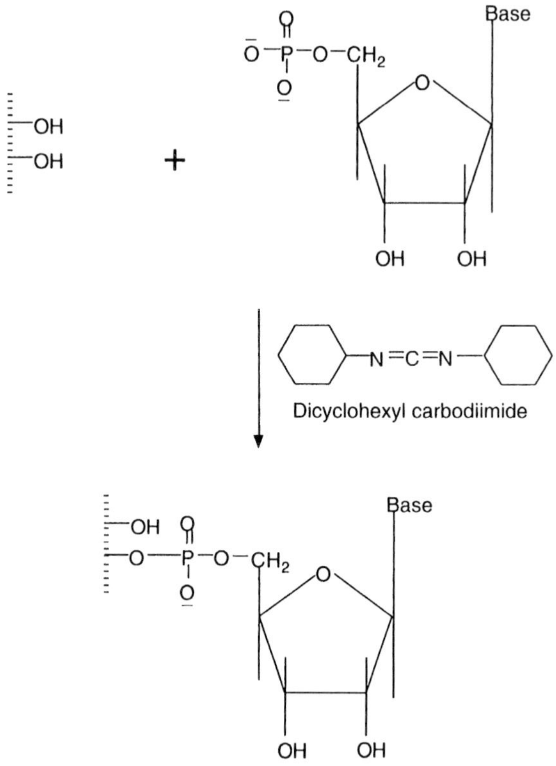

FIGURE 3.54
Coupling of the 5-phosphate end of a nucleotide with the hydroxyl group of a solid support using the carbodiimide reaction.

3.6.6.2 Preparation of Oligo(dT)–/ /–Cellulose Using Carbodiimide Coupling

In this method, thymidine-5′-phosphate is polymerized using dicyclohexyl carbodiimide (DCC) in pyridine; on addition of cellulose, the activated terminal phosphate groups of the oligo(dT) are condensed with the hydroxyl groups of cellulose (Gilham, 1971). Oligo(dT)–/ /–cellulose prepared by this method will consist of cellulose containing a mixture of the bound homopolymer chains (up to 10 nucleotides in length) linked by phosphodiester linkages at the 5′ terminals.

IMMOBILIZATION PROTOCOL

1. Add a twofold excess of DCC to a thymidine-5′-phosphate (2 mmol) solution in 3 ml dry pyridine. Add a few glass beads and shake the thick gum that forms for 5 days at room temperature. Nonporous glass beads of $1/8$-in. diameter are suitable to add as an agitation aid.
2. Add 5 gm washed and dried cellulose (see Section 3.6.1) to the reaction mixture along with 50 ml dry pyridine and 2 mmol of DCC.
3. Shake this mixture for an additional 5 days at room temperature.
4. Filter the reaction mixture in a Buchner funnel and wash with 500 ml pyridine.
5. Suspend oligo(dT)–/ /–cellulose in 200 ml 50% aqueous pyridine for 10 hr and wash with 3 L methanol to remove the dicyclohexyl urea that formed during the synthesis.
6. Finally, wash oligo(dT)–/ /–cellulose with deionized water until the absorbance of the washing at 260 nm is essentially zero. The gel can be stored in 10 mM Tris-HCl buffer, pH 7.4, containing 0.15 M NaCl and 0.02% sodium azide at 4°C.

3.6.7 Cyanuric Chloride Activation

Cellulose activated with cyanuric chloride can be used to immobilize double-stranded DNA (Biagioni *et al.*, 1978). DNA is linked to cellulose through the amino groups of adenine, guanine, and cytosine (Fig. 3.55). Cyanuric chloride-activated cellulose provides a viable matrix for the preparation of DNA–cellulose. The support has the following desirable characteristics. (1) It can be prepared by an easy and reproducible method. (2) Double-stranded DNA is coupled in satisfactory quantities. (3) DNA–cellulose is very stable when exposed to urea and high pH. (See Section 2.2.1.10, Chapter 2 for a description of how to prepare a cyanuric chloride-activated matrix.)

3.6.7.1 Preparation of Double-Stranded DNA–Cellulose Using Cyanuric Chloride Activation

IMMOBILIZATION PROTOCOL

1. Wash cellulose and dry the matrix as described in Section 3.6.1.
2. Swell cellulose powder (5 gm) for 15 min by suspending in 50 ml 3.0 N NaOH.

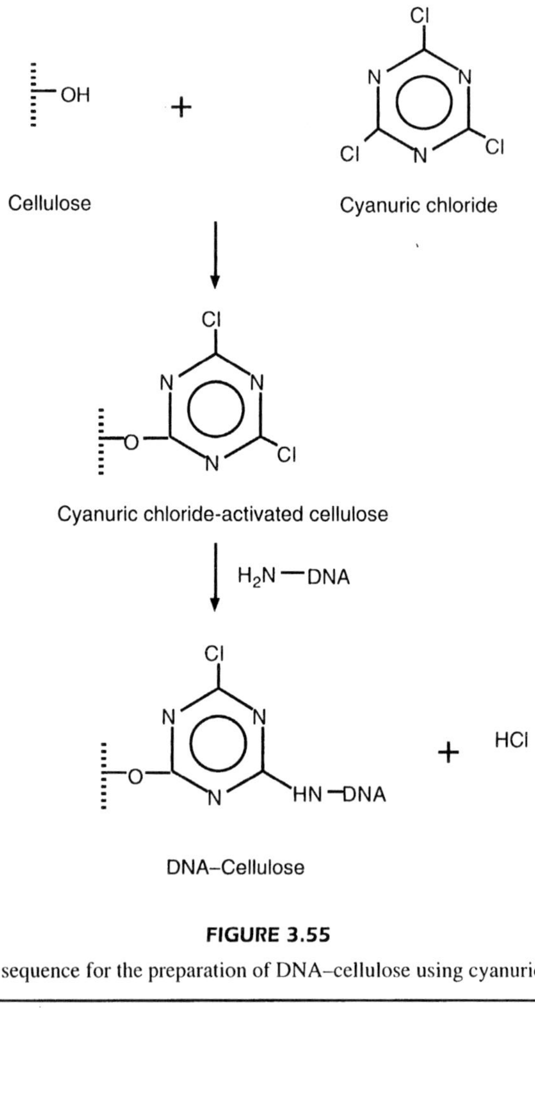

FIGURE 3.55
Reaction sequence for the preparation of DNA–cellulose using cyanuric chloride.

3. Filter the cellulose suspension in a Buchner funnel, suction dry to a moist cake, and add with stirring to a cyanuric chloride solution (5 gm cyanuric chloride dissolved in 100 ml of a 1:1 dioxane/xylene mixture).

NOTE: Use purified cyanuric chloride for activation. Commercial cyanuric chloride contains degradation products. It is purified by extraction with chloroform, using 10 ml/gm cyanuric chloride. Cyanuric chloride, but not its degradation product, is soluble in chloroform and can be obtained as a white solid by removal of the solvent with a rotary evaporator under reduced pressure.

4. After stirring for 30 min at room temperature, filter the reaction mixture and wash the cyanuric chloride-activated cellulose successively with 1 L each of dioxane, a 1:2:1 mixture of acetic acid/dioxane/water, water, and acetone.
5. Dry cyanuric chloride-activated cellulose under vaccuum and use immediately or store in a desiccator.
6. Couple DNA to the cyanuric chloride-activated cellulose as follows. Add cyanuric chloride-activated cellulose (1 gm) to a calf thymus DNA solution (20 mg DNA dissolved in 10 ml water).
7. Stir the suspension for 1 hr at 4°C and filter.
8. In order to abolish any unreacted chlorotriazinyl functions, mix the DNA–cellulose with 50 ml 0.1 M ethanolamine, pH 8.0, for 1 hr at room temperature.
9. Finally, collect the DNA–cellulose by filtration, wash extensively with deionized water, and store as a water suspension at 4°C.

3.6.8 Coupling with Carboxymethyl-Cellulose

Nucleic acids can be immobilized onto carboxymethyl-cellulose (CM-cellulose) easily and efficiently (Potuzak and Wintersberger, 1976). The coupling of calf thymus DNA to CM-cellulose is achieved by esterification of the carboxyl groups of CM-cellulose with terminal hydroxyl groups of DNA. Immobilized nucleic acids prepared by this method contain large amounts of bound ligand. The adsorbents are stable when exposed to elevated temperature and to high formamide concentrations, but only somewhat stable under alkaline conditions. As much as 15 mg DNA can be easily coupled to 1 gm CM-cellulose.

3.6.8.1 Preparation of Calf Thymus DNA–Carboxymethyl-Cellulose

IMMOBILIZATION PROTOCOL

1. Suspend CM-cellulose (1 gm; Whatman, Clifton, New Jersey) in 0.5 N NaOH (50 ml) and stir for 30 min. Filter the gel suspension on Whatman No. 1 filter paper in a Buchner funnel and wash extensively with deionized water until the wash reaches a pH of 8.0.
2. Suspend the CM-cellulose derivative in 100 ml water and adjust the pH of the gel suspension to 3.5 by addition of 0.01 N HCl.

3. Filter and wash the gel suspension with 50 ml ethanol/diethylether mixture (1:1) and 50 ml ether.
4. Finally, dry the gel for 60 min at 40°C.
5. Add the dried CM-cellulose to a calf thymus DNA solution (Sigma; Worthington) (20 mg DNA dissolved in 10 ml water). Spread the resulting gel suspension on a Petri dish in a thin layer and dry slowly over a period of 60 hr at 40°C, controlling the drying process by periodically covering the dish.
6. Scrape the resulting DNA–/ /–cellulose from the surface of the Petri dish, grind with a mortar and pestle, and suspend the powder in 50 mM sodium phosphate buffer, pH 7.0, containing 50% glycerol (50 ml).
7. Leave the DNA–/ /–cellulose suspension at room temperature for 24 hr.
8. Finally, wash the DNA–/ /–cellulose with 100 ml water and store in 1.0 M NaCl at 4°C.

3.6.9 Epoxy Activation

Epoxy activation is a simple and efficient method to covalently couple DNA to cellulose. It uses the bifunctional epoxide 1,4-butanediol diglycidyl ether to activate cellulose and subsequently link DNA to the cellulose (Fig. 3.56). DNA immobilized by the epoxy method is stable when exposed to elevated temperature, formamide, and alkaline conditions (Moss *et al.*, 1981).

3.6.9.1 Immobilization of Calf Thymus DNA Using Epoxy-Activated Cellulose

IMMOBILIZATION PROTOCOL

1. Swell dry cellulose powder (10 gm) in 1 L water, filter on Whatman No. 1 filter paper in a Buchner funnel, and wash successively with 1 L ethanol, 1 L water, and 200 ml 0.6 N NaOH. The ethanol wash is used to remove residual pyridine in commercial preparations.
2. Suction dry the washed cellulose to a moist cake and combine with 25 ml 1,4-butanediol diglycidyl ether (Aldrich) and 25 ml 0.6 N NaOH containing 2 mg/ml sodium borohydride.

NOTE: This epoxide is toxic and flammable. All reactions and subsequent steps are carried out in a well-ventilated hood and users should wear appropriate protective clothing.

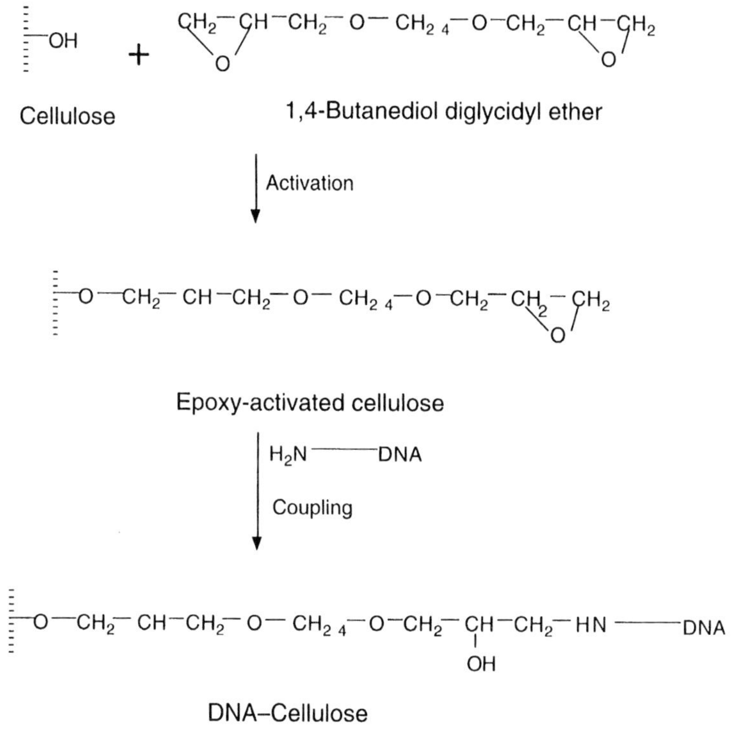

3.6 Immobilized Nucleic Acids

FIGURE 3.56
Reaction sequence for the preparation of DNA-cellulose using 1,4-butanediol diglycidyl ether.

---

3. Stir the reaction mixture with a paddle for 20 hr at room temperature.
4. Stop the reaction by filtering the gel suspension in a Buchner funnel. Extensively wash the activated gel with deionized water until the washings are neutral. Quickly wash the gel with 200 ml ethanol to remove any residual 1,4-butanediol diglycidyl ether. Follow this by a 200 ml wash of deionized water.
5. Lyophilize the epoxy-activated cellulose; store desiccated at −20° C.
6. Prepare a sonicated calf thymus DNA (Sigma) solution at a concentration of 1.5 mg/ml in water, heat to 100°C for 5 min in water, and quickly cool in ice.
7. Suspend epoxy-activated cellulose (50 mg) in 0.1 ml 0.1 N NaOH and mix in a 1.5-ml Eppendorf microfuge tube with 0.1 ml of the sonicated DNA solution.

8. Pipette the gel slurry onto a standard microscope slide and spread to half the slide surface area. Allow the spread gel slurry to sit at 21°C in 100% humidity for 4–8 hr and dry in room air for 2 hr.
9. Carefully scrape the mixture from the glass surface with a razor blade into a 1.5-ml Eppendorf tube and mix with 1 ml water by intermittent vortexing. Separate the DNA–cellulose from the aqueous solution by low speed centrifugation.
10. Extensively wash the DNA–cellulose with water by repeatedly suspending it in water and recovering it by low speed centrifugation.
11. Block unreacted epoxy groups on the resin by reacting the resin with 1.0 M ethanolamine, pH 9.0, for 1 hr at room temperature.
12. Finally wash the DNA–cellulose with water as before and store in 10 mM Tris-HCl, pH 7.2, containing 0.15 M NaCl and 0.02% sodium azide at 4°C.

3.6.9.2 Coupling of Nucleotide Homopolymers to Epoxy-Activated Cellulose

Nucleotide homopolymers (400–600 bases), for example, poly(dT), poly(dC), poly(dA), and poly(dG), can be successfully coupled to epoxy-activated cellulose essentially according to the procedure just given for calf thymus DNA. For coupling, 1 µg nucleotide homopolymer is used per mg epoxy-activated cellulose. Coupling efficiency varies in a range of 80–90%.

3.6.10 Periodate Oxidation of RNA

RNA can be immobilized easily onto amino group-containing matrices by first oxidizing the vicinal 2' and 3' hydroxyl groups of RNA with sodium *meta*-periodate and adding the resulting aldehydo-RNA to the amine matrix to form a Schiff base. The Schiff base is finally reduced with sodium borohydride or sodium cyanoborohydride to form a stable linkage between the RNA and the matrix. Aldehydo-RNA can also be added to hydrazide-containing matrices to form a stable hydrazone bond (Fig 3.57).

3.6.10.1 Preparation of Immobilized RNA by Periodate Oxidation

OXIDATION OF RNA

1. Add sodium *meta*-periodate (10 mg) to an RNA solution (10 mg RNA dissolved in 1.5 ml ice-cold deionized water).

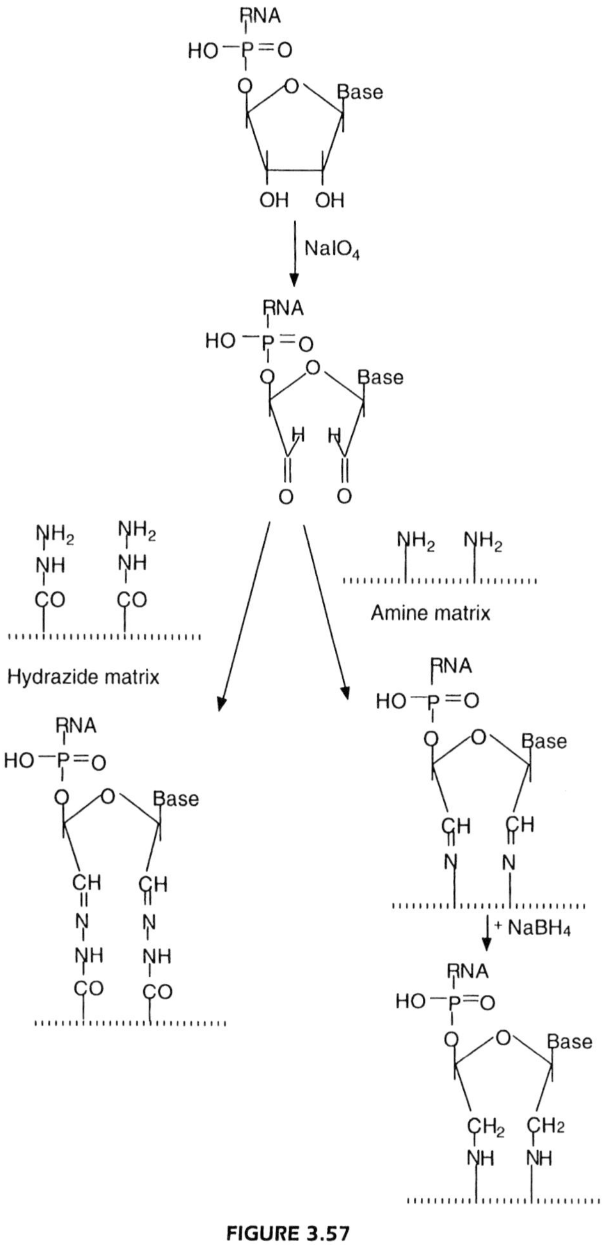

FIGURE 3.57

Reaction sequence for the preparation of immobilized RNA by the periodate oxidation/reductive amination method.

2. Stir the reaction mixture in an ice bath for 30 min. Immediately use the periodate-oxidized RNA to couple to aminoethyl-cellulose.

Coupling to Aminoethyl-Cellulose

1. Swell aminoethyl-cellulose (500 mg) in 2.0 M NaCl, 0.2 M Tricine, pH 8.2 (coupling buffer), and thoroughly wash with coupling buffer.
2. Suspend the gel in 1.5 ml coupling buffer and add to the periodate-oxidized RNA solution prepared as just described.
3. Stir the reaction mixture at room temperature for 6 hr.
4. Add 50 mg sodium borohydride (dissolved in 2 ml coupling buffer) to the reaction mixture and stir for 30 min. This step reduces the Schiff bases that are formed between the oxidized RNA and the aminoethyl-cellulose to stable secondary amine linkages.
5. Finally, extensively wash the RNA–cellulose with coupling buffer and water. RNA–cellulose can be stored in 10 mM Tris-HCl buffer, pH 7.2, containing 0.15 M NaCl and 0.02% sodium azide.

COUPLING TO HYDRAZIDE-ACTIVATED SUPPORTS

1. Wash hydrazide-agarose (2 ml settled gel; prepared as described in Chapter 2, Section 2.2.3.1) with 0.1 M sodium acetate buffer, pH 5.0 (coupling buffer), and then suspend in 2 ml coupling buffer.
2. Mix the hydrazide-agarose gel suspension with periodate-oxidized RNA solution prepared as described earlier. Adjust the pH of the reaction to 5.0 and stir the mixture for 4 hr at 4°C.
3. Add 2.0 M NaCl solution (5 ml) to the reaction mixture and stir for an additional 30 min.
4. Wash the gel successively with 20 ml each of coupling buffer, 2.0 M NaCl, and water. RNA-agarose can be stored in 10 mM Tris-HCl, pH 7.2, containing 0.15 M NaCl and 0.02% sodium azide.

3.6.11 Reversible Complex Formation with Immobilized Boronic Acid

Immobilized boronic acid can be used to immobilize oligo- and polyribonucleotides in a reversible fashion (Weith *et al.*, 1970; Rosenberg and Gilham, 1971; Rosenberg *et al.*, 1972; Schott *et al.*, 1973; Moore *et al.*, 1974; Rosenberg, 1974; Duncan and Gilham, 1975; McCutchan *et al.*,

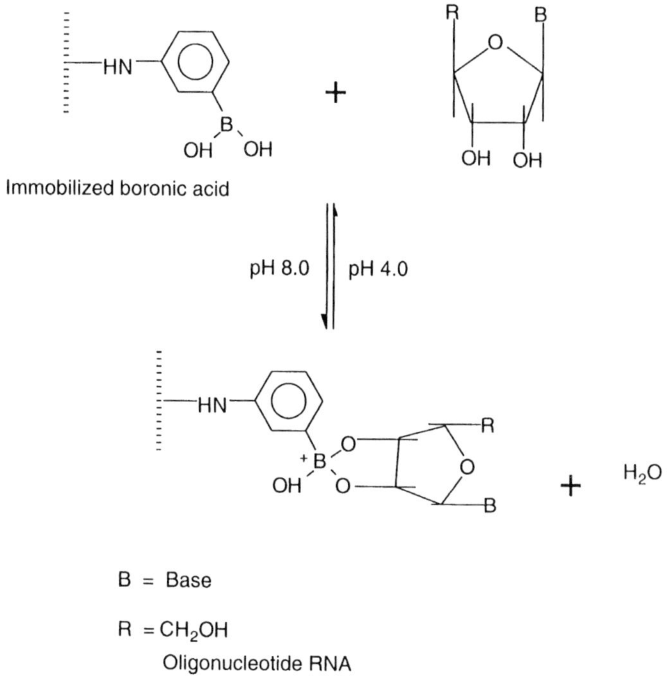

3.6 Immobilized Nucleic Acids

FIGURE 3.58
Reversible complex formation of ribonucleotides with immobilized boronic acid.

1975; Hecht, 1977; Majumder et al., 1979; Pace and Pace, 1980; Singhal et al., 1980; Hohnson, 1981). Vicinal diol groups present in these molecules can form specific cyclic complexes with boronic acid groups (Fig. 3.58). These cyclic boronate complexes are formed at pH 8–9 and are dissociated at pH 3–4. The smaller pored boronate-polyacrylamide supports (Pierce, BioRad, Aldrich) are useful for the affinity purification of mono- and oligoribonucleotides, whereas boronate-agarose supports (Pierce) are used for the purification of higher molecular weight polyribonucleotides.

Figure 3.59 shows the separation of AMP from cAMP using boronate-polyacrylamide. AMP binds to boronate-acrylamide since it contains the necessary vicinal diol groups to interact with the ligand. Since cAMP lacks vicinal diol groups, it is not capable of binding to the boronate support and therefore elutes from the column unretarded.

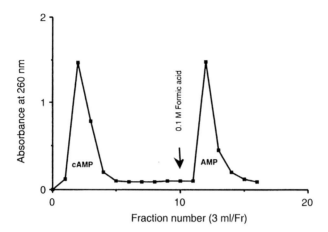

FIGURE 3.59

Separation of AMP and cAMP on a boronate–acrylamide column (2.5 ml gel). The column is equilibrated with 0.25 M ammonium acetate, pH 9.5, containing 0.05 M $MgCl_2$ (binding buffer). AMP and cAMP (200 mg each; free acids; Sigma) are dissolved in 0.5 ml binding buffer and applied to the boronate–acrylamide column. The column is washed with 30 ml binding buffer while 3.0-ml fractions are collected. Bound AMP is eluted with 0.1 M formic acid. Chromatography is conducted at room temperature.

3.7 THIOPHILIC ADSORBENTS

Thiophilic adsorption chromatography is a newly recognized type of protein–ligand interaction (Porath *et al.,* 1985; Hutchens and Porath, 1986, 1987a,b; Belew *et al.,* 1987; Porath, 1987; Nopper *et al.,* 1989; Lihme and Heegaard, 1990). A thiophilic adsorbent has broad specificity for immunoglobulins based on their recognition of a sulfone group in close proximity to a thioether group. Thiophilic interaction chromatography provides an alternative method for the purification of immunoglobulins under mild conditions that preserve biological activity.

A thiophilic gel is prepared by coupling β-mercaptoethanol to DVS-activated Sepharose 4B according to the following reaction sequence (Fig. 3.60).

IMMOBILIZATION PROTOCOL

1. Wash DVS-activated Sepharose 4B (100 ml settled gel; prepared as described in Section 2.2.1.8, Chapter 2) in a sintered glass filter funnel with

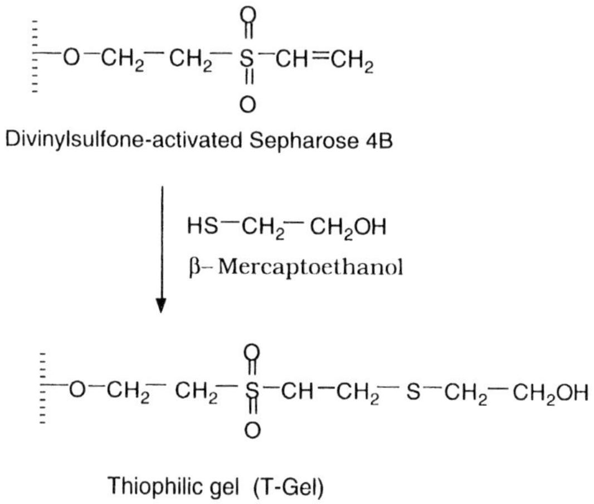

3.8 Immobilized Disulfide Reductants

$$\text{---O---CH}_2\text{---CH}_2\text{---}\underset{\underset{O}{\|}}{\overset{\overset{O}{\|}}{S}}\text{---CH=CH}_2$$

Divinylsulfone-activated Sepharose 4B

$$\Big| \quad \begin{array}{c} \text{HS---CH}_2\text{---CH}_2\text{OH} \\ \beta\text{-Mercaptoethanol} \end{array}$$

$$\text{---O---CH}_2\text{---CH}_2\text{---}\underset{\underset{O}{\|}}{\overset{\overset{O}{\|}}{S}}\text{---CH---CH}_2\text{---S---CH}_2\text{---CH}_2\text{OH}$$

Thiophilic gel (T-Gel)

FIGURE 3.60
Reaction sequence for the preparation of a thiophilic gel (T-gel).

---

200 ml 0.1 M sodium carbonate buffer, pH 9.0 (coupling buffer), suction dry to a moist cake, and add to 75 ml coupling buffer.

2. Add β-mercaptoethanol (10 ml) to the gel suspension while stirring and continue the reaction for 24 hr at room temperature. **Caution:** Coupling and subsequent steps should be carried out in a well-ventilated hood. Filter the reaction mixture and extensively wash the gel with water. The thiophilic gel can be stored in 0.02% sodium azide at 4°C.

3.8 IMMOBILIZED DISULFIDE REDUCTANTS

Many extracellular proteins such as immunoglobulins, protein hormones, serum albumin, pepsin, trypsin, and ribonuclease contain one or more disulfide bonds. For functional and structural studies of proteins, it is often necessary to cleave these disulfide bridges. Disulfide bonds in proteins are commonly reduced with mercaptans, such as mercaptoethanol, thioglycolic acid, and cysteine. High concentrations of mercaptans (molar excess of 20- to 1000-fold) are usually required to drive the reaction to completion. The reaction sequence for this reduction process is shown in Figure 3.61.

Cleland (1964) showed that DTT and dithioerythritol (DTE) are superior reagents in reducing disulfide bonds in proteins. DTT and DTE have low

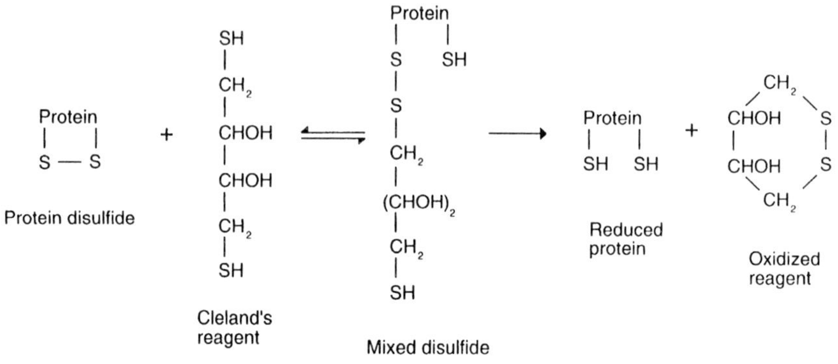

FIGURE 3.61
Reaction sequence for the reduction of disulfide bonds by 2-mercaptoethanol.

oxidation-reduction potential and are capable of reducing protein disulfides at reagent concentrations far below those required for mercaptoethanol. However, even these reagents must be used in approximately 20-fold molar excess to achieve nearly 100% reduction of a protein. The reaction sequence for the reduction of protein disulfides with Cleland's reagent is shown in Figure 3.62.

The use of immobilized disulfide reductants has many advantages over solution phase agents such as β-mercaptoethanol, cysteine, DTT, and DTE.

1. Immobilized disulfide reductants can be used to reduce all types of biological disulfides without producing product or by-product contaminants.
2. Soluble components that interfere with the assay of free thiol groups are not present when immobilized disulfide reductants are used.

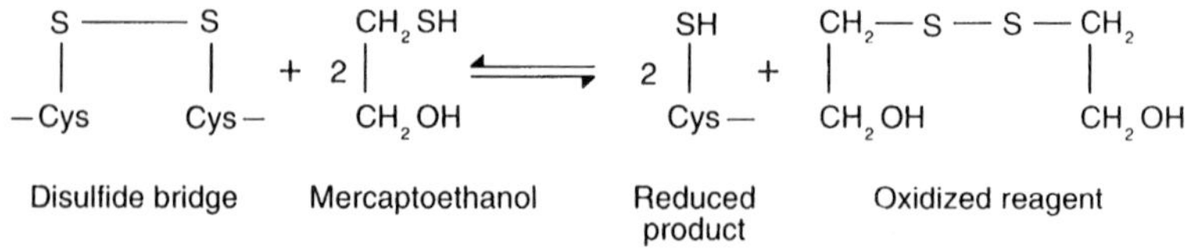

FIGURE 3.62
Reaction sequence for the reduction of protein disulfides by Cleland's reagent.

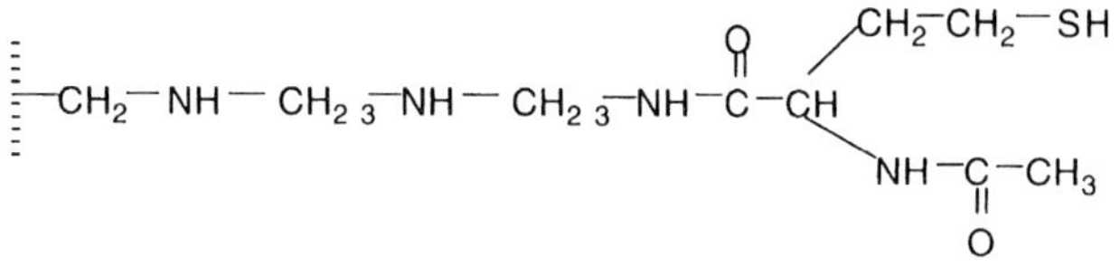

3.8 Immobilized Disulfide Reductants

$$\equiv\!-CH_2\!-NH\!-CH_{2\ \overline{3}}\!-NH\!-CH_{2\ \overline{3}}\!-NH\!-\overset{O}{\underset{\parallel}{C}}\!-\underset{\underset{\underset{O}{\parallel}}{NH-\underset{\parallel}{C}-CH_3}}{\overset{CH_2-CH_2-SH}{CH}}$$

Immobilized
N-acetyl homocysteine

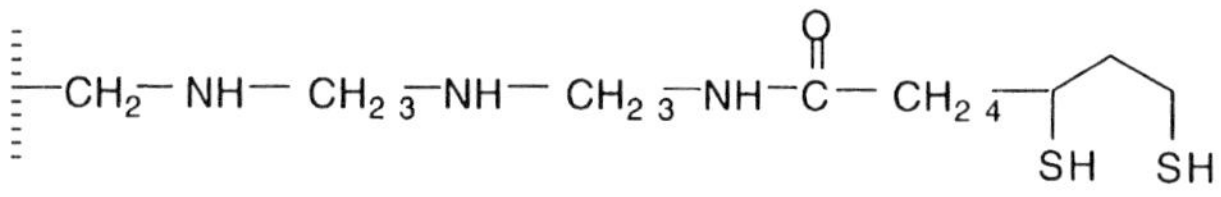

$$\equiv\!-CH_2\!-NH\!-CH_{2\ \overline{3}}\!-NH\!-CH_{2\ \overline{3}}\!-NH\!-\overset{O}{\underset{\parallel}{C}}\!-CH_{2\ \overline{4}}\underset{SH\ \ SH}{\overbrace{\qquad}}$$

Immobilized dihydrolipoamide

FIGURE 3.63
Partial chemical structure of sulfhydryl agarose and immobilized dihydrolipoamide.

---

3. Immobilized disulfide reductants can be regenerated easily and reused several times.

Immobilized dihydrolipoamide (thioctic acid) (Gorecki and Patchornick, 1973; Gorecki and Patchornick, 1975) and immobilized N-acetyl-homocysteine thiolactone (Eldjarn and Jellum, 1963; Jellum, 1964) are the two most commonly used immobilized disulfide reductants (Fig. 3.63). The authors have used immobilized reductants successfully to reduce many types of biological disulfides (oxidized glutathione, bovine insulin, ribonuclease, and lysozyme).

IMMOBILIZATION PROTOCOL

Immobilized N-acetyl-homocysteine is prepared by the following reaction sequence (Fig. 3.64).

1. Wash immobilized DADPA (100 ml settled gel), prepared as described earlier (Section 3.1.1.1), successively with 1 L ice-cold water and 1 L ice-cold 1.0 M sodium bicarbonate in a sintered glass funnel, and suction dry to a moist cake.
2. Suspend the gel in 100 ml ice-cold 1.0 M sodium bicarbonate and stir with a paddle in a cold room at 4°C.

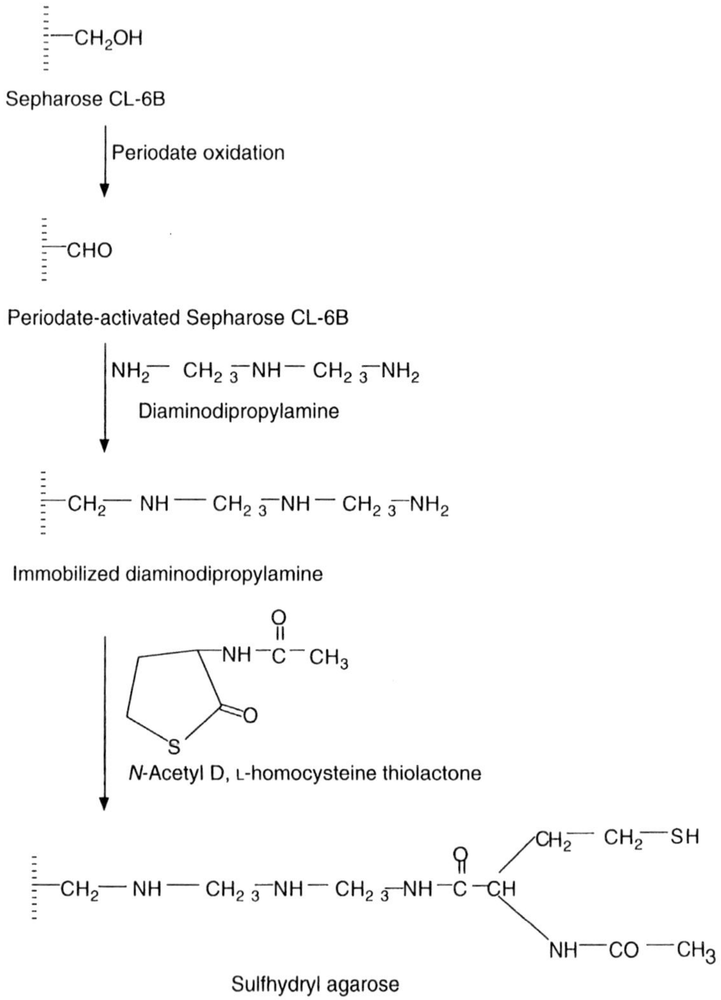

3. Immobilization of Ligands

≡—CH$_2$OH

Sepharose CL-6B

↓ Periodate oxidation

≡—CHO

Periodate-activated Sepharose CL-6B

↓ NH$_2$—CH$_2$$_3$—NH—CH$_2$$_3$—NH$_2$
Diaminodipropylamine

≡—CH$_2$—NH—CH$_2$$_3$—NH—CH$_2$$_3$—NH$_2$

Immobilized diaminodipropylamine

↓ N-Acetyl D, L-homocysteine thiolactone

≡—CH$_2$—NH—CH$_2$$_3$—NH—CH$_2$$_3$—NH—C(=O)—CH(CH$_2$—CH$_2$—SH)(NH—CO—CH$_3$)

Sulfhydryl agarose

FIGURE 3.64
Reaction sequence for the preparation of sulfhydryl agarose.

3.8 Immobilized Disulfide Reductants

3. Add N-acetyl-D,L-homocysteine thiolactone (20 gm; Sigma) to the gel suspension and stir the reaction mixture at 4°C for 20 hr.
4. Filter and wash the reaction mixture in a sintered glass funnel successively with 1 L each of water, 1.0 M NaCl, and water.
5. Immobilized N-acetyl-homocysteine can be stored in 10 mM Tris-HCl buffer, pH 7.6, containing 100 mM NaCl, 1 mM EDTA, and 0.02% sodium azide at 4°C.

Immobilized lipoamide can be prepared as follows.

1. Wash immobilized DADPA (100 ml settled gel), prepared as described earlier (Section 3.1.1.1), with 200 ml 0.1 M sodium phosphate, pH 7.3 (reaction buffer). Suction the gel free of excess buffer until it is left as a wet cake.
2. Dissolve 1.0 gm lipoic acid (thioctic acid; Aldrich) in 100 ml reaction buffer. The compound is only sparingly soluble at pH 7.3, but raising the pH to 10–11 with addition of 6 N NaOH will bring the compound into solution. Readjust the pH to 7.3 with 6 N HCl.
3. Add the wet gel to the lipoic acid solution and stir the slurry with a paddle. To initiate the coupling reaction, add 3.0 gm water-soluble carbodiimide EDC to the mixture.
4. React overnight at room temperature.
5. Wash the gel extensively with reaction buffer, 1 M NaCl, and water. The support may be stored as an aqueous slurry containing 0.02% sodium azide as preservative.

4

TECHNIQUES OF THE TRADE

4.1 MEASUREMENT OF ACTIVATION LEVEL AND IMMOBILIZED LIGAND DENSITY

Activated matrices designed to immobilize specific target ligands may have widely varying levels of coupling potential. The degree to which a support is chemically modified to produce active coupling groups governs the amount of ligand that can be immobilized per unit volume or unit area of matrix material. The activation level may extend from only a few micromoles of coupling groups per bead for a support such as a nonporous polystyrene particle to several hundred micromoles per ml for a support such as cross-linked agarose. The ligand density that can be expected from coupling to such widely differing materials may vary even more than the activation level.

The successful application of immobilized ligands to particular biotechnology problems is often predicated by selecting the optimal ligand density range. Too low a density of ligand molecules can produce a weakly interacting affinity matrix that will not perform well in its intended application. Conversely, too high a ligand loading may produce nonspecific binding effects or create conditions under which the elution of target molecules is difficult. These considerations are especially true when the interaction between the target molecule and the ligand is either extremely weak or extraordinarily strong. A weak affinity interaction between ligand and target molecule may actually benefit from a high ligand loading, whereas a strong affinity pairing could respond well to average or low ligand loadings (Lowe *et al.*, 1973).

Two important parameters in insuring the production of an optimal immobilized affinity ligand are an awareness of the activation level for the chosen chemistry and matrix material and a knowledge of the resulting density of the coupled ligand. The broad selection of coupling chemistries and matrices available differ considerably in their activation and coupling potentials. The following sections are designed to present some viable assay methods for determining these parameters. Some of the methods are designed to measure only the activation level by coupling a nonrelevant small molecule that can be easily monitored to the support. Other procedures are useful for measuring the density of the intended ligand after immobilization.

4.1.1 Measurement of Activation Levels Using Ninhydrin

Ninhydrin-based monitoring systems are among the most widely used for the quantitative determination of the amino acid content of proteins. Ninhydrin reacts with primary amines to form a colored complex known as Ruhemann's purple (Fig. 4.1). This same reaction also can be used to measure the amount of free primary amino groups attached to an insoluble support. Indirectly, the activation level of the matrix can be determined from the quantity of primary amine detected. A small molecule is needed with a functional group on one end that will react with the activated matrix and a terminal primary amino group on the other end. For amine-reactive immobilization chemistries, diamines such as ethylene diamine (EDA) or diaminodipropylamine (DADPA) work well. The small size of these compounds insures that the maximal amount of ligand will be able to couple to the activated matrix, giving a good indication of the total amount of activity originally present.

The following protocol illustrates the measurement of activity for a carbonyl diimidazole (CDI)-activated agarose support. Other amine-reactive matrices may be substituted for this one without changing the basic assay procedure. To couple the diamine to other activated supports, follow the recommended coupling protocol given in the appropriate section of Chapter 2. The ninhydrin solution used in the detection step is Nin-Sol AF, a preformulated reagent from Pierce Chemical.

PROTOCOL

1. React 1 ml CDI-activated agarose in an equal volume of acetone with 0.5 ml DADPA in a test tube. (Use a fume hood and be careful handling the diamine.)
2. Mix by rocking for at least 4 hr at room temperature.
3. Wash the gel extensively with acetone and then with water.

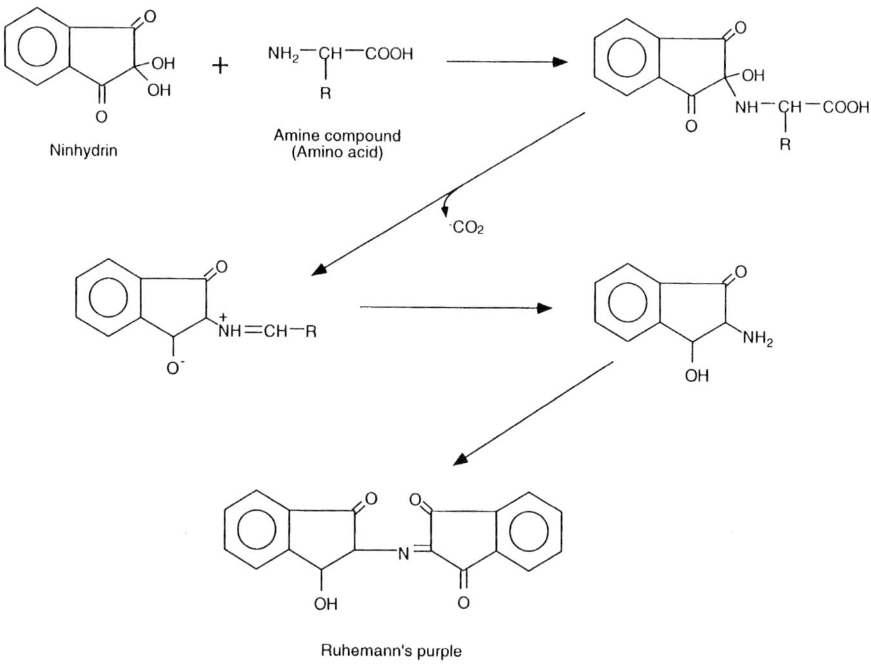

4.1 Measurement of Activation Level

FIGURE 4.1

The reaction of ninhydrin with a primary amine produces the colored compound Ruhemann's purple. Immobilized amines will also react with this reagent and yield color in direct proportion to the density of amine substitution.

---

4. Make a gel slurry consisting of about 50% gel in water (by volume). The exact volume of gel relative to the total slurry volume should be determined by centrifugation in a graduated tube. Record the total volume of slurry and the volume of gel in it. These values will be necessary for the final calculation (see subsequent text).
5. Add 200 µl gel slurry, 2 ml water, and 1 ml ninhydrin reagent to each of two test tubes.
6. Heat in a boiling water bath for 15 min.
7. Cool and transfer the contents of each tube to a 100-ml volumetric flask.
8. Dilute to the mark with a 50:50 (v/v) mixture of ethanol:water. Mix well and let the gel settle. The ethanolic solution fully solubilizes the Ruhemann's purple color from the gel.
9. Measure the absorbance of the solution at 570 nm.

Calculate the quantity of amines coupled to the support using the following equation, which takes into account the extinction coefficient for the reaction between ninhydrin and one primary amine on DADPA.

$$\mu\text{mol amine/ml gel} = (A_{570})(100/0.2)(\text{ml slurry/ml gel})(1/8750)(1000)$$

4.1.2 Determination of Immobilized Protein Using Bicinchoninic Acid

The reaction of protein with alkaline Cu^{2+} to form Cu^{1+} can be monitored using the intense purple color formed with the specific Cu^+ chelating compound bicinchoninic acid (BCA; Pierce, Sigma), sodium salt (Fig. 4.2). The color produced by this complex is stable and increases linearly over a broad range of protein concentrations (Smith *et al.*, 1985). The common use of the BCA reagent to measure protein in solution can be extended to assaying protein coupled to insoluble support materials. Immobilized protein samples can be reacted with the BCA reagent; the color produced is directly proportional to the amount of protein present (Stich, 1991).

The following protocol describes the use of BCA for the determination of protein immobilized on chromatography matrices. However, the basic principles of the assay can be applied to the measurement of protein coupled to other support materials. For instance, polystyrene balls or microplates can be assayed for immobilized protein in a similar manner (Sorensen and Brodbeck, 1986). In these cases, instead of adding a gel slurry to the BCA working reagent, add a few polystyrene balls to the BCA or add the working reagent directly to the wells of a coated microplate.

PROTOCOL

1. Prepare a 50% (v/v) slurry of the immobilized protein in water. Accurate slurry measurements can be made by centrifugation in a graduated test tube. A control should be prepared using uncoupled gel slurry.
2. Prepare a set of protein standard solutions using either the soluble form of the protein that was immobilized or BSA as a representative protein. Make at least five concentrations of the standard protein, ranging from 0.1 mg/ml to 2.0 mg/ml.
3. Using test tubes as the assay vessels, mix 100 µl sample (either the gel slurry or the standards) with 2 ml BCA working reagent in each tube (Pierce Chemical; prepared according to manufacturer instructions). Each sample should be assayed in duplicate.

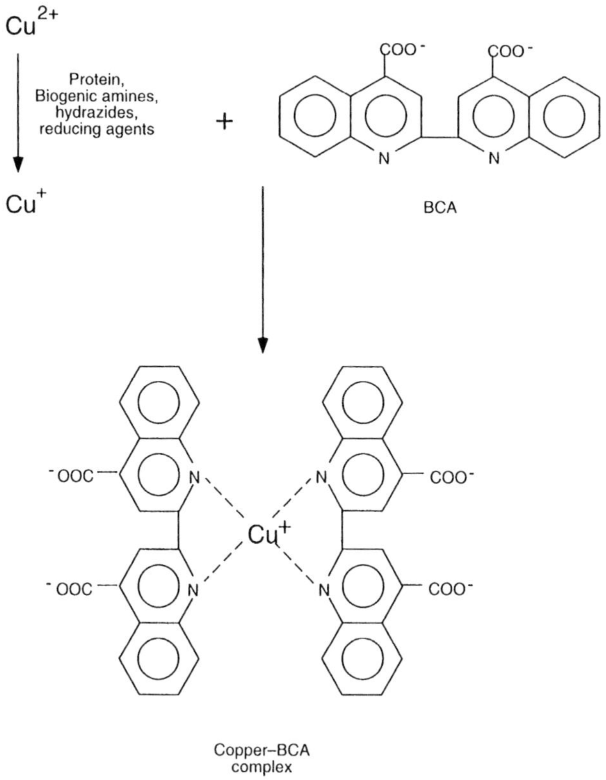

4.1 Measurement of Activation Level

FIGURE 4.2
Bicinchoninic acid (BCA) can be used as a detection reagent for the formation of copper I from copper II as a result of reduction by proteins, biogenic amines, hydrazides, reducing sugars, or other reducing agents. Immobilized proteins can be measured by the colored product that results from the BCA–Cu^+ chelate.

---

4. Seal the tubes with Parafilm and incubate at 37°C for 30 min. Alternatively, the tubes may be incubated at room temperature for 2 hr. Periodically mix the tubes containing the gel slurry.
5. Cool the tubes to room temperature. Separate the gel from each tube by using a serum separator tube or by centrifugation.
6. Measure the absorbance of each supernatant at 562 nm.
7. For the standards, plot the concentration against the absorbance. Determine the concentration in the gel sample tubes by comparison with the standard curve or by linear regression. The amount of protein

immobilized then can be calculated from the fact that each tube contained 50 µl gel in 100 µl slurry.

4.1.3 Qualitative Assay for Amine, Hydrazide, or Sulfhydryl Functionalities Using 2,4,6-Trinitrobenzenesulfonate

Compounds containing primary aliphatic or aromatic amines or hydrazide functionalities will react with 2,4,6-trinitrobenzenesulfonate (TNBS) to form a highly colored complex (Fig. 4.3). This reaction provides an excellent visual check for the presence, or absence, of these groups during the preparation of immobilized ligands (Inman and Dintzis, 1969). The color formed varies slightly depending on the type of functional group present. Aliphatic primary amines will render a bright orange color, whereas aro-

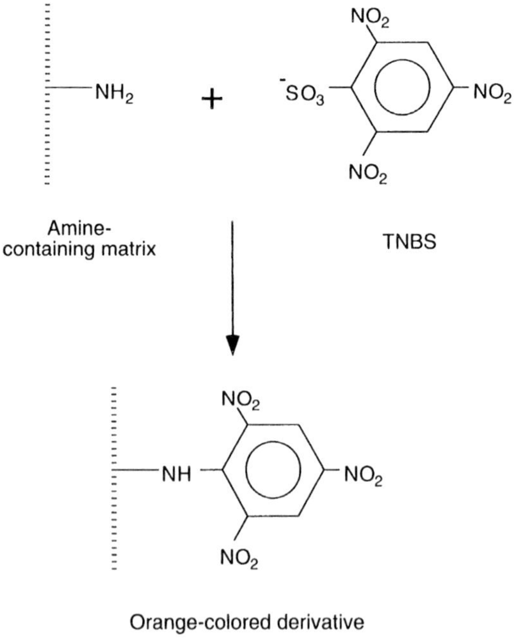

FIGURE 4.3

Trinitrobenzene sulfonate (TNBS) reacts with amine-containing matrices to produce a bright orange covalent derivative. Sulfhydryl- and hydrazide-containing supports also yield colored complexes. The reaction forms the basis of an excellent qualitative check for the presence of these functionalities.

matic amines produce a reddish orange and sulfhydryls a strong red-brown color (Cuatrecasas, 1970). Hydrazides react with TNBS to produce a deep red shade.

When preparing various immobilized species, TNBS can be used to monitor the success of some of the reaction steps if an amine or hydrazide compound is being coupled or blocked. For instance, in the preparation of an immobilized succinylated amino spacer arm (Section 3.1.1.7, Chapter 3), TNBS can be used to confirm the presence of the initial terminal amine on the support. An orange color developing on the matrix surface indicates a positive result. In addition, the presence of any orange color in the solution phase indicates incomplete washing of the support to remove unreacted amino spacer. TNBS can also be used to monitor for the success of the succinylation step, since the reagent will give only a pale yellow color with the matrix if the amines are completely blocked. Other ligands coupled to amine spacers can be assessed similarly if they are expected to block the amines completely or block them enough to show a significant difference in color before and after coupling.

PROTOCOL

1. Add about 0.5 ml derivatized matrix to 2 ml saturated sodium borate solution.
2. Add 3–4 drops 1.5% TNBS solution in ethanol to the suspended matrix.
3. Mix the tube contents and observe the color changes on the surface of the support material. Complete color development may require 5–10 min. Centrifugation of the slurry may aid in visualization since the matrix will pellet at the bottom of the tube.

4.1.4 Assay of Hydrazide Functionalities Using Bicinchoninic Acid

Hydrazide groups efficiently reduce Cu^{2+} to Cu^+. The Cu^+ formed by this reaction can be monitored using the specific chelator BCA in alkaline solution (Pierce Chemical). The BCA response of an activated hydrazide-containing matrix is compared with a standard curve generated from serial dilutions of an adipic dihydrazide solution. After adjusting for the presence of two hydrazides per molecule of adipic dihydrazide, the activation level of the activated support can be calculated in μmol of hydrazide functionality (authors' unpublished results).

PROTOCOL

1. Prepare a standard curve using adipic dihydrazide dissolved in water. Make a series of dilutions of a stock solution of this compound ranging from 1 mg/ml to 25 µg/ml. Use at least five different concentrations to construct the standard curve.
2. Prepare a slurry of the hydrazide-activated support in water. Adjust the slurry concentration to 50% (v/v), measuring the gel volume by centrifugation in a graduated tube and adjusting the total slurry volume as necessary.
3. In separate test tubes, mix 50 µl each standard solution or gel slurry sample with 5 ml BCA working reagent (prepared according to manufacturer instructions). Each sample should be assayed in duplicate.
4. Let the tubes incubate at room temperature for 5 min.
5. Separate the gel from the solution in each tube by centrifugation or by using a serum separator tube.
6. Measure the absorbance of each tube at 562 nm.
7. Construct a standard curve from the response of the adipic dihydrazide standards. Take into account that the concentration of hydrazide functionalities is twice that of adipic dihydrazide.
8. From the standard curve, determine the hydrazide concentration represented in the gel sample. The level of hydrazide functionalities determined by this assay represents the activation level contained in 25 µl support material.

4.1.5 Assay of Biotin Binding Sites on Immobilized Avidin or Streptavidin

The utility of immobilized avidin or streptavidin in the purification of biotinylated proteins and other molecules is well documented in the literature (for review see Wilchek and Bayer, 1990). Theoretically, avidin has a maximum activity of 14.4 units/mg protein. One unit is defined as the binding of 1 µg biotin per mg avidin. Attaching avidin to a solid support will typically lower the maximal binding potential by active site blocking, denaturation, and steric hindrance. To determine the functional capacity of such a support, the compound biotin-p-nitrophenyl ester provides a biotin derivative that is easy to monitor because of its intense yellow color, which can be released in alkaline solution (Fig. 4.4). Its small size also provides an indication of the true level of biotin binding sites still active after immobilization. The capacity for a biotinylated protein with a molecular size much

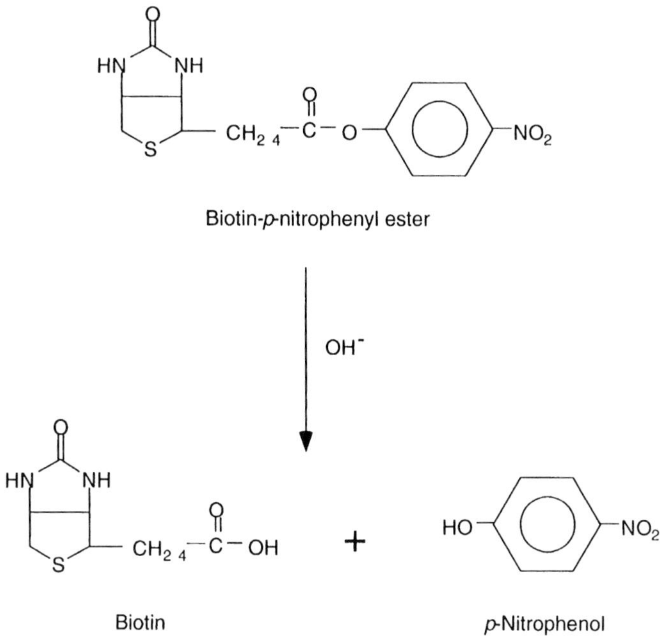

4.1 Measurement of Activation Level

FIGURE 4.4

Biotin-*p*-nitrophenyl ester can interact with immobilized avidin to form a tight complex. Subsequent cleavage of the ester under basic conditions liberates the colored *p*-nitrophenol group. This reaction can be used to determine the specific activity of immobilized avidin preparations.

greater than that of this molecule naturally will be much less on a molar basis.

PROTOCOL

1. Dissolve 10 mg biotin-*p*-nitrophenyl ester (Sigma) in 50 ml 0.5 M sodium acetate, pH 5.0 (buffer A). It may take 30 min of stirring for complete dissolution. Filter and measure the absorbance of the solution at 210 nm. Use immediately.
2. Prepare an immobilized avidin minicolumn containing 2–5 ml gel. Note the exact volume since it will be used in the calculations later. Equilibrate the column with 2–3 column volumes of buffer A.
3. Add 3-ml aliquots of the biotin-*p*-nitrophenyl ester solution until the eluted fractions have an absorbance close to that of the original solution. Collect 3-ml fractions.

4. Wash the column with 2–3 column volumes of water.
5. Elute the colored *p*-nitrophenol group off the column by applying a 0.1 N NaOH solution. A yellow band should be seen migrating through the column as the ester breaks down and elutes.
6. Measure the absorbance at 410 nm of the yellow eluted fractions using a cell with a 1-cm path length. If a dilution is necessary, use 0.1 N NaOH as the diluent. The extinction coefficient of the eluted *p*-nitrophenol is 18.3 ml/μmol (molar extinction coefficient = 18,300).

Calculate the biotin binding sites according to the following equation (the molecular weight of biotin is 244.3).

$$[(A_{410})(3 \text{ ml})(\text{number of fractions})(244.3)]/[(18.3)(\text{ml gel in column})]$$
$$= \mu g \text{ biotin bound/ml gel}$$

4.1.6 Qualitative Determination of Immobilized Protein Using Coomassie Dye

Under acidic conditions, the dye Coomassie Blue G-250 (Bradford reagent) will specifically bind to protein molecules, forming a blue complex that can be monitored at 595 nm (Bradford, 1976). This interaction forms the basis for a quantitative assay of protein concentration in solution. Immobilized protein will also interact strongly with this dye. The blue complex formed between an immobilized protein and the dye is an excellent qualitative indicator of successful protein coupling.

PROTOCOL

1. Wash a small quantity of immobilized protein with water. Place about 0.5 ml washed gel in a test tube.
2. Add 2–3 ml Coomassie protein assay reagent (Pierce Chemical) to the gel.
3. Mix and observe the color of the gel. Centrifugation may help visualize the color change since the gel will pellet on the bottom of the tube.

An intense blue color on the gel indicates the presence of protein molecules. The surrounding solution should be the brown color of the acidic dye reagent. If the solution becomes blue it could indicate incomplete washing of the support and/or leaking of protein into solution. The solution also may turn blue if the pH of the dye reagent is raised as a result of coming in contact with the immobilized protein support. Thus, it is recommended to remove any buffers from the affinity support with a water wash prior to adding the Coomassie reagent.

4.1.7 Measurement of Ligand Coupling by Difference Analysis

A common way to determine ligand loading is to monitor the decrease in absorbance of the ligand coupling solution after the immobilization reaction is complete. For instance, the absorbance at 280 nm of a protein solution can be recorded prior to mixing with an activated matrix. After coupling, the initial washes may be measured at the same wavelength and any difference can be related to the amount of protein immobilized. For a 1-cm cell path, the absorbance values can be directly multiplied by the total solution volume to obtain the total optical density (OD) for the two solutions. The difference in the total OD before coupling and the total OD after coupling equals the total OD of protein that was coupled. Dividing by the extinction coefficient for a 1 mg/ml solution of the protein yields the total number of milligrams immobilized.

For ligands with convenient absorptivities, this method is perhaps the easiest for estimating the amount of ligand coupled. However, care must be taken not to rely on this value too heavily, since many factors can cause great variation in the calculation. If the activation chemistry contributes a leaving group that also has absorptivity at the wavelength at which the ligand is monitored, the determination will be inaccurate unless this contaminant is removed by dialysis or some other method. CDI coupling chemistry is a good example of a reaction releasing such a contaminant. During the coupling reaction, CDI-activated supports release imidazole into solution. Imidazole has a broad absorbance contribution that affects absorbance readings at 280 nm. Therefore, the degree of protein immobilization can only be determined by the difference method if imidazole is first removed from the washes.

Another difficulty may be encountered when the uncoupled ligand does not wash off the support easily after the coupling reaction. This may occur frequently with hydrophobic molecules and certain dyes that have a tendency to bind noncovalently to one another. If the washes after immobilization are too dilute, accurate difference determinations will not be possible. Nevertheless, this method is used repeatedly and with some degree of reliability in countless examples of ligand immobilization.

4.1.8 Direct Absorbance Scan of the Immobilized Ligand

Another way to measure the amount of ligand immobilized to a chromatographic support is by performing a direct absorbance scan of the gel suspended in 50% glycerol in water (v/v). For ligands with known absorptivity in the uv or visible wavelengths, this is an easy method of monitoring successful ligand coupling and quantifying the amount of ligand immobilized (by using the appropriate extinction coefficients).

In one example of this technique, the dye phenol red was immobilized using the Mannich reaction (Section 3.1.4.3, Chapter 3) on DADPA–/ /-agarose. In this case, the success of the coupling reaction is visually evident by the intense red color of the matrix. The absorbance scan of the gel is illustrated in Figure 4.5. Even for colorless ligands, the same basic protocol is applicable.

PROTOCOL

1. Wash 1 ml of the affinity support with water to remove any buffer salts. Drain to remove excess water.
2. Suspend the washed gel in 2 ml 50% glycerol in water (v/v). Mix gently to suspend the gel in the solution fully and uniformly. Be careful not to introduce air bubbles during the mixing process.
3. Take a 0.5-ml aliquot of this suspension and further dilute it with 2.5 ml of the 50% glycerol solution. This concentration of gel in the glycerol solution usually gives acceptable absorbance readings with a strongly

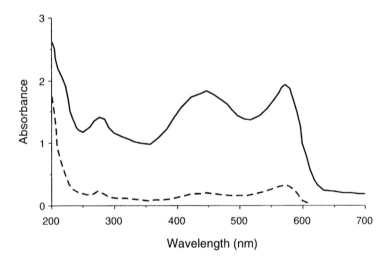

FIGURE 4.5

Direct absorbance scan of immobilized phenol red on agarose in 50% glycerol. The dye was immobilized using the Mannich coupling chemistry at two temperatures. The 57°C reaction (solid line) yielded a much higher density of immobilized dye than the 37°C protocol (dashed line).

absorbing ligand (such as the phenol red ligand described earlier). Adjusting the final slurry concentration may be necessary to obtain the optimal absorbance for a particular immobilized ligand.
4. Using the same optimal slurry concentration determined in Step 3, prepare a control gel sample suspended in glycerol using the support material from which the affinity gel was made (include spacer arm molecules attached if appropriate).
5. Zero a spectrophotometer at the wavelength of absorptivity for the ligand molecule using a 50% glycerol solution in both the sample and reference cells. Do a wavelength scan of the suspended derivatized support in the sample cell against the control support suspension in the reference cell. An estimate of the amount of bound ligand may be determined from the measured extinction coefficient of the ligand in solution.

4.1.9 Measurement of Sulfhydryl Groups Using Ellman's Reagent

Ellman's reagent, 5,5'-dithiobis(2-nitrobenzoic acid), can be used to assay the amount of sulfhydryl residues immobilized on a solid support (Ellman, 1958; Riddles *et al.*, 1979). In the presence of free sulfhydryl groups, Ellman's reagent is reduced to liberate the chromogenic substance 5-sulfido-2-nitrobenzoate (Fig. 4.6), also known as 5-thio-2-nitrobenzoic acid (TNB). The intense yellow color of the anion TNB can be quantified by measuring the absorbance at 412 nm.

Since each sulfhydryl functionality generates one chromogenic TNB anion from one Ellman's molecule, a support containing sulfhydryl groups can be assayed easily according to the following protocol.

PROTOCOL

1. Wash 2 ml of a sulfhydryl-containing gel with several volumes of 0.1 M sodium phosphate, 1 mM EDTA, pH 8.0 (assay buffer).
2. Dilute the gel slurry to a total volume of 8 ml with assay buffer to prepare a 25% slurry (v/v). Take 1 ml of this well-mixed slurry and further dilute with 1 ml assay buffer to create a 12.5% slurry.
3. Mix the gel suspension using a vortex mixer and aliquot 20 µl, 40 µl, 60 µl, 80 µl, and 100 µl slurry into separate test tubes. Add enough assay buffer to each tube to increase the total volume to 1 ml.

4. Add 100 μl Ellman's reagent [prepared by dissolving 5,5'-dithiobis(2-nitrobenzoic acid) at a concentration of 4 mg/ml in assay buffer] to each tube. Mix and incubate at room temperature for 15 min.
5. Separate the gel from the solution using a serum separator or by centrifugation. Measure the absorbance of each supernatant at 412 nm against a blank (prepared by mixing 1 ml assay buffer and 100 μl Ellman's reagent).
6. Plot the resulting absorbances against the amount of gel added. Use a point in the linear portion of the curve to determine the sulfhydryl content. For that sample, calculate the sulfhydryl content of the gel using the following formula. (The molar extinction coefficient of the reduced Ellman's reagent is 13,600 and the total volume of the reaction is 1.1 ml.)

$$\mu\text{mol -SH groups present} = (A_{412}/13{,}600)(1.1/1000)(10^6)(0.125)(\text{ml slurry added})$$

To determine the μmol sulfhydryl groups per ml gel, recalculate, allowing for the μl gel added to the particular tube used in the calculation.

4.2 AFFINITY TECHNIQUES

The manipulations involved in affinity chromatography can determine the success of an immobilized ligand. Preparing an affinity matrix—choosing the correct support, selecting the optimal activation chemistry, and coupling a suitable ligand—is only the first step in using the matrix to solve a biotechnology problem.

Before an immobilized ligand can be used to bind a specific target molecule, decisions must be made concerning how best to use the affinity support. Will it be used in a column format? Will the column be small, large, tall, or squat? Perhaps a batch technique should be employed. What buffer should be used to bind the target molecule? Once bound, how is the target eluted? How is nonspecific binding best prevented? How are minicolumns and larger-sized columns packed and used?

These questions must be considered each time an affinity matrix is used. Even if the support is not meant to be used in a chromatography format, the method of optimizing an interaction can be applied to any affinity problem. In the following sections, these concerns will be addressed with the goal of putting an affinity matrix to practical use.

4.2.1 Column Packing and Use

4.2.1.1 Disposable Plastic Minicolumns

Most affinity chromatography applications use less than 5 ml gel. These small-scale separations are performed easily using plastic minicolumns suspended in test tubes and run under gravity flow. Usually, the columns are made of polystyrene or polypropylene; the barrels are merely adaptations of commercially available plastic pipette tips with the addition of porous disks to hold and bracket the gel (Fig. 4.7). (**NOTE:** Pierce, BioRad, and Isolab market a wide range of small disposable columns and porous disks.) Most columns range in size to accommodate a bed volume of

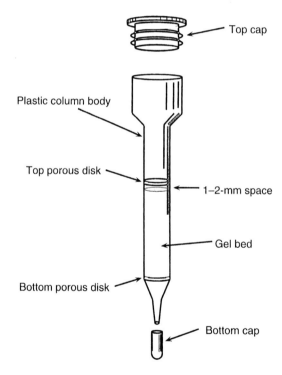

FIGURE 4.7

Plastic minicolumns can be used to hold convenient sizes of immobilized affinity gels for chromatographic separations. The columns come with top and bottom caps to seal the column, and top and bottom disks to hold and bracket the gel bed.

0.5–10 ml gel. Affinity purification operations can be completed in less than 1 hr using these columns, with minimal expense beyond the cost of the gel itself.

Another convenience of these minicolumns is the stop-flow characteristic of the top porous disk. When an aqueous buffer is applied to the column, it will flow through until it reaches the level of the top disk. At that point, capillary force in the porous disk stops the flow and prevents the gel from drying out. This effect allows accurate collection of fraction volume, since the amount of buffer applied to the top of the column will be same as that collected in the eluent up to the point at which the flow stops. This effect also allows making quantitative measurements based on the affinity separation (Section 5.4.1, Chapter 5).

PACKING PROCEDURE

1. Prepare a slurry of the affinity gel to be packed using either water or the binding buffer for the chromatography. The slurry concentration should be such that one addition of the mixture will contain enough gel to fill the column to the desired bed volume. In many cases, a 50% slurry (v/v) is ideal. Thoroughly remove air bubbles from the slurry by placing it in a small suction filter flask, sealing the top, and applying a vacuum for 5 min while gently swirling the gel.
2. Place the bottom cap on the end of the column.
3. Place the column in a 16 × 125-mm test tube or other suitable holder and add 2.75 ml degassed water to a 2-ml column, 6 ml to a 5-ml column, or 12 ml to a 10-ml column. (Column size is measured by the maximum amount of gel the column can hold.)
4. Float one of the porous polyethylene disks on top of the water in the column.

NOTE: These disks may have considerable hydrophobic character and often will not wet properly with aqueous buffers. Also this hydrophobicity may cause nonspecific binding of protein if the disks are not first treated with a dilute solution of 0.1% Tween 20. Before use, wash the disks thoroughly with this solution to block nonspecific sites and wet the inner pores. This process also eliminates much of the entrapped air in the disk, which markedly improves the flow rate of the resulting column.

5. Push the disk to the bottom of the column. Pushing it through the water will remove air entrapped in the disk and insure good flow through it during the chromatography. To push the disk to the bottom without tipping it, use the top end of a 10-ml plastic graduated pipette or the

4.2 Affinity Techniques

top end of a serum separator tube. The circular diameter of these devices will keep the disk straight all the way to the bottom.

6. Empty the column of water and add the appropriate volume of gel slurry to produce a final bed volume of the desired size. Add enough slurry, or extra water, to bring the total liquid level in the column up to the point of the neck (the point at which the plastic barrel begins to taper outward).
7. Let the column sit upright for at least 30 min to allow the gel to settle.
8. Place one of the porous disks on top of the liquid in the column and depress it, as before, to just above the settled gel level. Leave approximately 1–2 mm space between the top of the gel bed and the bottom of the top disk.
9. Wash the inside top part of the column to clean out any gel that may have remained along the sides during packing.
10. For storage, keep about 2 ml water or buffer over the top disk to prevent the gel from drying. Store in an upright position with the top cap fully inserted. A preservative such as sodium azide (0.02%) should be included to prevent microbial growth.

Prepacked affinity columns prepared and stored in this manner should remain stable for years. Prefilled columns with immobilized affinity ligands are commercially available (Pierce, BioRad, Isolab).

Using a Minicolumn for Affinity Chromatography

To use a packed minicolumn, remove the top cap first. This prevents air from being drawn into the gel from the bottom. Then remove the bottom cap, and place the column in a suitable holder such as a test tube. Allow the storage solution to drain through the column or pour it off the top.

Next, equilibrate the column with a binding buffer appropriate for the affinity separation. Let the final application of binding buffer flow through until the flow stops when the liquid level reaches the top disk. Transfer the column to a fresh test tube. The affinity column is now ready for the application of a sample solution. After each aliquot of sample is added, binding buffer for washing or elution buffer is added and completely drained through; transfer the column to fresh tubes to collect fractions for analysis (Fig. 4.8). Usually, the fraction sizes for minicolumn separations are between 1 ml and 3 ml.

Performing a Coupling Reaction in a Minicolumn

In addition to its chromatographic applications, the plastic minicolumn also provides a convenient chamber for performing various coupling reac-

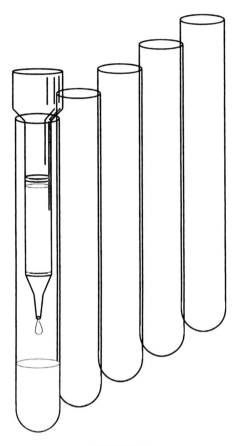

FIGURE 4.8
Collecting fractions using a minicolumn is as easy as moving the column to a fresh test tube after an aliquot of eluent has passed through.

tions. It has a top and bottom cap, so it can be sealed easily. Mixing can be done simply by rocking the column to keep the gel suspended. Most affinity support preparations use less than 5 ml gel, so these columns are the ideal size for reaction vessels. The gel can be mixed, reacted, washed, and used for affinity separations without being removed from the column. For reactions that use aqueous buffer systems or solvents that do not harm the polymeric structure of the columns, this is the easiest way to prepare small volumes of beaded affinity supports. Often, the gel can be activated in the columns, so the support never has to be transferred for washing or subsequent reactions. Preactivated supports already packed in minicolumns and ready for coupling reactions are commercially available (Pierce).

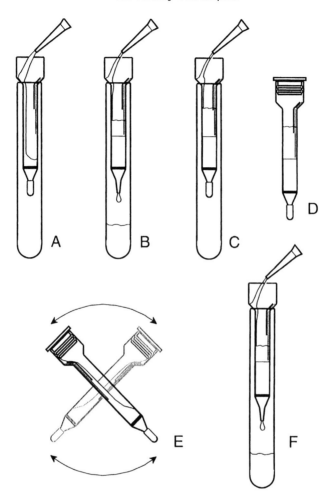

FIGURE 4.9

Using a minicolumn as a reaction chamber for performing immobilization procedures. (A) With the bottom cap on, add an aliquot of activated matrix as a 50% slurry in water or buffer. (B) Remove the bottom cap and wash the support into the appropriate coupling buffer. (C) Reapply the bottom cap and add the ligand to be coupled (dissolved in 1–2 gel volumes of coupling buffer). (D) Seal the column by inserting the top cap. (E) Mix by gentle rocking. (F) Wash the support free of uncoupled ligand and by-products of the reaction.

Figure 4.9 shows the manipulations that are common to using a minicolumn as the immobilization vessel. If the gel is not already preactivated and supplied in the columns, a sample of activated gel slurry is added first. The column is packed with a bottom disk according to the packing protocol, but no top disk is added. The activated matrix is then washed into the appropriate coupling buffer and the bottom cap is applied. Next, an equal volume of ligand solution is added and the top cap is inserted to seal the column. The gel is suspended in the ligand solution by gentle inversion and rocked for the duration of the reaction. A "blood rocker" (Aliquot Mixer, HemaTek, Miles) is ideal for this purpose. After the coupling is complete, the gel is washed to remove unreacted ligand and by-products of the reaction. Finally, the top disk can be placed in the column according to the packing procedure and the column can be used for affinity chromatography.

4.2.1.2 Larger Columns

The operation and use of medium- to large-scale chromatography columns is usually similar for those ranging in size from 10 ml to 10 L. Affinity separations, even in process-scale applications in industry, usually are not much bigger. Even the smallest of these columns is much more expensive than the disposable plastic minicolumns just discussed. Most consist of glass barrels with molded plastic bottom and top screw-on caps. Some less expensive varieties have plastic barrels. The caps incorporate a channel for buffer flow into and out of the column; the bottom cap typically has a supporting device for a membrane or porous plastic frit to hold the gel. Often these columns can be adapted for use with an adjustable plunger that is inserted into the top of the column. The plunger is positioned at the level of the gel so there is no buffer layer on the gel surface. This configuration insures minimal sample or buffer dilution during the chromatography and thus maximizes peak resolution.

Large columns are typically used in conjunction with fraction collectors and peristaltic pumps to create uniform flow rates and fraction sizes. A fully automated system can be constructed that incorporates a uv monitor and strip chart recorder; the monitor communicates with the fraction collector to isolate particular peaks as they elute off the column (Pharmacia-LKB; Isco; BioRad). Computerized chromatography systems are available that perform all these functions in one bench-sized instrument. The Pharmacia FPLC system and the Waters Protein Purification Unit are two examples of instruments that can perform small- to large-scale affinity purifications.

The dimensions of the columns used for large-scale affinity purifications are not as critical as they are in other types of chromatography. They may be tall and thin or short and squat. One way to help maintain the elution characteristics when scaling up from small to large columns is to use the same column height. In some cases, this may not be practical because columns with a diameter large enough to hold the desired amount of gel may be unavailable. In most cases, short squat columns may be used to advantage to increase flow rates without adversely affecting the quality of the affinity separation.

Packing an affinity matrix in such a column setup is relatively simple. The protocol here describes the steps to follow and the most important cautions.

PACKING PROCEDURE

1. Select a column that will hold slightly more gel than the bed volume desired. Shorter columns will provide higher flow rates and faster purifications. It is not necessary to use a plunger pressed down on the gel for affinity chromatography, but using one will eliminate the need to open the column to apply sample or the first aliquot of elution buffer.
2. Prepare a slurry of affinity gel in water or binding buffer. The slurry concentration should be between 50 and 80% (v/v) for the packing operation. Degas the slurry by putting it in a suction filter flask and pulling a vacuum on it with periodic swirling for about 10 min. Also degas all the buffers used for the chromatography in a similar fashion.
3. Assemble the column with the bottom cap and connect tubing that leads to a waste container. The tube should have a pinch clamp attached to stop the flow. Alternatively, the tube may be connected to a peristaltic pump to regulate flow during the packing and chromatography operations. Suspend the column in a vertical position using a clamp. For medium-sized columns it is not necessary to use a level to plumb the column.
4. Fill the column about $\frac{1}{3}$ full of water or buffer (degassed) and allow it to flow out through the bottom tube to eliminate air pockets. Stop the flow just before the liquid level reaches the bottom membrane.
5. Now add the thoroughly mixed and degassed gel slurry to the column. Ideally, the slurry containing the desired quantity of gel should be added in one step to prevent layering of different diameter beads with multiple additions. In practice, however, multiple additions are acceptable if care is taken not to disturb the top layer of settling gel when another addition of slurry is made. As soon as the first slurry addition is made, start the flow by opening the pinch clamp or by starting the peristaltic pump. The flow rate used to pack the column should be equal to or greater than the rate that will be used during the chromatography.
6. Once the gel slurry has been added, allow it to fully settle under flow by adding water or buffer to the top of the column. If a plunger is to be used, place it in position after the gel bed has formed completely (stop the flow during this operation). If no plunger is used, maintain a layer of buffer or water above the gel so the flow of additional liquid into the column will

not disturb the gel surface. To protect the gel surface, a circular piece of $1/8$-in. thick styrofoam packing material may be floated on the liquid layer so incoming drops of liquid cannot disturb the gel.

7. At this time, the flow may be stopped and the top cap screwed onto the column. Attach an inlet tube to the top cap, the other end of which is attached to a pipette submerged in the eluent buffer reservoir (which is placed at a level higher than the top of the column so siphon flow may be used to add buffer to the column). Prime the tube with buffer (with a syringe) before attaching it to the column. A pinch clamp attached to the incoming buffer line is used to prevent buffer flow when the top of the column is opened. The column is now ready for use. Figure 4.10 shows the completed column setup.

Using Larger Columns

Columns up to several liters in size are relatively simple to operate. Several facts should be remembered, however, to assure long-term affinity support stability. To maintain the integrity of the gel bed, all samples should be clarified by centrifugation or filtration before application to the gel. This practice will prevent column clogging with precipitated or aggregated components, especially in concentrated protein solutions. Cellular debris should be removed similarly from samples prior to application. In samples containing large amounts of lipids, for example, lipemic sera, chromatography at 4°C may be slowed because of lipid layering at the top of the gel bed. Delipidizing these solutions prior to affinity separation will avoid this problem and may even help increase purification yields of the desired protein. Samples containing protease activity should be inactivated if the affinity ligand is a protein. This may seem obvious, but we have traced many stability problems with immobilized proteins to contamination of samples with proteolytic activity. If an affinity support does seem clogged by lipids or precipitated matter, unpack the column and thoroughly wash the gel batchwise using a Buchner or fritted glass filter funnel. Such supports often can be restored completely to their original chromatographic potential by periodic cleanups.

Clarified samples may be applied to medium-size affinity columns either directly through the inlet tube or by disassembling the top cap and manually layering the solution on the gel. For large sample volumes, the first approach is definitely the easiest. However, when working with small sample solutions, it may be more prudent to layer the sample manually. When applying the sample by hand, allow the buffer level to drain through the gel to the gel surface; then stop the flow. Using a Pasteur pipette, carefully apply the sample without disturbing the gel bed. Running the sample solution

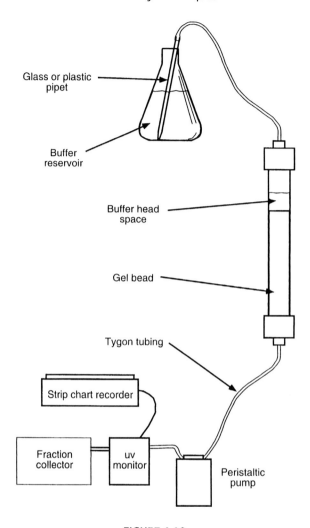

FIGURE 4.10

A typical column setup using larger sized columns. The buffer reservoir is usually placed above the gel bed to encourage siphoning of buffer into the column. A peristaltic pump, uv monitor, and strip chart recorder are often optional equipment. The use of a fraction collector, however, is recommended to obtain reproducible fraction sizes.

---

slowly down the sides of the column usually works well. After the sample if fully applied, start the flow again, and allow the solution to enter the gel. Follow addition of the sample solution with addition of several aliquots of manually added buffer to wash the sides of the column and drive the sample into the affinity gel. Finally, add enough additional buffer to the top of the column to create the buffer head needed to prevent gel surface disturbances from incoming buffer dripping from the inlet port. Reassemble the column and resume flow with the binding/wash buffer.

These larger columns can be stored for long periods packed with gel. The column should remain in a vertical position to maintain the integrity of the gel bed. A preservative also should be added to the storage solution (for example, 0.02% sodium azide) to prevent microbial growth. Alternatively, unpack the column and store the gel in bulk. Storage in either form should be at 4°C.

4.2.2 Buffer and Flow Rate Considerations

4.2.2.1 Binding Buffers

The correct choice of binding conditions can have a dramatic effect on the separation potential of an affinity column. Even when the support is not used in a chromatographic operation, the binding conditions chosen are often critical to establishing a favorable affinity environment. The binding buffer establishes this environment and should be used to initially wash and equilibrate the immobilized support in preparation for binding the target molecule. If the binding buffer does not have the optimal components, molarity, and pH, the affinity interaction may not occur at all or, if it does, it may be much weaker than expected.

Unless the affinity interaction is well documented in the literature, it is always a good practice to optimize the binding buffer by performing several affinity experiments using different buffer compositions. A series of separations can be done using buffers of different pH, buffer salt composition, and molarity. In some interactions, the presence of metals is also an important factor in creating the proper binding site configuration for the affinity pair. For instance, mannan binding protein will only bind IgM if calcium is present in the binding buffer (see Section 5.1.13, Chapter 5). Remove calcium with EDTA and the affinity interaction is broken. An example of the effect of pH on the binding capacity of an affinity gel is shown in Figure 4.11 for the interaction of immobilized protein A, protein G, and protein A/G with immunoglobulins. The affinity interaction is optimal at about pH 8 for protein A, whereas protein G and protein A/G have maximal capacity over a much broader range of pH values. The relative salt strength can also affect binding constants. In the affinity system using immunoglobulin binding proteins, NaCl concentration drastically affects the binding of mouse immunoglobulins. At high salt strength (>1 M) mouse IgG_1 binds well to immobilized protein A, but at low salt concentrations it binds

4.2 Affinity Techniques

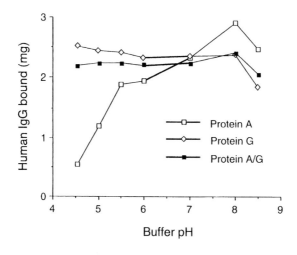

FIGURE 4.11
The effect of pH on the binding of immunoglobulins to immobilized protein A, immobilized protein G, and immobilized protein A/G.

poorly. Other IgG molecules from various species also experience changes in their binding to protein A, depending on the relative salt concentration. Only by optimizing the binding buffer variables will an affinity support perform at its full potential.

Some of the more common binding buffer formulations include those listed here. The list is by no means exhaustive, but provides a starting point for developing optimal affinity interactions. The pH of most of the formulations may be varied somewhat, depending on the buffering range of the buffer salts used. Sodium chloride is usually included at physiological concentration (0.15 M) to prevent nonspecific interactions caused by ion-exchange effects and to stabilize some proteins in solution. Higher salt concentrations often are used to reduce severe ionic interactions or to enhance hydrophobic interactions, as in the case of protein A–immunoglobulin binding. Buffers containing metal ions are meant to be used with proteins requiring chelated metal to initiate ligand binding or enzymatic activity. Potential elution buffers for these conditions can be found in Section 4.2.2.3; the formulation includes EDTA to remove the metals.

1. 0.01 M sodium phosphate, 0.15 M NaCl, pH 7.0–7.4 (PBS)
2. 0.008 M sodium phosphate, 0.002 M potassium phosphate, 0.14 M NaCl, 0.01 M KCl, pH 7.4 (modified Dulbecco's PBS)
3. 0.1 M sodium phosphate, 0.15 M NaCl, pH 6.8–7.4 (0.1 M PBS)

4. 0.1 M sodium phosphate, 0.5 M NaCl, pH 6.8–7.4 (high salt concentration to prevent severe ionic interactions)
5. 0.1 M sodium phosphate, 1.0–1.5 M NaCl, pH 6.8–7.4 (high salt to promote hydrophobic interactions, e.g., protein A–immunoglobulins)
6. 0.1 M sodium phosphate, 0.15 M NaCl, 10 mM EDTA, pH 7.4 (often used to stabilize sulfhydryl groups)
7. 25 mM Tris, 0.15 M NaCl, pH 7.6 (TBS; often used in immunoaffinity applications); note that Tris buffers may be formulated with 10 mM $CaCl_2$ or 1 mM $MgCl_2$ to stabilize metal binding proteins
8. 25 mM Tris, 192 mM glycine, pH 8.0 (Tris-glycine)
9. 50 mM sodium borate, pH 8.0–8.5
10. 50 mM sodium borate, 0.15 M NaCl, pH 8.0–8.5
11. 0.05 M sodium acetate, pH 4.5
12. 0.05 M sodium acetate, 0.15 M NaCl, pH 4.5
13. 0.2 M ammonium acetate, pH 8.8 (often used with boronic acid–sugar interactions)
14. 0.05 M sodium carbonate, 1 M NaCl, pH 11.0 (used with iminobiotin–α-vidin interactions)
15. 10 mM HEPES, 50 mM NaCl, pH 7.5
16. 0.2 M sodium carbonate, pH 8.7

4.2.2.2 Nonspecific Interactions

Nonspecific binding is the adsorption of sample components that do not have targeted sites of interaction with an immobilized ligand. Secondary nonspecific interactions of this type may be caused by ion exchange or hydrophobic character in the matrix material itself or by cross-reactivity inherent in the immobilized ligand. Nonspecificity originating in the ligand is often due to the use of substances that are not truly biospecific for the desired target molecule. Immobilized dyes, for example, typically will have some binding affinity for more than one component in a particular sample. Similarly, immobilized metal chelators may have affinity for more than one metal binding protein in a mixture. This type of nonspecificity may be overcome by more careful selection of potential affinity ligands or by developing an elution condition that will separate the desired target molecule from more than one binding component (see Section 4.2.2.3).

Nonspecific interactions originating in the matrix are often traceable to a hydrophobic or ionic character in the chemistry of the support material. Particularly, matrices made from glass, silica, polystyrene, and some of the other synthetic polymers such as Toyopearl have varying degrees of nonspecific effects associated with their structures. If the activation and coupling chemistry does not modify this character or if the ligand does not sufficiently block all these potential nonspecific sites, then significant adsorption of unwanted molecules may occur.

Some supports such as glass and silica routinely undergo a hydrophilic chemical coating step prior to the immobilization of an affinity ligand to effectively block their underlying nonspecific character (see Section 1.2.3, Chapter 1). Nonspecific ionic interactions may be overcome by the inclusion of salt in the binding buffer. Optimization of the binding buffer composition is often the key to eliminating nonspecific binding in affinity chromatography applications (see previous discussion on binding buffers).

Immobilized affinity ligands used in analytical applications such as immunoassays often require buffer-phase blocking agents to eliminate the nonspecific binding of immunological reagents (Towbin *et al.,* 1979; Johnson *et al.,* 1984; Hauri and Bucher, 1986; Vogt *et al.,* 1987; Douglas *et al.,* 1988; 1989; Harlow and Lane, 1988). Zimmerman and Van Regenmortel. Nonspecificity in this case is again caused by the surface characteristics of the matrix, usually consisting of exposed areas of hydrophobicity (as is common with polystyrene microplates and beads and many forms of membranes). If left unblocked, these surface interactions can obliterate completely any specific affinity interactions and ruin an immunoassay by creating high backgrounds. To prevent this type of nonspecificity, protein buffer additives that have no active part in the immunochemical reactions of an assay are typically used to noncovalently block and eliminate sites of adsorption. These protein blockers are usually added to the binding buffer and the affinity support is treated by incubation with the mixture.

Common blocking additives include BLOTTO (5% nonfat dry milk), bovine serum albumin (BSA; 1–3%), casein (1%), gelatin (0.25%), horse serum (1–3%), and the detergent Tween-20 (0.1%). Incubation of the affinity matrix with any of these additives dissolved in binding buffer for 1 hr at 37°C, for several hours at room temperature, or overnight at 4°C will effectively coat and block all sites of nonspecificity.

4.2.2.3 Methods of Elution

The correct choice of elution conditions to break an affinity interaction is often as important to successful purification as correct choice of binding conditions. The optimal elution buffer will liberate the bound substance(s) in a minimum volume and maintain activity or integrity of the purified material. This result may be accomplished in a number of ways.

The most common elution conditions employ a shift in the composition of the mobile phase so the optimal binding environment created by the binding buffer is lost. The most basic approach involves displacement of the bound material by competition with a counterligand. The counterligand is either the same as the ligand that is immobilized or of similar structure so it will effectively compete for the binding sites on the adsorbed molecules. Usually a counterligand dissolved in the binding buffer at a concentration of 0.05–0.1 M will break the affinity interaction quickly and elute the bound substance off the column in a sharp peak. Exceptions to this statement include the extraordinarily strong binding between native avidin (or streptavidin) and biotin. This interaction cannot be broken simply by including biotin in solution, but requires the harsh conditions of 6 M guanidine hydrochloride at pH 1.5 for full desorption. Any other affinity interactions with very high affinity constants may also be difficult to break simply by including a counterligand in the elution buffer.

When a counterligand is not available or does not work well on its own, several alternative methods can be used, either in conjunction with or in place of the competing substance. Sometimes the affinity interaction is pH dependent so the binding constant is very low at pH values that are not optimal. In these cases, elution can be effected with a buffer that has a pH different enough from the binding buffer to rapidly break the interaction (Collier and Kohlhaw, 1971). For instance, most antibody–antigen interactions can be eliminated at low or high pH values. Elution in immunoaffinity chromatography often is done by using buffers at pH 2–3 (0.1 M glycine·HCl) or at pH 10–11 (0.1 M glycine·NaOH). Many other types of affinity interactions can be broken similarly at such extremes of pH. One caution in using pH to effect elution: some labile proteins and other macromolecules cannot withstand the severe conditions without suffering denaturation. In particular, some monoclonal antibodies lose their antigen binding ability if they are exposed to pH values too far from neutral. In these cases, other methods of elution must be found to recover the bound molecule with high activity (see Section 5.1.9, Chapter 5).

Chaotropic agents may be used to effect elution by changing the structure of water in and around the site of the affinity interaction. These substances work especially well when the bond between the ligand and target molecule is primarily formed from hydrophobic interactions. A chaotropic elution buffer will break these affinity bonds by essentially dissolving the hydrophobic binding regions into the aqueous phase. A side effect of this type of eluent is the creation of structural changes in protein molecules. Many of these chaotropic agents are also known as protein denaturants. In most instances, removal of the chaotrope after elution will restore the native protein structure. In some sensitive proteins, however, activity may be lost or significantly reduced. Successful retention of activity often depends on the choice of the proper chaotropic agent at the right concentration. According to the Hofmeister series, chaotropic salts can be mild to severe, depending on their composition, in the order shown here (Green, 1931).

$$NH_4^+ < K^+ < Na^+ < Cs^+ < Li^+ < Mg^{2+} < Ca^{2+} < Ba^+$$

$$PO_4^{2-} < SO_4^{2-} < Acetate^- < Cl^- < Br^- < NO_3^- < ClO_4^- < I^- < SCN^-$$

Guanidine, urea, and ethylene glycol are also strongly chaotropic.

Therefore, salts consisting of ions chosen from the left side of the series will have little chaotropic character, whereas those from the right side will be powerful chaotropic agents. High concentrations of ammonium sulfate or sodium sulfate will actually enhance the structure of water and strengthen hydrophobic interactions. This phenomenon is sometimes called a "salting out" effect. Occasionally, a hydrophobic interaction can be created by high concentrations of these lyotropic agents and broken by simply removing the salt. Guanidine, urea, and ethylene glycol also can be used to break affinity interactions, both through a chaotropic effect and as protein denaturants. Ethylene glycol gradients (0–50%) are capable of resolving multiple binding components on hydrophobic supports.

Selecting the mildest chaotropic or denaturing agent at the lowest possible concentration that will just effect elution is the best strategy for insuring rapid elution with good retention of activity.

Eluting agents also can be created by removing metal ions essential for affinity binding. Many proteins depend on complexed metals to form the correct binding site orientation for interaction with a ligand or substrate. Lectins and enzymes are often particularly dependent on chelated metals such as calcium, magnesium, and zinc. The presence of these metals in the binding buffer is usually required to effect binding. In most cases, the affinity bond can be broken simply by including a chelating substance in the elution buffer, commonly EDTA or the calcium specific chelator EGTA. A concentration range of 1–10 mM is usually sufficient to remove the metals from the protein molecules and effect elution. These chelators also are used to cause elution from metal chelate affinity supports by removing the metal from the immobilized ligand.

NOTE: When using EDTA as an eluent, maintain the pH at levels greater than 7.0 to prevent precipitation of the free acid form of the molecule.

Resolution of multiple binding components is possible by choosing an elution condition that will break the affinity interactions differentially and cause each component to elute at a slightly different point. The eluting agent (whether counterligand, chaotrope, or chelator) can be present at a concentration high enough to break the affinity bonds, but not high enough to cause all the components to elute at once. Alternatively, the eluting agent can be used in a gradient elution from low to high concentration (e.g., low to high effectiveness), thus breaking the weakest affinity interactions first and the strongest last (Lowe et al., 1974). Finally, the immobilized ligand itself can be used as a weak affinity binding molecule to retard, but not bind, any interacting target molecules. In this approach (called "weak affinity chromatography"), none of the affinity interactions is strong enough to stop the progress of any component through the column (Ohlson et al., 1988, 1989). All interacting molecules are slowed in their passage through the affinity support, so molecules with different affinities will elute at different points. The best performance of any of these differential elution techniques will be realized only using an HPLAC (Section 5.4.2, Chapter 5) system, in which computerized automation and small particle support materials make the affinity separations mechanically reproducible with the highest resolution possible.

4.2.2.4 Flow Rates

The flow rate through an affinity chromatography support is an important parameter in generating an optimal separation. Under normal circumstances, porous beaded chromatography supports contain internal volume that is accessible to the mobile phase only by diffusional processes (Ackers, 1964; Little et al., 1971). The rate at which the eluent moves through the support governs the efficiency of this internal pore access. Too fast a flow will cause the mobile phase to move past the beads faster than the diffusion time necessary to reach the entire inner structure (Cuatrecasas, 1968). Too slow a flow will create a time of chromatography that is unreasonably long, or secondary diffusion effects may obscure some eluting peaks and decrease resolution. Optimizing the linear flow rate through an affinity support will result in maximal binding capacity and create the best possible separations.

In general, the faster the flow through an affinity support, the lower the apparent binding capacity for a particular target molecule (Cuatrecasas, 1968). As linear flow rates increase, diffusion and contact time limitations will allow fewer target molecules to find and bind the immobilized affinity ligand. For a given sample size, therefore, the faster the flow, the less efficient the removal of the target molecule from solution. For instance, in our laboratories an immobilized protein A affinity support with a total binding capacity for human IgG of 35 mg/ml gel at 1 cm/min linear flow fell to a capacity of about 12 mg/ml at 10 times that flow rate. The only support that is an extraordinary exception to this flow-versus-capacity phenomenon is

4.2 Affinity Techniques

the Poros matrix, which reportedly reaches its optimum binding only at elevated flow rates (Section 1.3.3.1, Chapter 1).

Most affinity supports will perform at or near optimal binding levels when operated under gravity flow (i.e., without auxiliary pumping systems). The actual linear flow (in cm/min through the gel bed) under gravity operation will vary depending on the type of matrix used and the physical dimensions of the column. Tall thin column configurations will produce slower linear flows than short squat columns. Finer support materials will also produce slower flow rates than larger diameter or coarser materials. For process operations, the larger particle diameter supports usually are selected to enhance linear flow rates when using large columns.

Another factor affecting flow is the total height of the buffer drop from the buffer reservoir (usually positioned above the column) to the bottom of the outlet tube below the column. The higher the buffer reservoir is placed above the column, the greater the head pressure at the top of the gel bed and the greater the flow rate.

Nascent gravity flow can be artificially increased by pumping systems up to 2–3-fold without a severe loss in total column binding capacity. When scaling up from small separations to process levels, flow rate differences can affect the yield of purified product significantly. Care should be taken, however, in setting a pumping system to increase throughput. Too fast a pumping rate will cause channeling or air bubbles may develop within the gel bed. This is particularly true when the pump is placed after the column in the chromatography setup. If the pump is placed before the column, the result will be an increase in head pressure and air will not be drawn through the matrix. Be careful, however, not to exceed the pressure limitations of the column that is being used. Most glass or plastic barrels, especially the larger columns, will only withstand 10–15 psi before breaking.

Occasionally, an affinity interaction will require more time for optimal binding than the normal column flow rate will allow. We have found this to be the case especially in some antigen–antibody interactions in immunoaffinity separations. More efficient binding of the target molecule often can be accomplished if the sample is incubated on the column for 10–30 min by stopping the flow after application. If the sample size is greater than the column volume, recycling the sample also can cause more target molecule to bind. For example, when eliminating endotoxins from solutions containing albumin, we have found that recycling the sample over a column of Detoxi-Gel (see Section 5.2.3, Chapter 5) dramatically improves endotoxin removal.

4.2.3 The Sample

The sample used in an affinity system may be of almost any origin imaginable. It could be a crude biological extract, a cell culture supernatant, serum, ascites, waste water, or an artificial control made from a purified substance dissolved in binding buffer. Several aspects of the sample must be considered to create a successful affinity interaction, regardless of the system being used.

The sample should not contain any chemical that could degrade or modify the affinity ligand or the matrix. Columns can easily become clogged if the sample is not free of precipitated material. Some precipitates are difficult to identify visually, but routine centrifugation or filtration of the sample through a 0.5-µm filter will eliminate even the finest particles. Sample clarification is a good practice and will extend the life of any affinity matrix.

Lipids can be a cause of column fouling, as well. Specifically if the sample is of serum origin, the lipid content may be high. Even if the sample appears clear with no lipemic haze, an affinity column may filter out the lipids at the top as the sample passes through, especially if the chromatography is done at 4°C. Over time, this effect can significantly slow the flow of buffer through the column. When working with large sample volumes that contain lipid, it is probably best to remove the lipid before performing the chromatography.

The sample also should be free of any competing substances that could interact with either the ligand or the target molecule. For instance, when purifying a lectin on an immobilized carbohydrate column, the sample must be free of any sugars that could compete in the interaction. Frequently, dialysis or gel filtration of a sample is sufficient to remove low molecular weight competing substances, but other fractionation methods (e.g., ammonium sulfate precipitation or ion-exchange separations) must be used for substances of greater molecular size.

When the immobilized ligand is sensitive to proteases in the sample (e.g., when the ligand is a protein), such enzymatic activity should be eliminated prior to application to the affinity support. In this case, the addition of a protease inhibitor or performing other prepurification steps to remove enzymatic activity will prevent degradation of the affinity column. Often, a bonus of removing protease activity is the ability to run the affinity separation at room temperature instead of 4°C. See Section 5.2.4, of Chapter 5, for a discussion of affinity supports designed to remove proteolytic activity.

Finally, it is best to dilute or dialyze the sample in binding buffer prior to its application to an affinity support. The inherent pH and salt composition of a sample is not always optimal for an affinity interaction. Dialysis or a 50% (1:1, v/v) dilution of the sample in binding buffer (or 2X binding buffer) will create the best conditions for binding. When working on a separation problem for the first time, we routinely add a 2X binding buffer concentrate to the sample to create a 1:1 dilution. The presence or absence of any precipitate is then noted, since it is far better to remove precipitates before adding the sample to the column than to have the precipitates clog the column during the purification process.

4.2.4 Measuring Capacity or Activity of Immobilized Ligands

The total capacity or activity of an affinity support is an important parameter for determining how much sample can be processed in a given time. This parameter also provides a reasonable comparison of similar immobilized ligands from different manufacturers. Capacity or activity measurements are sometimes thought to be applicable only to affinity supports that are to be used for purification. These measurements, however, can have a profound effect on the usefulness of any affinity support, not only those meant for purification purposes, but also supports used in scavenging, catalysis, modification, or analytical applications.

When the immobilized ligand functions simply as a specific binder of a target molecule, the total binding capacity measurement is the most useful way to describe how well the support performs under ideal conditions. When the matrix is catalytic or a chemical modifier of some kind, the specific activity per unit volume should be measured.

The density of an immobilized ligand usually directly affects the binding capacity or activity (Lowe and Dean, 1974). Therefore, within limits, the more ligand immobilized during the coupling reaction, the greater the capacity or activity of the support. The major limitation to this effect is that nonspecific interactions become substantial when too much ligand is immobilized. This is especially true with small ligands. At very high densities, some ligands will bind other components in a sample by hydrophobic or ionic interactions. Care should be taken when measuring or comparing capacity determinations to be certain that the amount bound is actually representative of a specific interaction with the intended target molecule. This assessment can best be made by measuring the purity of the bound fraction by electrophoresis or HPLC techniques.

The sample can affect the binding capacity measurement, especially if the target molecule is dissolved in a solution containing considerable other components. Sometimes target molecules are present in low concentration relative to other substances in samples originating from natural sources. Impure samples can radically affect the efficiency of the affinity interaction. For instance, immobilized protein A usually will have a much higher apparent capacity for purified human IgG dissolved in binding buffer than for human IgG present in human serum. The target molecule concentration as well as the concentrations of other sample constituents can dramatically alter the binding parameters of an affinity matrix. For the measurement of total binding capacity, always use a purified sample dissolved in the optimal binding buffer.

Flow rates through or past an affinity matrix also have a significant effect on the capacity. In general, if a sample containing fewer target molecules than an affinity support is capable of binding and is applied to a column (i.e., the column is not overloaded), then the faster the flow of the sample through the column the less efficient will be the capture of target molecules (Cuatrecasas, 1968). In other words, as the flow rate of sample through a column is increased, the apparent capacity of the affinity support will drop. When measuring the total binding capacity of a support, it is important to overload the matrix significantly with target molecule to negate this flow rate dependency. Usually, sample is applied until the absorbance of the breakthrough fractions (the sample that passes through unbound) equals the absorbance of the sample. At this point, the affinity sites are completely saturated with bound target molecules and no more can be taken up. The following protocol describes this overload procedure (sometimes called "frontal analysis") for a target molecule that is a protein.

MEASUREMENT OF TOTAL BINDING CAPACITY

1. Pack 1 ml affinity matrix in a minicolumn according to the procedure described in Section 4.2.1.1.
2. Equilibrate the column with a suitable binding buffer that will optimize the affinity interaction (see Section 4.2.2.1).
3. Prepare a solution of purified target molecule in binding buffer. For proteins, a 5–10 mg/ml solution works well to quickly overload most affinity columns. Prepare at least 10 ml of the sample solution.
4. Place the minicolumn in a test tube and add the sample to the affinity support in 1-ml aliquots, moving to fresh tubes after each addition of solution has completely entered the gel. Continue to add sample until the concentration of target molecule in the column effluent equals its concentration in the original solution. If the target molecule has an absorbance or activity that can be easily measured (e.g., A_{280}), compare these parameters in the sample solution and effluent fractions. Once the measurements indicate that as much target protein is breaking through the column as is being applied, then the affinity matrix is completely saturated. Stop applying sample at this point.

5. Wash the affinity matrix with binding buffer until any unbound protein still in the column has been removed completely. Usually, A_{280} in the range of 0.01–0.02 is considered baseline (measurement made against a blank consisting of binding buffer). Sometimes a saturated affinity matrix will continue to bleed bound protein, giving a baseline higher than this range. In this case, stop washing when the absorbance from fraction to fraction levels off.
6. Elute the bound protein with a suitable elution buffer (see Section 4.2.2.3). Pool the main fractions containing protein, and measure the absorbance of this pooled solution at 280 nm. Determine the amount of recovered protein in this bound fraction by using the original sample solution of known concentration for comparison. The amount of protein contained in the bound fraction is representative of the total capacity of the affinity matrix (expressed as mg/ml gel).

MEASUREMENT OF SPECIFIC ACTIVITY

The activity of an immobilized affinity ligand is important if the support is designed to be used as a catalyst or modifying reagent. In these cases, this measurement indicates how much target molecule the matrix is capable of acting on in a given time.

For immobilized enzymes, an activity determination is usually performed using synthetic substrates that give potentiometric or colorimetric products when enzymatically changed. The measurement itself is no different than determining enzymatic activity of a soluble enzyme, except a quantity of gel slurry (usually as μliter amounts of a 50% slurry) rather than an enzyme solution is added to the reaction. A change in pH or a change in the absorbance over time is then monitored to calculate the activity (per min) of the immobilized reagent.

5

SELECTED APPLICATIONS

5.1 PURIFICATION

Use of immobilized affinity ligands for the purification of target molecules is perhaps the most common of all affinity techniques. An affinity ligand coupled to an insoluble support can be a powerful tool in the isolation of a particular substance from a complex mixture. In many cases, affinity chromatography can reduce the number of chromatographic steps in a purification procedure to one or two. Exploiting a specific affinity interaction often can eliminate the need for auxiliary ammonium sulfate precipitations, ion-exchange, and gel filtration steps in an isolation protocol. If an affinity purification step can be used early in an isolation scheme, dramatic increases in the yield of final product and substantial cost savings can result.

In this section, we describe some of the most important purification procedures commonly used in isolating biologically useful molecules. In particular, several parts of this chapter are dedicated to the technique of antibody or antigen purification. Immunoaffinity chromatography is one of the most rapidly developing divisions of immobilized affinity ligand technology.

5.1.1 Purification of Trypsin Using Immobilized p-Aminobenzamidine

Immobilized p-aminobenzamidine can be used to purify trypsin and trypsin-like enzymes (e.g., urokinase, enterokinase, or plasminogen activators). Commercially available trypsin or TPCK-trypsin is frequently contaminated with autolytic products of trypsin digestion. These autolytic products must be removed before trypsin can be used in analytical assays based on the enzymatic digestion of proteins and peptides. Affinity-purified trypsin generates well-defined tryptic peptides free of endogenous contamination. The HPLC analysis of these fragments then becomes an easy task. A protocol on a laboratory scale for the purification of commercially available trypsin, suitable for sequence analysis, is given here. The protocol can be easily adapted to large-scale purifications.

The chromatography is performed at 4°C. Absorbance of all fractions is measured at 280 nm. If necessary, enzyme activity of all fractions is measured using a suitable synthetic substrate for trypsin (e.g., BAEE). A typical affinity chromatographic purification profile of trypsin on an immobilized *p*-aminobenzamidine column is shown in Figure 5.1.

PROTOCOL

1. Pack immobilized *p*-aminobenzamidine (1.0 ml gel; prepared as described in Section 3.1.3.1, Chapter 3) in a disposable polypropylene column (see Section 4.2.1.1, Chapter 4, for column packing procedure).

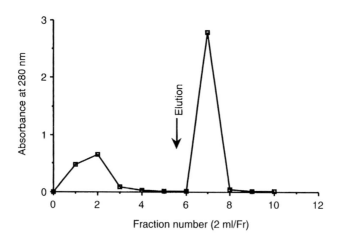

FIGURE 5.1

Affinity chromatography of commercial trypsin on an immobilized *p*-aminobenzamidine column (1.0 ml gel). The column is equilibrated with 50 m*M* Tris–HCl, pH 7.2, containing 150 m*M* NaCl and 10 m*M* CaCl$_2$ (binding buffer). Trypsin (7 mg) is dissolved in 1.0 ml binding buffer and applied to the column. The column is washed with 12 ml binding buffer while 2.0-ml fractions are collected. Bound trypsin is eluted with 0.1 *M* acetic acid containing 10 m*M* CaCl$_2$. Chromatography is conducted at 4°C.

2. Equilibrate the column with 10 ml 50 mM Tris-HCl buffer, pH 7.2, containing 0.1 M NaCl and 10 mM CaCl$_2$ (binding buffer).
3. Dissolve trypsin (7.0 mg) in 1.0 ml binding buffer and apply to the benzamidine column.
4. After the sample has entered the gel, wash the column with 12 ml binding buffer and collect 2-ml fractions. Elute bound trypsin with 0.1 M acetic acid containing 10 mM CaCl$_2$.

Fractions 1–6 (peak 1, Fig. 5.1) are devoid of any trypsin activity and contain autolytic products present in commercial trypsin preparations. Fractions 7 and 8 (peak 2) contain active trypsin of high specific activity, without any contaminating autolytic products. This entire purification procedure requires less than 1 hr at this scale. Fraction 7 should be used immediately for tryptic digestion studies.

5.1.2 Purification of Avidin or Streptavidin Using Immobilized Iminobiotin

The extraordinary affinity ($K_a = 10^{15}$ M^{-1}) between biotin and avidin (or streptavidin) has been used for countless applications in both research and biotechnology. Refer to Wilchek and Bayer (1990) for an excellent review of avidin–biotin technology. Although the remarkable nature of the avidin-biotin complex makes it an ideal choice for many applications, it has drawbacks in the affinity chromatographic purification of avidin, streptavidin, and their conjugates. For instance, immobilized biotin has been used for the purification of avidin. However, the harsh conditions (6 M guanidine-HCl, pH 1.5) that are required for the elution of bound avidin make it impractical for many applications. Although the avidin is not permanently inactivated by these denaturing conditions, many avidin conjugates (e.g., ferritin–avidin, protein A–avidin, alkaline phosphatase–avidin) cannot tolerate such harsh conditions and will lose their biological activity.

This problem has been overcome by replacing biotin with its cyclic guanidino analog, 2-iminobiotin. At high pH (9 and above) 2-iminobiotin exists primarily in its free base form and retains the tight specific binding to avidin characteristic of biotin. At low pH (4.0), at which it exists primarily in a charged form, it interacts weakly with avidin (Fig. 5.2). This interesting property of iminobiotin has been used to isolate avidin, streptavidin, and their conjugates under mild conditions and with complete retention of biological activity.

A protocol for the purification of commercially available avidin on laboratory scale is given here. The protocol can be adapted for large scale as

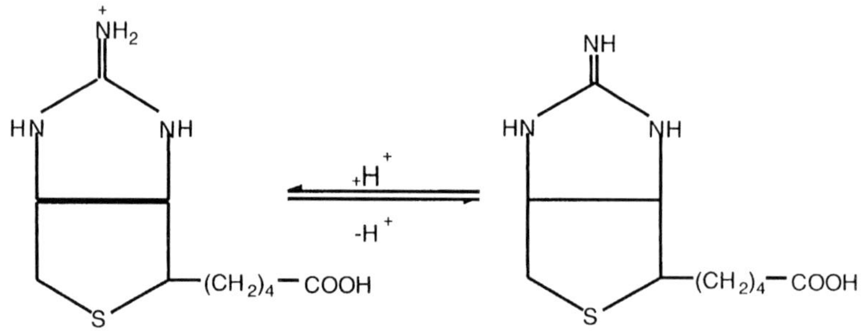

FIGURE 5.2
Structure of the biotin analog iminobiotin.

well as for the purification of streptavidin and avidin conjugates. Some commercial sources of bulk avidin, streptavidin, and their conjugates are frequently contaminated with undesired proteins or inactivated components. Therefore, if highly purified preparations are required, affinity chromatography can be performed easily on an immobilized iminobiotin column.

PROTOCOL

1. Pack immobilized iminobiotin (1.0 ml gel; prepared as described in Section 3.1.3.3, Chapter 3) in a disposable polypropylene column (see Section 4.2.1.1, Chapter 4, for the proper column packing method using these minicolumns).
2. Equilibrate the column by washing with 10 ml 50 mM sodium carbonate, pH 11.0, containing 0.5 M NaCl (binding buffer).
3. Dissolve 5 mg avidin in 1.0 ml binding buffer and apply to the iminobiotin column.
4. After the sample has entered the gel, wash the column with 12 ml binding buffer and collect 2-ml fractions.
5. Elute bound avidin with 50 mM ammonium acetate buffer, pH 4.0, containing 0.5 M NaCl.

The chromatography can be performed at room temperature or at 4°C. Absorbance of all fractions was measured at 280 nm. A typical profile of affinity chromatographic purification of commercial avidin on an immobilized iminobiotin column is shown in Figure 5.3.

Fractions 1–3 (peak 1, Fig. 5.3) contain protein contaminants present in commercial avidin preparations. These fractions also represent any inactive avidin no longer able to bind the immobilized iminobiotin. Fractions

5.1 Purification

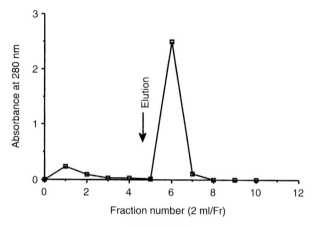

FIGURE 5.3
Affinity chromatography of commercial avidin on an immobilized iminobiotin column (1.0 ml gel). The column is equilibrated with 50 mM sodium carbonate buffer, pH 11.0, containing 0.5 M NaCl (binding buffer). Avidin (5.0 mg) is dissolved in 1.0 ml binding buffer and applied to the column. The column is washed with 12 ml binding buffer while 2-ml fractions are collected. Bound avidin is eluted with 50 mM ammonium acetate buffer, pH 4.0, containing 0.5 M NaCl. Chromatography is conducted at room temperature.

---

6 and 7 contain highly purified avidin that can be dialyzed against water and lyophilized. This purification procedure often results in an avidin preparation with the highest possible biotin binding activity as measured by the HABA (hydroxyazobenzoic acid) binding assay (Green, 1965).

5.1.3 PURIFICATION OF LECTINS

Lectins are carbohydrate binding proteins of nonimmune origin that agglutinate cells or precipitate glycoconjugates. They are powerful reagents for the study of carbohydrates and their analogs, both in solution and on cell surfaces. Lectins are used widely for preparative and analytical applications in biochemistry, cell biology, immunology, and related fields. The readers are referred to a book on lectins (Liener *et al.*, 1986) that deals with the properties, functions, and applications of lectins in biology and medicine.

In this section the procedures we have used successfully for the purification of lectins are described; special emphasis is given to wheat germ lectin, castor bean lectins, jacalin, and *Griffonia simplicifolia* lectin I. Plant seeds are the source of many lectin proteins. For the isolation of lectins, these seeds must be finely ground to a powder and extracted with an organic solvent (e.g., acetone, diethyl ether, hexane, or methanol) to remove lipids and pigment molecules. The delipidized seed meal is then extracted with a suitable buffer. Ammonium sulfate fractionation, centrifugation, and solubilization of the precipitate yields an extract containing the lectin(s). Most lectin purification methods use affinity chromatography that exploits the specific sugar-binding ability of the lectin. In principle, a sugar ligand with which the lectin interacts is immobilized to a solid support, the lectin is bound as the seed extract is passed slowly over the affinity column, and the bound lectin usually is eluted with a sugar that competes for lectin sites on the affinity support.

5.1.3.1 Purification of Wheat Germ Lectin Using Immobilized *N*-Acetyl-D-Glucosamine

Among several *N*-acetylglucosamine-binding lectins reported in the literature, wheat germ lectin is one of the best studied lectins and is used widely in glycoconjugate research. The protocol given here for the isolation of wheat germ lectin can be easily adapted for the purification of other *N*-acetylglucosamine-binding lectins.

PROTOCOL

NOTE: All operations are performed at 0–4°C unless otherwise indicated.

1. Grind unprocessed whole wheat germ (200 gm; ICN) to a coarse powder using a blender, and mix with 1 L dry-ice cooled acetone (−78°C) for 30 min. Filter the suspension using Whatman No. 1 filter paper on a Buchner funnel and wash twice with 500 ml dry-ice cooled acetone. Air-dry the defatted wheat germ powder at room temperature.
2. Mix the dried and defatted wheat germ powder (100 gm) with 500 ml 50 mM sodium acetate buffer, pH 4.5, and stir the suspension at 4°C for 24 hr. Centrifuge the suspension at 12,000 g for 20 min and discard the solid pellet.
3. Stir the extract and bring to 40% saturation with ammonium sulfate by the slow addition of solid ammonium sulfate. Continue the stirring for 1 hr. Collect the precipitate by centrifugation at 12,000 g for 20 min, suspend it in 60 ml 0.05 M sodium acetate buffer, pH 4.5. Dialyze the crude lectin solution extensively against the same buffer.
4. To remove any insoluble material, centrifuge the dialysate at 12,000 g for 20 min, and apply the supernatant to an immobilized *N*-acetyl-D-glu-

cosamine column (see Section 3.1.2.3, Chapter 3, for preparation of the affinity support) (size: 0.9×16.5 cm; 10.5 ml gel) that is equilibrated with 50 mM sodium acetate buffer, pH 4.5.
5. After sample application, wash the column with 50 mM sodium acetate buffer, pH 4.8, until the absorbance at 280 nm of the column effluent is less than 0.01 (about 400 ml buffer is required for washing).
6. Finally, elute the bound wheat germ lectin with 50 mM sodium acetate buffer, pH 4.5, containing N-acetyl-D-glucosamine (50 mg/ml). Collect fractions of 6.0 ml and measure the absorbance of all fractions at 280 nm. A typical profile of affinity chromatographic purification of wheat germ lectin on immobilized N-acetyl-D-glucosamine column is shown in Figure 5.4.
7. Fractions 85–88 are pooled, dialyzed extensively against water, and lyophilized. Total yield of wheat germ lectin is 52 mg.

5.1.3.2 Purification of Castor Bean Lectins

Aqueous extracts of castor beans (*Ricinus communis*) contain two distinct but structurally related lectins: a highly cytotoxic weak hemagglutinin (ricin, also known as ricin D and RCA_{II}) and a strong erythroagglutinin that is a weak inhibitor of protein synthesis (*Ricinus communis* agglutinin, also known as RCA_I). Ricin is a dimeric protein ($M_r = 63,000$ daltons) composed of two different polypeptide chains, A and B. Ricin A chain is widely

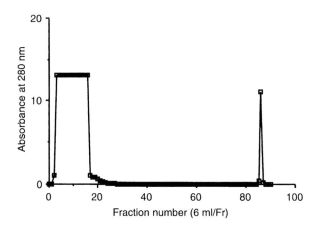

FIGURE 5.4

Affinity chromatography of wheat germ extract on an immobilized N-acetyl-D-glucosamine column.

used in immunotoxin preparation. *Ricinus communis* agglutinin (RCA_I) is a tetramer ($M_r = 120,000$ daltons) composed of two A' and two B' polypeptides and is believed to exist as two noncovalently associated heterodimers, each with a structure analogous to that of ricin.

Castor bean lectins specifically bind to galactose and N-acetylgalactosamine residues of glycoconjugates. The affinity chromatographic method described here for the purification of castor bean lectins is based on the specific binding of these lectins to galactose residues in an immobilized lactose matrix.

PROTOCOL

CAUTION: Castor bean extracts are extremely toxic and great care must be exercised when isolating the lectins. Extreme caution must be taken even when handling the dust of ground beans.

1. Blend and mix castor beans (25 gm) with 500 ml diethyl ether for 30 min. (Caution: Use a fume hood and an explosion-proof blender, and wear a dust mask.) Filter the suspension on Whatman No. 1 filter paper in a Buchner funnel and wash twice with 500 ml diethyl ether. Air-dry the defatted castor bean powder at room temperature. (Caution: The dry dust is extremely dangerous!)
2. Mix the dried and defatted castor bean powder with 300 ml PBS, pH 7.2 (20 mM sodium phosphate buffer, pH 7.2, containing 150 mM NaCl). Stir the suspension at 4°C for 24 hr. Centrifuge the suspension at 12,000 g for 20 min and discard the solid residue.
3. Stir and bring the castor bean extract to 60% saturation with ammonium sulfate by the slow addition of solid ammonium sulfate. Continue the stirring for 1 hr. Collect the precipitate by centrifugation at 12,000 g for 20 min, suspend the pellet in 70 ml PBS, pH 7.2, and dialyze extensively against PBS, pH 7.2.
4. Centrifuge the dialysate at 12,000 g for 20 min to remove any insoluble material and apply the clarified supernatant to an immobilized lactose column (Section 3.1.2.2, Chapter 3) (size: 0.9 × 16.5; 10 ml gel) that has been equilibrated previously with PBS, pH 7.2. After the sample application, wash the column with PBS, pH 7.2, until the absorbance at 280 nm of the column effluent is less than 0.01 (at least 100 ml PBS, pH 7.2, is required for washing).
5. Elute the bound castor bean lectins with 0.2 M lactose in PBS, pH 7.2. Collect 6.0-ml fractions and measure the absorbance of all fractions at

280 nm. A typical profile of affinity chromatographic purification of castor bean lectins on immobilized lactose is shown in Figure 5.5.
6. Pool fractions 32–35 and dialyze extensively against PBS, pH 7.2, to remove lactose. Total yield of castor bean lectins is 98 mg.

NOTE: Do not lyophilize castor bean lectins since they are extremely toxic.

The castor bean lectin preparation at this stage consists of a mixture of RCA_I and RCA_{II}. The two proteins can be separated by gel filtration on BioGel P-150 column (Nicolson *et al.*, 1974). RCA_I (agglutinin) and RCA_{II} (toxin) also can be separated on a Sepharose 4B column. Nicolson *et al.* (1974) noted the ability of RCA_{II} but not RCA_I to bind *N*-acetylgalactosamine and purified both lectins by sequential elution from Sepharose 4B of the toxin with *N*-acetylgalactosamine and the agglutinin with galactose.

5.1.3.3 Purification of Jacalin Using Immobilized Melibiose

Jacalin is an α-D-galactose-binding lectin isolated from jack fruit (*Artocarpus integrifolia*) seeds. Because of its binding and precipitating specificities for the heavy chain of human immunoglobulins, jacalin has been used as a new laboratory tool in immunochemistry and cellular immunology. Jacalin has been used for IgA subclass typing and as a preparative tool

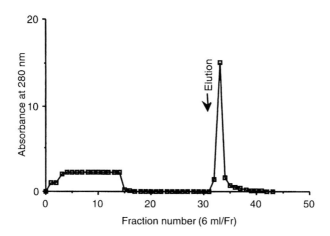

FIGURE 5.5

Affinity chromatography of castor bean extract on an immobilized lactose column.

(purification of IgA and IgD, removal of IgA from biological samples and preparations). For a review on jacalin, see Aucouturier *et al.* (1989).

PROTOCOL

NOTE: All operations are performed at 0–4°C unless otherwise indicated.

1. Grind jack fruit (*Artocarpus integrifolia*) seeds (20 gm) to a coarse powder using a blender and homogenize with 200 ml PBS, pH 7.4 (20 mM sodium phosphate buffer, pH 7.4, containing 0.15 M NaCl). Stir the homogenate at 4°C for 24 hr, centrifuge at 12,000 g for 20 min, and discard the solid residue.
2. Stir the jack fruit seed extract and bring it to 70% saturation with ammonium sulfate by the slow addition of solid ammonium sulfate. Continue the stirring for 1 hr. Collect the precipitate by centrifugation at 12,000 g for 20 min, suspend in 50 ml PBS, pH 7.4, and dialyze extensively against PBS, pH 7.4.
3. Centrifuge the dialysate at 12,000 g for 20 min to remove any insoluble material and apply the supernatant to an immobilized melibiose column (Section 3.1.2.1, Chapter 3) (50 ml gel packed in a 2.5-cm diameter column) that has been equilibrated previously with PBS, pH 7.4. After the sample application, wash the column with PBS, pH 7.4, until the absorbance at 280 nm of the column effluent is less than 0.005 (about 325 ml buffer is required for washing).
4. Elute the bound jacalin with 0.1 M melibiose in PBS, pH 7.4. Collect fractions of 10 ml and measure the absorbance of all fractions at 280 nm. A typical profile of affinity chromatographic purification of jacalin on an immobilized melibiose column is shown in Figure 5.6.
5. Pool and dialyze fractions 41–46 against PBS, pH 7.4, to remove melibiose. Total yield of jacalin is 70 mg.

5.1.3.4 Purification of *Griffonia (Bandeiraea) simplicifolia* Lectin I Using an Immobilized Melibiose Column

Griffonia simplicifolia seed extract contains two lectins designated *G. simplicifolia* lectin I (GS I) and *G. simplicifolia* lectin II (GS II). GS I has specificity for α-galactose residues, whereas GS II is specific for *N*-acetylglucosamine residues. GS I and GS II can be purified by affinity chromatography on immobilized melibiose and chitin, respectively. A protocol is given here for the purification of GS I lectin using an immobilized melibiose column (Hayes and Goldstein, 1974).

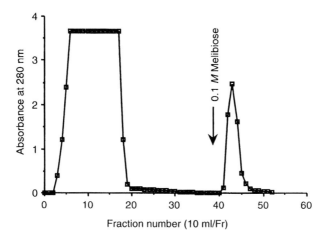

FIGURE 5.6
Affinity chromatography of jack fruit seed extract on an immobilized melibiose column.

PROTOCOL

NOTE: All operations are performed at 0–4°C unless otherwise indicated.

1. Grind *G. simplicifolia* seeds (200 gm) to a fine meal using a blender, and mix with 1 L methanol for 1 hr. Filter the suspension using Whatman No. 1 filter paper on a Buchner funnel. Repeat the extraction with methanol three times and air-dry the methanol-extracted seed meal at room temperature.
2. Add 12 gm polyvinylpyrrolidone (average MW 360,000, Aldrich) to the dry methanol-extracted seed meal and mix with 2 L PBS-Ca (20 mM sodium phosphate buffer, pH 7.2, containing 150 mM NaCl and 0.1 mM $CaCl_2$); then add 36 gm D-galactose. Stir the mixture at 4°C for 24 hr. Centrifuge the suspension at 12,000 g for 1 hr and discard the solid residue.
3. Slowly add solid ammonium sulfate to the crude extract to a 55% saturation level. Stir 1 hr at 4°C. Remove the precipitate by centrifugation and discard. Bring the supernatant solution to 75% saturation with ammonium sulfate. Stir for 1 hr at 4°C. Collect the precipitate by centrifugation, and dissolve it in 150 ml 0.15 M NaCl. Extensively dialyze the crude lectin isolate against PBS-Ca.

4. Centrifuge the dialysate (200 ml) at 12,000 g for 30 min to remove any insoluble material, and apply it to an immobilized melibiose column (Section 3.1.2.1, Chapter 3) (size: 2.5 × 18 cm; 90 ml gel) that has been equilibrated previously with PBS-Ca. After the sample application, wash the column with PBS-Ca until the absorbance at 280 nm of the column effluent is less than 0.01 (about 3 L PBS-Ca is required for washing).
5. Elute the bound lectin with 0.1 M melibiose in PBS-Ca. A typical profile of affinity chromatographic purification of *G. simplicifolia* lectin I is shown in Figure 5.7.
6. Dialyze peak 2 (the bound lectin) against PBS-Ca to remove melibiose and store as a frozen solution at –20°C. Total yield of GS I lectin is 235 mg.

5.1.4 Purification of Human Serum Albumin on an Immobilized Anti-HSA Immunoaffinity Column

Immunoaffinity chromatography is a widely used technique in basic biochemical research and in the biotechnology industry for the purification of complex mixtures. The protocol given here for the purification of HSA on

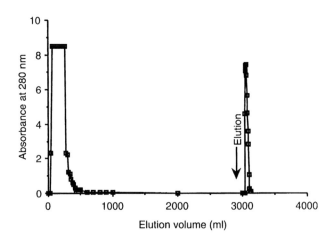

FIGURE 5.7

Elution profile of *G. simplicifolia* extract on an immobilized melibiose column. The ammonium sulfate fraction is applied to the column and the column is washed free of unbound proteins. Bound GS lectin I is eluted with 0.1 M melibiose.

an immobilized anti-HSA immunoaffinity column can be adapted easily for the purification of many antigens if the appropriate antibody immunoaffinity column is properly prepared. (See Section 3.4.2, Chapter 3, for the preparation of immunoaffinity matrices.)

PROTOCOL

1. Couple 6 mg affinity purified anti-HSA antibody to 2 ml periodate-activated Sepharose CL-6B as described in Section 3.4.2.1, Chapter 3.
2. Pack the immobilized anti-HSA gel in a disposable polypropylene minicolumn. (See Section 4.2.1.1, Chapter 4, for the protocol of packing a minicolumn.)
3. Equilibrate the column by washing with 10 ml binding buffer (0.1 M sodium borate buffer, pH 8.0). Perform the chromatography at 4°C.
4. Dissolve HSA (5 mg) in 1.0 ml binding buffer and apply to the immunoaffinity column. After the sample has entered the gel, wash the column with 18 ml binding buffer and collect 2-ml fractions.
5. Elute bound HSA with 0.1 M glycine-HCl, pH 2.8. Measure the absorbance of all fractions at 280 nm.

A typical affinity chromatographic profile of the purification of HSA on anti-HSA immunoaffinity column is shown in Figure 5.8.

5.1.5 Purification of α_2-Macroglobulin Using Immobilized Anti-α_2-Macroglobulin Immunoaffinity Column

Proteinase-binding α-macroglobulins are large glycoproteins found in circulation that function as "molecular traps" for proteinases. When bound to α-macroglobulins, a proteinase is "protected" from reaction with large proteinase inhibitors and substrates, but readily reacts with small substrates and inhibitors. The level of α_2-macroglobulin in human plasma varies from 2 to 4 mg per ml plasma. α_2-Macroglobulin can be isolated easily by immunoaffinity chromatography or by metal chelate affinity chromatography (discussed in Section 5.1.9). The immunoaffinity chromatographic purification of α_2-macroglobulin is discussed here and illustrated in Figure 5.9.

PROTOCOL

1. Prepare 2 ml immunoaffinity matrix by coupling the polyclonal antibody anti-α_2-macroglobulin to a suitably activated support material. Suggested coupling chemistries include reductive amination (Section

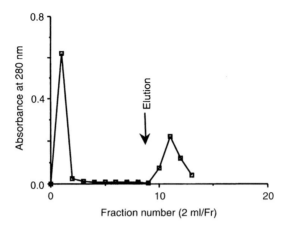

FIGURE 5.8

Affinity chromatography of human serum albumin (HSA) on an immobilized anti-HSA column (2 ml gel). The column is equilibrated with 0.1 M sodium borate buffer, pH 8.0 (binding buffer). HSA (5 mg) is dissolved in 1.0 ml binding buffer and applied to the column. The column is washed with 18 ml binding buffer while 2.0-ml fractions are collected. Bound HSA is eluted with 0.1 M glycine–HCl, pH 2.8. Chromatography is performed at 4°C.

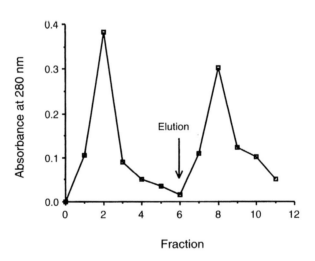

FIGURE 5.9

Elution profile for the immunoaffinity purification of α_2-macroglobulin on an immobilized polyclonal anti-α_2-macroglobulin column, performed according to the procedure discussed in the text.

2.2.1.4, Chapter 2), CNBr (Section 2.2.1.1, Chapter 2), iodoacetyl (Section 2.2.2.1, Chapter 2), or hydrazide (Section 2.2.3.1, Chapter 2). Also, see the discussion of antibody immobilization in Section 3.4.2.3, Chapter 3, for more details on how to best couple immunoglobulins. A cross-linked agarose support material makes an excellent immunoaffinity matrix.
2. Pack 2 ml antibody-coupled gel in a minicolumn according to the protocol outlined in Chapter 4 (Section 4.2.1.1).
3. Equilibrate the column with 10 ml 50 mM Tris-HCl, 0.15 M NaCl, pH 8.0 (binding buffer).
4. Dilute 2 ml human serum with an equal volume of binding buffer, and filter or centrifuge the solution to remove any cellular debris or precipitate. Apply the clarified sample to the immunoaffinity column, taking 2-ml fractions.
5. After the sample has fully entered the gel, wash the column with 20 ml binding buffer to completely remove unbound serum proteins.
6. Elute the bound α_2-macroglobulin with 0.1 M glycine, 0.15 M NaCl, pH 2.8. Immediately neutralize the eluting fractions by adding 0.1 ml 1 M Tris-HCl, pH 8.0. Dialyze the purified α_2-macroglobulin against 0.01 M sodium phosphate, 0.15 M NaCl, 0.02% sodium azide, pH 7.2. The purified protein may be stored at 4°C.

5.1.6 Purification of Immunoglobulin G Using Protein A, Protein G, or Protein A/G

Immunoglobulin binding proteins have been used extensively for the purification of IgG molecules from serum, ascites, or cell culture supernatants. The affinity of protein A, protein G, or the chimeric protein A/G for the Fc region of immunoglobulins makes them natural affinity ligands for IgG isolation. Section 3.4.4, Chapter 3, describes the immobilization of these proteins to solid supports. The use of these affinity matrices for the purification of antibody molecules is illustrated by the following general protocol. A typical elution profile for the isolation of mouse IgG on immobilized protein G is shown in Figure 5.10.

PROTOCOL

1. Pack a 2-ml minicolumn containing immobilized protein A, immobilized protein G, or immobilized protein A/G. (See Section 4.2.1.1, Chapter 4, for the column packing procedure.)

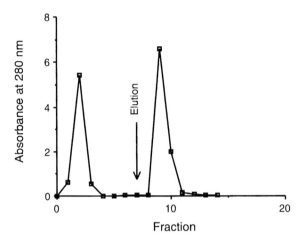

FIGURE 5.10

Elution profile for the purification of mouse IgG from mouse serum on an immobilized protein G affinity matrix.

---

2. Equilibrate the affinity column with an appropriate binding buffer by washing it with at least 10 ml. All three immunoglobulin binding proteins will give optimal binding performance using a buffer system at a pH between 7.5 and 8.0. Suggested buffers include 0.1 M Tris, 0.15 M NaCl, pH 7.5; 50 mM sodium borate, 0.15 M NaCl, pH 8.0; 0.1 M sodium phosphate, 0.15 M NaCl, pH 7.5; or any of these buffers containing a high salt environment (>1 M NaCl) to enhance the hydrophobic interactions between IgGs and the immobilized proteins. Such high salt buffers are commercially available (Pierce, BioRad), and typically increase the total binding capacity of the columns by about 50%. For immobilized protein G, a binding buffer at a lower pH, consisting of 50 mM sodium acetate, 0.15 M NaCl, pH 5.0, also works well.
3. Prepare a sample containing immunoglobulins by diluting it at least 1:1 in binding buffer. The sample may be of serum, ascites, or cell-culture origin and contain either polyclonal or monoclonal IgG antibodies. Apply the sample to the affinity column, collecting 2-ml fractions. The amount of sample applied to the column is governed by the column capacity for the particular immunoglobulin source. For mouse IgGs, apply no more than 10 mg immunoglobulin per 2-ml column. For human IgG, the column capacity may be almost twice this amount.
4. After the sample has entered the gel, wash the column with binding buffer until the absorbance at 280 nm is less than 0.02.

5. Elute the bound immunoglobulins with 0.1 M glycine, 0.15 M NaCl, pH 2.8. Immediately neutralize the eluting antibody by adding 0.1 ml 1 M Tris-HCl, pH 8.0, to each 2-ml fraction. Dialyze against 0.01 M sodium phosphate, 0.15 M NaCl, pH 7.2, and store at 4°C with a preservative (0.02% sodium azide) or lyophilize, depending on the stability of the antibody.

For antibodies that are sensitive to a low pH environment, an alternative elution buffer can be used. (See Section 5.1.7 for a discussion of gentler elution buffer formulations.)

5.1.7 Gentle Immunoaffinity Chromatography

In immunoaffinity chromatography it is critical to determine optimal conditions for the elution of bound antigen or antibody from the immunoaffinity column. Low dissociation constants of many antigen–antibody complexes require drastic conditions for desorption of the adsorbed component from the immunoaffinity matrix. Table 5.1 lists some of the elution buffers used by many investigators in immunoaffinity chromatography.

Table 5.1 shows that powerful desorption agents are frequently required to break the antigen–antibody interaction. Since exposure to such harsh

TABLE 5.1
Eluents Used in Immunoaffinity Chromatography

0.1 M Glycine–HCl, pH 2.3
0.1 M Glycine–NaOH, pH 10.0
0.1 M Tris–acetate, pH 7.7 containing 2.0 M NaCl
0.1 M Glycine–NaOH, pH 10.0 containing 50% ethyleneglycol
6.0 M Guanidine–HCl, pH 3.0
2.0 M Trichloroacetic acid–NaOH, pH 7.0
3.0 M Potassium chloride
0.5 M Acetic acid + 10% dioxane
5.0 M Potassium iodide
3.5 M Magnesium chloride
1.0 M Ammonium thiocyanate
1.0 M Lithium diiodosalicylate
6.0 M Urea
1% SDS
Deionized water
0.75 M Ammonium sulfate containing 40% ethylene glycol

agents can lead to the loss of the immunological activity of the antigen or antibody, elution under mild conditions is highly desirable and is of paramount importance in immunoaffinity chromatography.

Buffers suitable for the elution of bound antigen or antibody from immunoaffinity columns under mild conditions are commercially available. Sterogene Biochemicals and Pierce market elution buffers that are reported to be nonchaotropic and neutral in pH. Sterogene markets the elution buffer under the tradename AFC Elution Medium and Pierce markets under the tradename Immunopure Gentle Ag/Ab Elution Buffer.

Pierce claims that Immunopure Gentle Ag/Ab Elution Buffer can be used to elute bound antigen or antibody from immunoaffinity columns. The neutral pH elution buffer insures that fragile antibodies or antigens will be immunologically reactive when recovered from an immunoaffinity column.

5.1.8 Isotype Elution Buffers for Purification of Mouse IgG Subclasses

Immunoglobulin G from most species consists of several subclasses with different biological properties. Four subclasses of IgG have been identified in human (IgG_1, IgG_2, IgG_3, and IgG_4) and in mouse (IgG_1, IgG_{2a}, IgG_{2b}, and IgG_3). For immunological studies, it is often necessary to isolate one particular subclass of IgG from the other subclasses.

Although protein A interacts with all mouse IgG subclasses, the affinity of the interaction varies. Mouse IgG_2 binds to protein A very firmly, whereas IgG_1 interacts very weakly. It is also known that the binding of mouse IgGs to protein A is dependent on the pH and salt concentration of the buffer. This knowledge proves very useful in the fractionation of mouse IgG subclasses. The protocol given here for the separation mouse IgG subclasses using immobilized protein A is adapted from Ey *et al.* (1978) and Seppala *et al.* (1981).

BUFFERS FOR THE PURIFICATION OF MOUSE IgG SUBCLASSES

Binding buffer	*0.1 M potassium phosphate, pH 8.0, with 0.1% sodium azide*
Isotype elution buffer 1	0.1 M potassium phosphate, pH 6.0
Isotype elution buffer 2	0.1 M sodium citrate, pH 5.5
Isotype elution buffer 3	0.1 M sodium citrate, pH 5.0
Isotype elution buffer 4	0.1 M sodium citrate, pH 4.5
Isotype elution buffer 5	0.1 M sodium citrate, pH 3.5
Isotype elution buffer 6	0.1 M sodium citrate, pH 3.0

Fractionation of Mouse IgG Subclasses Using Immobilized Protein A

NOTE: Chromatography is performed at 4°C.

1. Pack immobilized Protein A (5 ml; prepared as described in Chapter 3, Section 3.4.4.1) in a disposable polypropylene column (for column packing, see Section 4.2.1.1, Chapter 4) and equilibrate by washing with 50 ml binding buffer.
2. Mix mouse serum (1.2 ml) with 1.2 ml 0.5 M potassium phosphate buffer, pH 8.1, and apply to the immobilized protein A column.
3. After the sample has entered the gel, wash the column with 40 ml binding buffer. Collect 3-ml fractions.
4. Elute the bound IgG subclasses sequentially with 30 ml each of isotype elution buffers 1 to 6 (formulation just listed).
5. Neutralize the fractions by collecting them into tubes that contain 0.8 ml 1.0 M Tris-HCl buffer, pH 8.5.
6. Wash the immobilized protein A column with 50 ml binding buffer and store at 4°C.
7. Measure the absorbance of all fractions at 280 nm. The elution profile for the affinity chromatographic fractionation of mouse serum on an immobilized protein A column is shown in Figure 5.11.

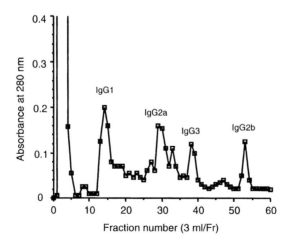

FIGURE 5.11

Fractionation of mouse IgG subclasses from mouse serum on an immobilized protein A column. For experimental details, see the text.

As evident from the profile, all four subclasses of mouse IgG can be separated by affinity chromatography on an immobilized protein A column if the bound IgGs are sequentially eluted using buffers of decreasing pH.

5.1.9 Purification of Immunoglobulin M

5.1.9.1 Purification of IgM Using Immobilized C1q

The complement protein C1q binds specifically to the Fc portion of IgG and IgM. The affinity of C1q for pentameric IgM is 18-fold higher than for monomeric IgG. Nethery *et al.* (1990) have used this selectivity for affinity chromatographic purification of IgM from myeloma serum and ascites fluid. The method is based on a temperature-dependent interaction of immobilized C1q with IgM. Immobilized C1q selectively binds IgM from crude samples at 5°C and the bound IgM is eluted simply and isocratically (using the same buffer) by bringing the gel to room temperature for 2 hr. Highly purified IgM is obtained using immobilized C1q affinity chromatography.

PROTOCOL

NOTE: The chromatography is performed at 5°C unless indicated otherwise.

1. Complement protein C1q is isolated from 1.5 L human plasma according to the method of Tenner *et al.* (1981). Total yield of C1q is 174 mg. Couple the purified C1q (174 mg) to 50 ml CNBr-activated Sepharose 4B as described (B.1.a, chapter 2). Approximately 2.8 mg of C1q are coupled per ml gel.
2. Pack the immobilized C1q (5 ml gel) in a disposable polypropylene column (for column packing, see Section 4.2.1.1, Chapter 4) and equilibrate by washing with 25 ml of 10 mM Tris-HCl buffer, pH 7.4, containing 65 mM NaCl (binding buffer).
3. Apply human myeloma IgM solution (1.0 ml) to the immobilized C1q column.
4. After the sample has entered the gel, wash the column with 20 ml binding buffer while 2-ml fractions are collected.
5. For elution of bound IgM, bring the column to room temperature, incubate at room temperature for 2 hr, and elute with 50 mM sodium phosphate buffer, pH 7.4, containing 150 mM NaCl, 2 mM EDTA, and 0.02% sodium azide (elution buffer).
6. Measure the absorbance of all fractions at 280 nm.

A typical affinity chromatographic profile for the purification of human myeloma IgM on immobilized C1q column is shown Figure 5.12.

5.1.9.2 Purification of IgM Using Immobilized Mannan-Binding Protein

Various mammalian sera contain a binding protein that binds mannose and N-acetylglucosamine residues (Kawasaki *et al.*, 1989). This binding protein, known as mannan-binding protein (MBP), has been shown to activate the complement system through the classical pathway. Nevens *et al.* (1992) have shown that immobilized MBP specifically binds mouse monoclonal IgM from mouse ascites fluid (U. S. Patent No. 5, 112, 952). Binding of IgM to MBP is calcium dependent and IgM bound to immobilized MBP can be eluted with EDTA.

PROTOCOL

NOTE: Chromatography is performed at 4°C unless indicated otherwise.

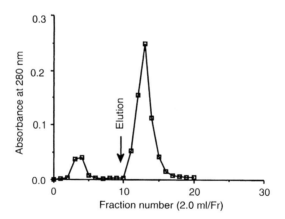

FIGURE 5.12

Affinity chromatography of human myeloma IgM on an immobilized C1q column (5 ml gel). The column is equilibrated by washing with 25 ml 10 mM Tris–HCl buffer, pH 7.4, containing 65 mM NaCl (binding buffer). Human myeloma IgM solution (1.0 mg IgM in 1.0 ml binding buffer) is applied to the column. The column is washed with 20 ml binding buffer while 2.0-ml fractions are collected. Chromatography is conducted at 5°C unless indicated otherwise. Elution of bound IgM is carried out at room temperature with 50 mM sodium phosphate buffer, pH 7.4, containing 150 mM NaCl, 2 mM EDTA, and 0.02% sodium azide.

1. Prepare a mannan–/ /–Sepharose 6B (1 L) affinity matrix by coupling 20 gm yeast mannan (Sigma) to 1 L CNBr-activated Sepharose 6B. (See Section 2.2.1.1, Chapter 2, for CNBr activation and coupling.)
2. MBP is purified from 10 L rabbit serum using affinity chromatography on immobilized mannan according to the method of Kozutsumi *et al.* (1980). The resulting 385 ml dialysate containing the purified protein (yield of MBP is 388 mg) is used directly to couple to 110 ml CNBr-activated Sepharose 4B according to the method of Cuatrecasas (1970) (see Section 3.4.4.4, Chapter 3). The MBP elution profile resulting from this mannan–/ /–agarose affinity separation is shown in Figure 5.13.
3. Pack the MBP–/ /–Sepharose 4B (5.0 ml gel) in a disposable polypropylene column (see Section 4.2.1.1, Chapter 4, for column packing) and equilibrate by washing with 25 ml 10 mM Tris-HCl buffer, pH 7.4, containing 1.25 M NaCl, 20 mM CaCl$_2$, and 0.02% sodium azide (binding buffer).
4. Dialyze mouse IgM, λ (MOPC 104E) clarified ascites fluid (0.5 ml; Sigma) against 200 ml 10 mM Tris-HCl buffer, pH 7.4, containing 1.25 M NaCl for 20 hr at 4°C.
5. After dialysis, mix the dialyzed ascites fluid with 0.5 ml binding buffer and apply to MBP–/ /–Sepharose 4B column.
6. After the sample has entered the gel, wash the column with 40 ml binding buffer, collecting 3-ml fractions.

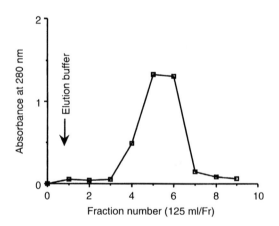

FIGURE 5.13

Affinity chromatographic purification of mannan-binding protein (MBP) from rabbit serum on a mannan–/ /–Sepharose 6B column. For experimental details, see the text. The elution profile of MBP from the mannan–/ /–Sepharose 6B column is shown.

7. For the elution of bound IgM, bring the column to room temperature and incubate at that temperature for 2 hr.
8. Elute bound IgM at room temperature using 10 mM Tris-HCl buffer, pH 7.4, containing 1.25 M NaCl, 2 mM EDTA, and 0.02% sodium azide.
9. Measure the absorbance of all fractions at 280 nm. A typical profile of the affinity chromatographic purification of mouse monoclonal IgM on an MBP–/ /–Sepharose 4B column is shown in Figure 5.14.
10. Pool the eluted mouse IgM fractions (16–19), dialyze against PBS, pH 7.4 (20 mM sodium phosphate containing 150 mM NaCl), and concentrate to 2.0 ml using Centricon 30 concentrators (Amicon).

The yield of IgM is 2.6 mg using an A_{280} value of 1.18 for a 1 mg/ml solution. Affinity-purified IgM is analyzed by SDS-PAGE under reducing conditions and by HPLC on a Superose 6 column. Only two bands corresponding to heavy and light chains of IgM are observed on an SDS-PAGE gel. HPLC analysis on the Superose 6 column indicates that mouse monoclonal IgM purified by affinity chromatography on MBP–/ /–Sepharose 4B column is at least 90% pure (Fig. 5.15). MBP–/ /–Sepharose 4B also binds human IgM and can be used to purify human IgM, although the capacity for human IgM of the column is less than that for mouse IgM.

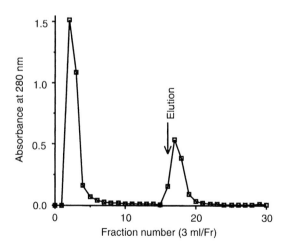

FIGURE 5.14

Affinity chromatographic purification of monoclonal mouse IgM from mouse ascites fluid on an immobilized MBP column (5.0 ml gel). For experimental details, see the text.

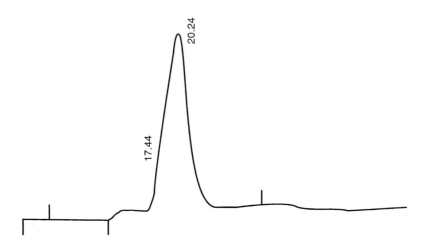

FIGURE 5.15

HPLC analysis of monoclonal mouse IgM purified from mouse ascites fluid using an immobilized MBP column. A Superose 6 column (10× 300 mm; Pharmacia-LKB) is used for the analysis. A 10 mM Tris–HCl buffer, pH 7.4, containing 1.25 M NaCl, 2 mM EDTA, and 0.02% sodium azide is used as the mobile phase. The HPLC separation is run for 40 min at a flow rate of 0.5 ml/min. Absorbance at 280 nm is used for detection.

5.1.10 Purification of Mouse IgG$_1$ from Mouse Ascites Fluid Using Thiophilic Gel Chromatography

Thiophilic adsorption is a newly recognized and highly selective type of salt-promoted protein–ligand interaction (Porath *et al.*, 1985; Hutchens and Porath, 1986, 1987a,b; Belew *et al.*, 1987; Porath, 1987; Nopper *et al.*, 1989; Lihme and Heegaard, 1990). This new interaction is termed thiophilic since it is distinguished by proteins that recognize a sulfone group in close proximity to a thioether. The essential structural features of an immobilized thiophilic ligand are represented in Figure 5.16. Thiophilic adsorption differs from hydrophobic interaction in some respects. Hydrophobic

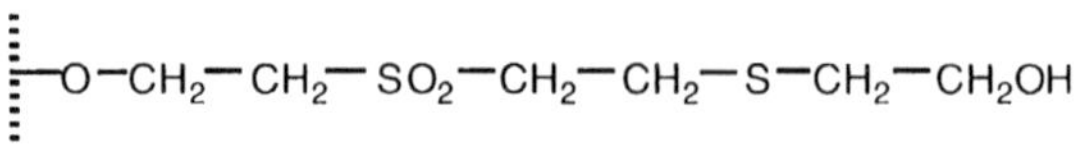

$$\vdots\!-\!O-CH_2-CH_2-SO_2-CH_2-CH_2-S-CH_2-CH_2OH$$

FIGURE 5.16

Partial chemical structure of a thiophilic gel (T-Gel).

interaction chromatography is strongly promoted by high concentrations of sodium chloride, whereas thiophilic adsorption is not. Also, a thiophilic adsorbent has a broad specificity for immunoglobulins. Thiophilic interaction chromatography provides an alternative method for the purification of immunoglobulins under conditions that preserve biological activity. A protocol is given here for the purification of mouse IgG_1 from mouse ascites fluid, that can be adapted easily for purifying all subclasses of mouse immunoglobulins (Belew, 1987; Nopper *et al.*, 1989).

PROTOCOL

1. Pack the thiophilic gel (4.0 ml; prepared as described in Section 3.7, Chapter 3) in a disposable polypropylene column. (See Section 4.4.1.1, Chapter 4, for column packing.) Equilibrate the column by washing with 25 ml 0.1 M Tris-HCl buffer, pH 7.6, containing 0.5 M K_2SO_4 (binding buffer).
2. Mix 1 ml mouse IgG_1, κ (MOPC) clarified ascites fluid (Sigma) with 2.0 ml binding buffer and apply to the column. After the sample has entered the gel, wash the column with 20 ml binding buffer while 4-ml fractions are collected.
3. Elute bound mouse IgG_1 with 0.1 M ammonium bicarbonate.

The chromatography is performed at 4°C. The absorbance of all fractions is measured at 280 nm. A typical profile of thiophilic gel chromatography of mouse IgG_1 ascites fluid is shown in Figure 5.17.

Pool fractions 8 and 9 (peak 2) and dialyze against 20 mM sodium phosphate buffer, pH 7.4, containing 0.15 M NaCl. Analyze by SDS-PAGE under reducing conditions. Only two bands corresponding to the heavy and light chains IgG are observed on an SDS-PAGE gel. These results demonstrate that mouse IgG_1 isolated by thiophilic gel chromatography is of very high purity.

5.1.11 Purification of Human Serum Albumin Using Immobilized Cibacron Blue F3GA

Immobilized dye chromatography has been useful in the purification of numerous proteins. An immobilized dye can bind to a sizable number of different proteins. Selective elution of the protein of interest is achieved using a competitive ligand, a salt gradient, or chaotropic agents. Many proteins have been purified to homogeneity in a single step using immobilized dye chromatography.

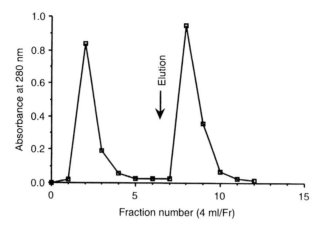

FIGURE 5.17

Chromatography of mouse IgG1 ascites fluid on a thiophilic affinity support (4.0 ml gel). The column is equilibrated with 0.1 M Tris–HCl buffer, pH 7.6, containing 0.5 M K_2SO_4 (binding buffer). Mouse IgG1, k (MOPC 21) clarified ascites (1.0 ml) is mixed with 2.0 ml binding buffer and applied to the column. The column is washed with 21 ml of binding buffer while 4.0-ml fractions are collected. Bound mouse IgG1 is eluted with 0.1 M ammonium bicarbonate. Chromatography is conducted at 4°C.

Immobilized Cibacron Blue F3GA is a widely used support for the purification of a vast array of proteins. The method given here for the purification of albumin from human plasma is based on Travis *et al.* (1976).

PROTOCOL

1. Pack 1 L immobilized Cibacron Blue F3GA (prepared as described in Section 3.1.4.1, Chapter 3) in a 5 cm×75 cm column and equilibrate by washing with 5 L of 50 mM Tris-HCl buffer, pH 8.0, containing 50 mM NaCl and 0.02% sodium azide (binding buffer). The washings should be colorless and free of leaching dye molecules. If not, wash the column with additional binding buffer until the washings are colorless.
2. Apply undiluted human plasma or serum (200 ml) to the column. After the sample has entered the gel, wash the column with 6 L binding buffer.
3. Elute bound HSA with 0.2 M NaSCN in the binding buffer. Collect eluted HSA as a single fraction until the absorbance at 280 nm of the column effluent is less than 0.1.
4. Extensively dialyze the eluent against water and lyophilized. The yield of albumin is 6.5 gm.

5.1 Purification

If required, the column can be re-equilibrated with the binding buffer and a new sample of plasma or serum can be applied. The column can be regenerated and reused at least 10 times with complete reproducibility. After 10 regenerations, wash the column with 6.0 M guanidine·HCl to remove traces of lipoproteins that are precipitated at the top of the column bed and other nonspecifically bound proteins. Exposure to guanidine·HCl has no deleterious effect on the performance of the immobilized Cibacron column.

5.1.12 Purification of Phosphoamino Acids, Phosphopeptides, and Phosphoproteins Using Iron Chelate Affinity Chromatography

Anderson and Porath (1986) have reported that phosphoproteins and phosphoamino acids bind to ferric ions immobilized on iminodiacetic acid–/ /–Sepharose CL-6B and can be eluted by increasing pH or by including phosphate ions in the eluent. A protocol is given here for the separation of L-serine from O-phospho-L-serine that can be adapted for the isolation of other phosphoamino acids, phosphopeptides, and phosphoproteins.

PROTOCOL

NOTE: Chromatography is performed at room temperature.

1. Pack an iminodiacetic acid–/ /–Sepharose CL-4B (2 ml gel; see Section 3.1.5.1, Chapter 3, for preparation of the affinity matrix) in a disposable polypropylene column. (For column packing, see Section 4.2.1.1, Chapter 4).
2. Wash the column with 10 ml deionized water; then apply 4 ml of 0.1 M ferric chloride solution.
3. Wash the column with 20 ml water to remove any unbound metal ions. Equilibrate the column by washing with 20 ml 0.1 M acetic acid/NaOH, pH 3.1 (binding buffer).
4. Dissolve L-serine and O-phosphoserine (3 mg each) in 1.0 ml binding buffer and apply the sample solution to the column.
5. After the sample has entered the gel, wash the column with 7 ml binding buffer and collect 1.0-ml fractions. L-Serine will not be retained by the iron-chelated column and therefore will be eluted in breakthrough fractions.
6. Elute bound O-phosphoserine by applying 8 ml 20 mM sodium phosphate buffer, pH 7.0. Collect fractions of 1.0 ml.

7. Elution of L-serine and O-phosphoserine can be monitored by the TNBS (trinitrobenzene sulfonic acid; see Section 4.1.3, Chapter 4) assay as follows. Add 2.0 ml saturated sodium borate solution to 100 µl of each fraction and mix with 100 µl freshly prepared 1.5% ethanolic solution of TNBS. After incubating at room temperature for 15 min, measure the absorbance at 340 nm using a control.

5.1.13 Purification of Human Plasma α_2-Macroglobulin Using Immobilized Metal Chelate Affinity Chromatography

α_2-Macroglobulin, a tetrameric glycoprotein, is a major component of human plasma, present in concentrations of 2–4 mg/ml. α_2-Macroglobulin plays an important role in regulating the activity and concentration of endoproteinases because of its ability to sequester endoproteinases of all four major classes. A protocol is given here for the purification of human plasma α_2-macroglobulin using zinc chelate affinity chromatography. This protocol is adapted from the method of Kurecki et al. (1979). It is based on the finding that α_2-macroglobulin binds tightly to zinc-chelated immobilized iminodiacetic acid at pH 6.0, allowing removal of contaminating proteins. At pH 5.0, α_2-macroglobulin can be eluted from the affinity column as a sharp concentrated peak. An electrophoretically homogeneous protein is obtained in good yield by this mild procedure.

PROTOCOL

1. Mix 1 L pooled citrated human plasma with an equal volume of saturated ammonium sulfate solution, stir for 30 min, and leave overnight at 4°C. Centrifuge the suspension at 15,000 g for 20 min and discard the supernatant. Suspend the pellet in 500 ml 50% saturated ammonium sulfate solution, stir for 1 hr, and centrifuge at 15,000 g for 20 min. Discard the supernatant and dissolve the pellet in 300 ml water. Dialyze for 72 hr against three changes (7 L each) of water. Centrifuge the dialyzed solution as before and save the decant supernatant. Bring the recovered supernatant to 40% saturation by the addition of solid ammonium sulfate (231 gm/L). Stir the solution for 1 hr and centrifuge as before. Decant the supernatant and bring it to 55% saturation by adding solid ammonium sulfate (102 gm/L). Stir the solution for 1 hr and centrifuge. Discard the supernatant and dissolve the precipitate in 60 ml 0.02 M sodium phosphate, 0.15 M NaCl, pH 6.0. Dialyze the solution for 48 hr against three changes (3 L each) of the same buffer. Clarify the crude dialyzed α_2-macroglobulin solution by centrifugation and purify it on a zinc-chelated immobilized iminodiacetic acid column as described next.

5.1 Purification

2. Pack a column containing immobilized iminodiacetic acid (75 ml settled gel; prepared as described in Section 3.1.5.1, Chapter 3) (size: 2.5 cm × 15 cm; see Section 4.2.1.2, Chapter 4, for column packing procedure). Wash the column with 500 ml of 0.05 M EDTA, 0.5 M NaCl, pH 7.0, followed by 500 ml water. Convert the gel to the zinc chelate form by equilibrating with aqueous $ZnCl_2$ (3 mg/ml, adjusted to pH 6.0). Monitor the column effluent until the pH value returns to 6.0 and Zn^{2+} breakthrough is indicated with the addition of 2 M sodium carbonate. Wash the column with 200 ml 0.25 M sodium acetate, 0.15 M NaCl, pH 5.0, and equilibrate it by washing with 500 ml 0.02 M sodium phosphate, 0.15 M NaCl, pH 6.0.
3. Apply the dialyzed α_2-macroglobulin solution (150 ml; total absorbance at 280 nm = 3,500; prepared as described in Step 1) to the zinc-chelated column, which has been equilibrated previously with 0.02 M sodium phosphate, 0.15 M NaCl, pH 6.0 (equilibration buffer), at a flow rate of 50 ml/hr. After sample application, wash the column with 400 ml equilibration buffer, collecting 20-ml fractions. Elute the bound α_2-macroglobulin with 0.02 M sodium cacodylate, 0.15 M NaCl, pH 5.0, collecting 11-ml fractions. Measure the absorbance of all fractions at 280 nm. A typical profile of the affinity chromatographic purification of α_2-macroglobulin is shown in Figure 5.18.
4. Pool the eluted α_2-macroglobulin fractions, adjust the pH to 6.0, and store at –20°C in 50% glycerol. Total yield of α_2-macroglobulin from 1 L human plasma is 320 mg. Homogeneous α_2-macroglobulin is

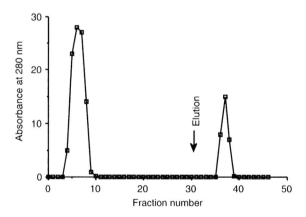

FIGURE 5.18

Affinity chromatographic purification of α_2-macroglobulin from human plasma on a zinc-chelated immobilized iminodiacetic acid column (75 ml gel). For experimental details, see the text.

obtained as indicated by the single band observed during electrophoresis at pH 9.4, and also after treatment with SDS and mercaptoethanol.

5.2 SCAVENGING

Affinity chromatography has been exploited successfully in the removal of unwanted trace amounts of contaminants from biological preparations. This highly attractive and useful application of affinity chromatography is called scavenging. Among the many contaminants removed using affinity chromatography are:

1. foreign immunogens
2. pyrogens (endotoxins, lipopolysaccharides)
3. proteolytic enzymes
4. detergents
5. lipids
6. heavy metals
7. viruses

Table 5.2 lists various affinity chromatographic supports that have been successfully used for scavenging.

TABLE 5.2
Affinity Chromatographic Supports Used for Scavenging

Affinity chromatographic support	Contaminant removed
Immobilized polymyxin B	Endotoxins
	Lipopolysaccharides
Immobilized Histidine	Endotoxins
	Lipopolysaccharides
Extracti-Gel D	Detergents
Bio-Beads SM-2	Detergents
Lipidex 1000	Lipids
Cab-O-Sil	Lipids
	Lipoproteins
Immobilized aprotinin	Proteolytic enzymes
Immobilized soybean trypsin inhibitor	Proteolytic enzymes
Immobilized p-aminobenzamidine	Trypsin-like enzymes
α_2-Macroglobulin, carrier fixed	Endoproteases
Immobilized iminodiacetic acid	Metals
Immobilized TED	Metals
Sulfhydryl agarose	Mercury
Immobilized octanoic acid hydrazide	Hepatitis B virus

5.2.1 Removal of Detergents from Protein Solutions

Detergents frequently are used for solubilization of membrane proteins. For many biological studies, after membrane solubilization it is necessary to remove detergents from protein solutions to restore the native conformation or activity of the protein. Dialysis, ion-exchange, and hydrophobic interaction chromatography have been used in the past to remove detergents from protein. In many cases, dialysis is not suitable to remove detergents, because many have low critical micelle concentrations. Although ion-exchange and hydrophobic interaction chromatography remove detergents from protein solutions, in many instances the final recovery of protein is poor. Two commercially available supports, Extracti-Gel D from Pierce and Bio-Beads SM-2 from BioRad, are claimed to remove detergents from protein solutions with excellent protein recovery. The use of these two commercially available supports to remove detergents from solutions containing proteins is described next.

5.2.1.1 Removal of Detergents from Protein Solutions Using Extracti-Gel D

Extracti-Gel D support is a versatile affinity matrix for the extraction of unwanted detergents from solutions containing proteins or other macromolecules. This unique affinity support allows relatively small detergent molecules to enter the gel matrix, where they can interact with a specially developed ligand capable of removing them from solution (Fig 5.19). Unlike the chromatographic schemes that have been used in the past to remove detergent, the low exclusion limit of the Extracti-Gel D base support over-

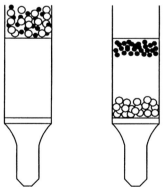

FIGURE 5.19
A pictorial representation of the removal of detergent on an Extracti-Gel D column.

comes the possibility of nonspecific binding by not permitting larger macromolecules to enter the gel. The result of this unique combination of properties insures that detergent molecules can be extracted from a solution with high efficiency, without losing valuable protein in the process. The Extracti-Gel D support material will remove detergents from solutions with a broad range of buffer compositions, salt concentrations, and pH values.

PROTOCOL FOR REMOVAL OF DETERGENTS FROM PROTEIN SOLUTIONS USING EXTRACTI-GEL D

NOTE: The column format rather than the batch format is best suited for optimal performance of Extracti-Gel D.

1. Pack a column of Extracti-Gel D (2.0 ml settled gel) in a disposable polypropylene minicolumn (for column packing see Section 4.2.1.1, Chapter 4).
2. Equilibrate the column by washing with 20 ml desired buffer.
3. Apply the protein solution containing the detergent to the column. The protein will be eluted in the void volume (0.7–1.0 ml per ml gel using the minicolumns). Since the exclusion limit of the gel is 10,000 MW, components that have molecular weights lower than this value will be able to enter the pores and interact with the affinity ligand.
4. After the sample has entered the gel bed, wash the column with the buffer and collect 2-ml fractions. The eluted protein fractions should be free from the detergent (provided that the maximum detergent-binding capacity of Extracti-Gel has not been reached). (See Table 5.3 for the binding capacity of Extracti-Gel for various detergents.)

TABLE 5.3

Binding Capacity of Extracti-Gel D for Various Detergents at Room Temperature

Detergent	Binding capacity (mg/ml gel)
CHAPS	50
SDS	80
Triton X-100	55
Triton X-114	85 [a]
Triton X-405	17
Brij 35	80
Tween 20	75
Lubrol PX	105
NP-40	75

[a] Performed at 4°C.

Extracti-Gel D can be regenerated successfully at least three times by washing sequentially with 5.0 ml each of deionized water, ethanol, butanol, ethanol, and deionized water. The column is finally equilibrated by washing with 20 ml desired buffer.

NOTE: Extracti-Gel D chromatography can be performed at room temperature or at 4°C.

In Figure 5.20, the detergent removing capability of Extracti-Gel D is shown. The absorbance at 275 nm of Triton X-100 eluting from the control support (contains no affinity ligand) can be followed as it rises sharply to a constant value. Each aliquot of Triton X-100 passes through the control column unchanged with respect to the detergent concentration. In contrast to the blank support, the Extracti-Gel D support is capable of removing the detergent from solution. Triton X-100 binding ability of Extracti-Gel D is seen as the difference between the profiles obtained for the blank support and Extracti-Gel D. This difference represents more than 98% of the Triton X-100 removed based on absorbance at 275 nm.

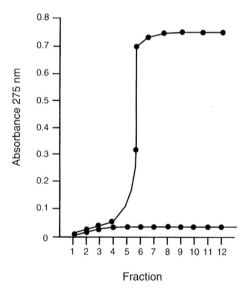

FIGURE 5.20

Removal of Triton X-100 using an Extracti-Gel D column. Extracti-Gel D (1.3 ml gel) and a blank control support (1.3 ml, containing no affinity ligand) are packed in disposable polypropylene minicolumns and equilibrated with 50 mM Tris–HCl buffer, pH 9.0, containing 0.2 M NaCl (equilibration buffer). Fractions of 1.0 ml are collected as 1.0 ml aliquots of 0.1% Triton X-100 solution (in equilibration buffer) are applied to the columns. All fractions are diluted to 3.0 ml with water and the absorbance is measured at 275 nm. The chromatography is performed at room temperature.

5.2.1.2 Removal of Detergents from Protein Solutions Using Bio-Beads SM-2

Bio-Beads SM-2 (BioRad Laboratories, Richmond, California) are neutral macroporous styrene–divinylbenzene copolymer beads of high surface area and are used for adsorbing detergents from protein solutions (Holloway, 1973). Table 5.4 shows some of the detergents that have been removed successfully from protein solutions using Bio-Beads SM-2.

PROTOCOL FOR REMOVAL OF TRITON X-100 FROM PROTEIN SOLUTIONS USING BIO-BEADS SM-2

1. Mix Bio-Beads SM-2 (30 gm) with 200 ml methanol and stir the mixture for 15 min. Filter and wash the copolymer beads on a sintered glass funnel with 500 ml methanol and 1 L water. Slowly wash the moist beads in a chromatography column with 2 L water. Washed Bio-Beads SM-2 can be stored in water at 4°C until required.
2. Add moist Bio-Beads SM-2 (0.6 gm) to 2 ml protein solution (1–10 mg protein per ml in 10 mM potassium phosphate buffer, pH 7.2) containing 1% Triton X-100. Gently mix the sample at 4°C on a blood tube rotator for 2 hr. Triton X-100 is completely adsorbed by Bio-Beads SM-2 in 2 hr. No protein is adsorbed by the copolymer beads.
3. Pack the moist Bio-Beads SM-2 (5 gm) in a column (1 × 8 cm) and equilibrate with 10 mM potassium phosphate buffer, pH 7.2, at 4°C. Apply a protein solution (1.0 ml, 5 mg protein) containing 1% Triton X-100 to the column and wash the column with 10 mM potassium phosphate buffer, pH 7.2, at a rate of 0.3 ml/min. Collect fractions every minute. A 1.0-ml sample [1% Triton X-100 and 0.5% bovine serum albumin (BSA)] yields a 2.1-ml sample containing 90% recovery of BSA and 0.01% Triton X-100.

Although the removal of Triton X-100 can be accomplished in 2 hr with no dilution of sample by batch format, the column format takes only 15 min with some dilution of sample. The column format is well suited to many

TABLE 5.4

Detergents That Adsorb to Bio-Beads SM-2

Triton X-100
Sodium cholate
Sodium deoxycholate
NP-40
Emulgen 911
Emulphogene BC-720

applications. The capacity of Bio-Beads SM-2 is found to be approximately 70 mg Triton X-100 per gm moist beads.

5.2.2 Removal of Lipids from Protein Solutions

Many proteins (e.g., enzymes, plasma carrier proteins, or membrane proteins) interact with lipids. A simple method for the removal of lipids from protein solutions that is nondestructive for proteins is described here.

5.2.2.1 Removal of Fatty Acids from Serum Albumins by Lipidex 1000 Chromatography

Removal of fatty acids from serum albumin is usually accomplished by treatment with activated charcoal at pH 3.0 and 0°C. Although activated charcoal treatment has been shown to have little effect on the native structure of albumin, it is possible that some deamidation of albumin can occur during such treatment at low pH.

Lipidex 1000 (Packard Instruments, Downers Grove, Illinois) is a 10% (w/w) substituted hydroxyalkoxypropyl derivative of Sephadex G-25. Fatty acids can be removed effectively from serum albumin by a single passage through a column of Lipidex 1000 at 37°C (Glatz and Veerkamp, 1983). The procedure is simpler and milder than activated charcoal treatment method and, after lipid removal, albumin is recovered in good yield.

PROTOCOL

1. Pack a column of Lipidex 1000 (100 ml gel; supplied in methanol) in a water-jacketed thermostated column (size: 3.0 × 15 cm) and equilibrate by washing with 1 L water at 37°C at a flow rate of 150 ml/hr.
2. Dissolve serum albumin (0.5 gm) in 5.0 ml water and apply to the Lipidex 1000 column. Elute the albumin with water. Albumin is recovered in the void volume and can be lyophilized. The column can be reused after thorough washing with methanol.

After Lipidex 1000 chromatography, at least 90% of the fatty acids are removed and 90–94% albumin is recovered, indicating that the support is well suited for defatting serum albumins. The fatty acid binding capacity of Lipidex 1000 is found to be 40–50 nmol/ml gel. Phospholipids, cholesterol, and steroids also can be removed by Lipidex 1000 chromatography.

5.2.2.2 Removal of Lipoproteins Using Cab-O-Sil

Cab-O-Sil is a tradename for a hydrated colloidal silica that can be obtained from Research Products International (Elk Grove, Illinois).

Cab-O-Sil has a high capacity for binding serum lipoproteins (Weinstein, 1979). Vance *et al.* (1984) have developed a procedure for the small-scale isolation and characterization of lipoproteins secreted by cultured rat liver hepatocytes. The lipoproteins are first adsorbed onto Cab-O-Sil. The lipid components are extracted from Cab-O-Sil with chloroform/methanol and the apoproteins solubilized in a buffer that contains 2% SDS and 6 M urea. Cab-O-Sil is shown to bind 90–95% of HDL and VLDL. The recovery of lipid components is essentially quantitative. The recovery of the apolipoproteins is only about 60% but with good precision.

PROTOCOL FOR REMOVAL OF LIPOPROTEINS USING CAB-O-SIL

1. Add 10 mg Cab-O-Sil as a freshly prepared slurry in water (50 mg/ml) to 1.0 ml lipoprotein solution.
2. Stir the Cab-O-Sil suspension at room temperature for 15 min.
3. Pellet the Cab-O-Sil by centrifugation at 20,000 g for 15 min. Remove the supernatant liquid with a Pasteur pipette. Lipoproteins are now adsorbed onto the Cab-O-Sil pellet.
4. Extract the lipids from the absorbed lipoproteins by suspending the pellet in 5 ml chloroform/methanol (2:1) using a Teflon stirring rod.
5. After 15 min, centrifuge the tube at 20,000 g for 15 min. Carefully decant the organic solvent into a glass screw-cap tube and transfer any remaining solvent with a Pasteur pipette. Add 1 ml water and vortex the tube, followed by centrifugation at 1500 g for 5 min. Remove the upper phase and wash the lower phase once with 2 ml methanol/water (1:1). Evaporate the organic solvent in a vacuum oven and analyze the lipid samples.
6. Solubilize the apoproteins still adsorbed onto the Cab-O-Sil pellet by adding 1 ml extraction buffer (2% SDS, 50 mM Tris, pH 9.0, 6 M urea, 0.1% EDTA, 0.1% DTT, 0.13% 6-aminocaproic acid, 0.05% glutathione) to the pellet in the centrifuge tube. Place the tube, with a marble on top, in a 95°C water bath for 10 min. Subsequently, cool the tube and centrifuge for 15 min at 20,000 g. The supernatant liquid contains apolipoproteins that can be analyzed by electrophoresis.

5.2.3 Removal of Endotoxins from Protein Solutions

Endotoxin is an integral component of the cell wall of gram-negative bacteria. Because gram-negative bacteria are ubiquitous, bacterial endotoxin frequently contaminates biological preparations. Endotoxin is actually a complex that consists of lipid, carbohydrate, and protein. Highly purified endotoxin does not contain protein and is thus referred to as a lipopolysaccharide (Fig. 5.21).

Endotoxin is known to cause febrile reactions in animals, causing symptoms of high fever, vasodilation, diarrhea, and, in extreme cases, fatal shock. Maintaining a low endotoxin concentration is extremely important for *in vivo* studies. Even nanogram levels of endotoxin have been shown to exert potent biological effects in human, animals, and cell cultures.

A number of known methods remove or reduce the level of endotoxin in fluids, including:

1. ultrafiltration
2. reverse osmosis
3. hydrophobic interaction chromatography
4. anion-exchange chromatography
5. adsorption to activated charcoal
6. affinity chromatography

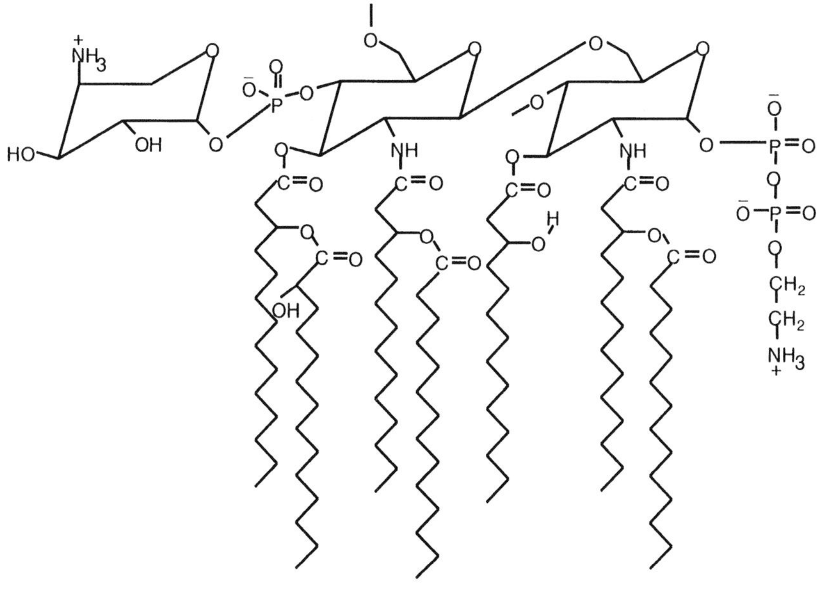

FIGURE 5.21
The proposed chemical structure of the lipid A component of *Salmonella minnesota* lipopolysaccharide. The endotoxic properties are anchored in the lipid A portion of the lipopolysaccharide.

5.2.3.1 Removal of Endotoxin Using Immobilized Polymyxin B

Affinity chromatography can be used successfully to remove endotoxin from biological preparations. Immobilized polymyxin B is a widely used affinity matrix for endotoxin removal. Immobilized histidine also can be used to remove pyrogens.

Polymyxin B is a cationic antibiotic that can neutralize the biological activity of lipopolysaccharide, presumably by binding with the lipid A component (Fig. 5.22). This property of polymyxin B has been exploited to remove endotoxin from solutions by affinity chromatography on an immobilized polymyxin B column (Issekutz, 1983).

PREPARATION OF IMMOBILIZED POLYMYXIN B

Couple polymyxin B sulfate (100 mg; Sigma) to CNBr-activated Sepharose 4B (10 ml gel; see Section 2.2.1.1, Chapter 2, for the preparation of CNBr-activated Sepharose 4B) or periodate-activated Sepharose CL-6B (10 ml gel; see Section 2.2.1.4, Chapter 2, for the preparation of periodate-activated Sepharose CL-6B). Pyrogen-free water is used for all buffer preparations. Immobilized polymyxin B can be stored in 25% ethanol.

REMOVAL OF ENDOTOXIN FROM BOVINE SERUM ALBUMIN USING IMMOBILIZED POLYMYXIN B

1. Pack immobilized polymyxin B (10 ml gel) in a disposable polypropylene column. (For column packing, see Section 4.2.1.1, Chapter 4.)

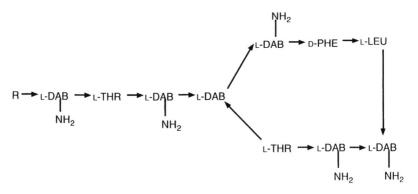

FIGURE 5.22
Schematic diagram of the chemical structure of polymyxin B. Abbreviations: DAB, α,τ-diaminobutyric acid; THR, threonine; PHE, phenylalanine; LEU, leucine; R, 6-methyloctanoic acid.

2. Equilibrate the column by washing with 100 ml endotoxin-free 0.1 M ammonium bicarbonate buffer, pH 8.0.
3. Dissolve BSA (400 mg; 80 E.U./mg) in 2.0 ml endotoxin-free 0.1 M ammonium bicarbonate buffer, pH 8.0, and apply to immobilized polymyxin B column.
4. After the protein solution has completely entered the gel, elute BSA with endotoxin-free 0.1 M ammonium bicarbonate buffer, pH 7.8, at a flow rate of 1.0 ml per hr, while 1.0-ml fractions are collected.
5. Collect the fractions containing BSA and immediately lyophilize. Endotoxin testing of the purified BSA should indicate a level of less than 1 E.U./mg protein. If the actual level is higher than this, recycle the protein through the affinity matrix to remove further quantities of endotoxin.

The immobilized polymyxin B column can be regenerated at least 10 times. For regeneration, wash the column with 50 ml 1% sodium deoxycholate solution followed by 50 ml endotoxin-free water. The column can be stored in 25% ethanol at 4°C.

5.2.3.2 Removal of Endotoxin from Protein Solutions Using Immobilized Histidine

Matsumae *et al.* (1990) have described a method for reducing endotoxin contamination in various solutions using immobilized histidine. Immobilized histidine binds endotoxin with a high affinity over a wide range of pH and temperature, and at low ionic strength. The affinity support is able to remove various types of endotoxin originating from gram-negative bacteria. The concentration of endotoxin can be typically reduced from 1000 ng/ml to less than 0.01 ng/ml in water.

5.2.4 Removal of Proteases

For many biological studies it is essential to completely remove undesirable proteases from biological solutions. These four affinity supports can be used successfully for this purpose:

1. immobilized aprotinin (trasylol)
2. immobilized soybean trypsin inhibitor (STI)
3. immobilized *p*-aminobenzamidine
4. immobilized α_2-macroglobulin, carrier-fixed

5.2.4.1 Immobilized Aprotinin

Aprotinin is a single-chain basic 58-amino-acid polypeptide with a molecular mass of 6512 daltons that has polyvalent inhibitory action on proteinases (for a review, see Hewlett, 1990). It inhibits a variety of enzymes belonging to the family of serine proteases (Table 5.5) by binding to the active site of the enzyme, forming tight complexes. A number of proteases are not inhibited by aprotinin (Table 5.6)

Aprotinin has proved to be a valuable tool in many branches of biomedical research. Aprotinin is very efficient in cell and tissue culture for the inhibition of proteases produced by the cell line or inadequately inactivated during trypsin neutralization. This inhibitory function has great potential to stabilize the cell membrane and cell-secreted proteins against the proteolytic effects of undesirable enzymes. We have found that immobilized aprotinin is highly efficient in removing or purifying trypsin, chymotrypsin, and urokinase.

PROTOCOL FOR THE REMOVAL OF TRYPSIN FROM SOLUTIONS

NOTE: Chromatography is conducted in a cold room at 4°C.

1. Pack 5.0 ml immobilized aprotinin [prepared by periodate activation method (Section 2.2.1.4, Chapter 2); 2 mg aprotinin coupled per ml gel] in a disposable polypropylene minicolumn. (See Section 4.2.1.1, Chapter 4, for a description of how to pack these minicolumns.)
2. Equilibrate the column by washing with 25 ml 50 mM Tris-HCl buffer, pH 8.0, containing 1 M NaCl and 10 mM CaCl$_2$ (binding buffer).
3. Apply the sample containing trypsin..

NOTE: The sample must be dialyzed against the binding buffer for the optimal removal of trypsin. Binding capacity of immobilized aprotinin should be approximately 1.0 mg trypsin per ml gel.

TABLE 5.5
Proteases Inhibited by Aprotinin

Trypsin
Trypsinogen
Chymotrypsin
Plasmin
Kallikrein
Elastase
Urokinase

5.2 Scavenging

TABLE 5.6
Proteases Not Inhibited by Aprotinin

Thrombin
Subtilisin
Papain
Pepsin
Rennin
Carboxypeptidase A, B

Collect 3.0-ml fractions.

4. After the sample application, wash the column with binding buffer until the absorbance at 280 nm of the column effluent is less than 0.01. The breakthrough (unbound) fractions will be free of any trypsin activity, however, these fractions will contain autolytic products of trypsin.
5. Elute the trypsin bound to immobilized aprotinin using 0.1 M acetic acid containing 1 M NaCl and 10 mM CaCl$_2$.
6. Regenerate the column by washing with 25 ml binding buffer. After regeneration, the column is ready for the next run. For storage, the column is washed with 25 ml 0.02% sodium azide and stored at 4°C.

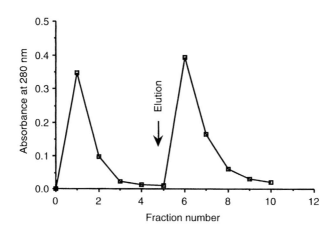

FIGURE 5.23

Removal of trypsin from a solution using an immobilized aprotinin column (5 ml gel). A solution (1 ml) containing trypsin (3 mg) is applied to an immobilized aprotinin column that has been equilibrated with 50 mM Tris–HCl buffer containing 1 M NaCl and 10 mM CaCl$_2$ (binding buffer). The column is washed with 25 ml binding buffer. Bound trypsin is eluted with 0.1 M acetic acid containing 1 M NaCl and 10 mM CaCl$_2$. The chromatography is performed at 4°C.

5.2.4.2 Immobilized Soybean Trypsin Inhibitor

Soybean trypsin inhibitor (STI) first crystallized by Kunitz (1945) is one of several inhibitors found in soybeans. The best known preparation is that of Kunitz (1947). STI consists of a single polypeptide chain (molecular mass 21,500 daltons) cross-linked by two disulfide bridges. It inhibits trypsin mole-for-mole and, to a lesser extent, chymotrypsin. STI also inhibits Factor Xa, plasmin, and plasma kallikrein.

We have used immobilized STI successfully to remove and purify trypsin from biological samples. Immobilized STI has been used to stabilize proteins in crude extracts containing potent proteases (Alhanaty *et al.*, 1979).

PROTOCOL FOR THE REMOVAL OF TRYPSIN FROM SOLUTIONS

NOTE: Chromatography is conducted in a cold room at 4°C.

1. Pack 3.0 ml immobilized STI [prepared by CNBr activation method (Section 2.2.1.1, Chapter 2); 15 mg STI coupled per ml gel] in a disposable polypropylene minicolumn (see Section 4.2.1.1, Chapter 4).
2. Equilibrate the column by washing with 15 ml of 50 mM Tris-HCl, pH 7.2, containing 0.1 M NaCl and 10 mM CaCl$_2$ (binding buffer).
3. Apply the sample containing trypsin.

NOTE: The sample must be dialyzed against the binding buffer for the optimal removal of trypsin. Binding capacity of immobilized STI is found to be 6.0 mg trypsin per ml gel.).

Collect 3.0-ml fractions.

4. After the sample application, wash the column with binding buffer until the absorbance at 280 nm of the column effluent is less than 0.01. The breakthrough fractions will be free of any trypsin activity; however, these fractions will contain autolytic products of trypsin.
5. Elute the trypsin bound to immobilized STI using 0.1 M acetic acid containing 10 mM CaCl$_2$.
6. Regenerate the column by washing with 20 ml binding buffer. The column is ready for the next run. For storage, the column is washed with 20 ml of 0.02% sodium azide and stored at 4°C.

A typical chromatographic profile for the removal of trypsin from a solution using an immobilized STI column is shown in Figure 5.24. As evident from experimental results, enzymatically active trypsin is completely

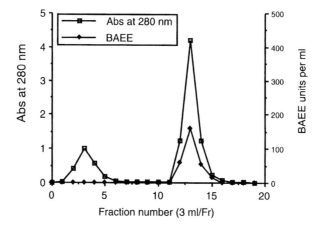

FIGURE 5.24
Removal of trypsin from a solution using an immobilized soybean trypsin inhibitor (STI) column (3 ml gel). A solution (3 ml) containing trypsin (21 mg) is applied to an immobilized STI column that has been equilibrated by washing with 15 ml 50 mM Tris–HCl buffer, pH 7.2, containing 100 mM NaCl and 10 mM CaCl$_2$ (binding buffer). The column is washed with 30 ml binding buffer while 3-ml fractions are collected. Bound trypsin is eluted with 0.1 M acetic acid containing 10 mM CaCl$_2$. Each fraction is assayed for trypsin activity using 10 mM BAEE (Na-benzoyl-L-arginine ethyl ester) as substrate at pH 8.0 and 25°C in a pH stat. The chromatography is conducted at 4°C.

---

bound onto the immobilized STI column and is eluted using 0.1 M acetic acid containing 10 mM CaCl$_2$ (peak 2), whereas autolytic products of trypsin are not retained on the column and elute unretarded (peak 1).

5.2.4.3 Immobilized p-Aminobenzamidine

p-Aminobenzamidine is a competitive inhibitor of trypsin. Immobilized p-aminobenzamidine has been used extensively to purify trypsin and trypsin-like enzymes (e.g., urokinase, enterokinase, or plasminogen activators). We have used immobilized p-aminobenzamidine to remove trypsin from biological preparations. (For, details, see Section 5.1.1.)

5.2.4.4 α$_2$-Macroglobulin, Carrier-Fixed

α$_2$-Macroglobulin is the most universal endoprotease inhibitor known. Boehringer Mannheim Corporation has developed a new method for the removal of endoproteases from biological solutions using "carrier-fixed"

α_2-macroglobulin. The protein has a high affinity for zinc and binds very tightly to a zinc-chelated immobilized iminodiacetic acid support. (See Section 5.1.13 for details on the purification of α_2-macroglobulin using metal chelate affinity chromatography.) Using α_2-macroglobulin fixed to such a metal chelate carrier support creates an excellent affinity adsorbent for the binding and removal of proteases from biological solutions.

5.3 CATALYSIS AND MODIFICATION

Immobilized ligands also may be used as insoluble catalysts or modification reagents to effect a number of biochemical transformations. For instance, immobilized enzymes can function as reusable reagents in the preparation of protein fragments or in the analysis of protein purity. Immobilized reductants can be used to reduce specific disulfide residues in proteins, freeing subunit structure for further analysis or conjugation. Immobilized oxidants can function as mild but efficient mediators in preparing ^{125}I-labeled reagents or in oxidizing specific sugar residues to create formyl coupling groups. Whatever the application, immobilized catalysts and modification reagents can effect the creation of useful biochemical derivatives without contaminating the resulting preparations. In most cases, the immobilized reagent is also reusable after a regeneration step.

5.3.1 Immobilized Enzymes

The technology of immobilized enzymes provides powerful tools for the development of new chemical and biochemical processes. The great potential of immobilized enzymes for biotechnology, medical treatment, analysis, or studies of fundamental biochemistry has been documented. (For reviews, see Gutcho, 1974; Mosbach, 1976, 1987.) The uses of immobilized enzymes are numerous:

1. isolation of labeled peptides
2. isolation of protease inhibitors
3. limited enzymic degradation of proteins
4. isolation of products of enzyme action
5. assaying inhibitors and activators

In this section, the uses of immobilized proteases for the analysis of peptides and proteins and for the preparation of protein fragments are described. The use of immobilized enzymes for peptide and protein analysis has many advantages over the use of soluble enzymes: immobilized enzymes can be reused if desired, no autolysis occurs, backgrounds are lower, the digest is readily removable from the enzyme, high enzyme concentrations may be used, and, finally, a process based on bound enzymes can be automated.

5.3.1.1 Immobilized Trypsin

Trypsin is a widely used protease for amino acid sequence determination of peptides. At pH 7–8, trypsin cleaves the peptide bond at the carboxy terminus of arginine, lysine, and S-2-aminoethylcysteine residues. If the ε-amino group of lysyl and ω-amino group of S-2-aminoethylcysteinyl residues are blocked by acylation, specific hydrolysis at arginine by trypsin can be achieved. When acylation is carried out using maleic anhydride, subsequent cleavage at both lysyl and S-2-aminoethylcysteinyl residues can be achieved by a second treatment with trypsin at pH 7–8 after demaleylation. Thus, it is possible to obtain overlapping peptides using only one protease.

Trypsin hydrolysis at pH 11.0 permits selective cleavage of lysyl residues in the presence of S-2-aminoethylcysteinyl peptide bonds. However, at pH 11.0 the discrimination between arginyl and lysyl bonds is not very precise.

Digestion of Oxidized Insulin B Chain Using Immobilized TPCK-Trypsin

PROTOCOL

1. Dissolve oxidized insulin B chain (1.0 mg) in 1.0 ml 0.1 M ammonium bicarbonate buffer, pH 8.0, and mix with immobilized TPCK-trypsin (0.25 ml gel; see Section 3.3.1, Chapter 3, for the preparation of immobilized TPCK-trypsin).
2. Mix the gel suspension for 18 hr at 37°C.
3. Centrifuge the digest suspension at 2000 rpm for 10 min. Extract the gel pellet twice with 0.5 ml 0.1 M ammonium bicarbonate, pH 8.0. Combine the three supernatants and lyophilize.
4. Dissolve the residue in 1 ml 1.0 M acetic acid and analyze 0.1 ml digest by HPLC using a Waters DeltaPak C_{18} reverse-phase column (15cm × 3.9 mm) with a 5–47.8% gradient of acetonitrile in water, both containing 0.1% TFA, over 15 min at a flow rate of 1 ml/min. Results of HPLC analysis are shown in Figure 5.25.

Digestion of Oxidized Ribonuclease A Using Immobilized TPCK-Trypsin

The protocol for the digestion of oxidized ribonuclease A is the same as that just given for the digestion of oxidized insulin B chain. Results of HPLC analysis of the digest are shown in Figure 5.26.

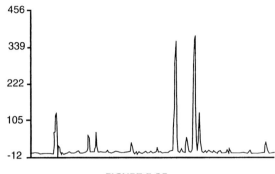

FIGURE 5.25

HPLC analysis and peptide maps of oxidized insulin B chain digested with immobilized TPCK-trypsin. For the experimental details, see the text.

Fragmentation of Human IgM to Fab and Fc5μ Using Immobilized TPCK-Trypsin

Enzymatically generated fragments of immunoglobulins are invaluable tools in many fields of research. Fragmentation becomes particularly important when dealing with immunoglobulins as large as pentameric IgM (M_r, 900,000 daltons). Some common uses of these immunoglobulin fragments are in:

1. histochemistry
2. tumor imaging
3. preparation of conjugates in diagnostics and therapeutics

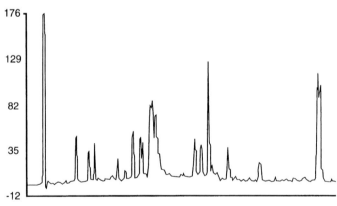

FIGURE 5.26

HPLC analysis and peptide maps of performic acid-oxidized ribonuclease A digested with immobilized TPCK-trypsin. For the experimental details, see the text.

4. anti-antibody production
5. molecular studies
6. genetic engineering
7. antigen binding studies

Studies using histochemistry or tumor imaging require the labeled immunoglobulin to cross capillaries and diffuse through tissue surfaces. When fragments are used instead of the whole immunoglobulin, better labeling occurs and backgrounds are reduced. Use of fragments also provides more rapid circulatory clearance, which is very important in tumor imaging and therapy.

Removal of the Fc fragment from the immunoglobulin allows studies to be carried out on antigen binding sites without interference from effector functions. Immunoglobulin fragments lacking Fc are less susceptible to catabolism by phagocytic cells. Fv, Fab, and F(ab')$_2$ all contain the antigen binding site. Table 5.7 lists the major fragments of IgM with a definition for each one.

Figure 5.27 shows the gross structural differences between human and mouse IgM. Compared with human IgM, mouse IgM disulfides are arranged around the molecule in series rather than in parallel and are attached at different points. These structural differences affect the selective cleavage sites for enzymatic digestion of the immunoglobulins (Fig. 5.28). When human IgM is digested with trypsin at 60°C, Fab (45 kDa) and Fc5µ (350–450 kDa) are generated as depicted in Figure 5.29 (Plaut and Tomasi, 1970; Beale, 1987). Theoretically, 45% Fc5µ and 50% Fab can be obtained from trypsin digestion of human IgM at 60°C. Trypsin at 60°C cleaves human IgM at Arg_{214} and results in the loss of a few peptides and a major glycopeptide from the hinge region.

Mouse IgM digestion with trypsin, however, proceeds to yield only F(ab')$_2$ and Fab fragments, and no Fc5µ. Unfortunately, trypsin digestion of mouse IgM is extremely inefficient and the yield of recovered fragments is very low, making the method impractical.

PROTOCOL FOR THE GENERATION OF FAB AND FC5µ FROM HUMAN IGM

1. Heat a water bath to 60°C.
2. Pack immobilized TPCK-trypsin (2 ml gel; for preparation of immobilized TPCK-trypsin, see Section 3.3.1, Chapter 3) in a disposable polypropylene column (for column packing, see Section 4.2.1.1, Chapter 4) and equilibrate by washing with 10 ml 50 mM Tris-HCl buffer, pH 8.0, containing 150 mM NaCl and 10 mM CaCl$_2$ (Tris-Ca buffer).

TABLE 5.7
Nomenclature for IgM and Its Fragments

Abbreviation	Definition
L	22,000 dalton peptide chain folded into two globular domains designated VL and CL
μ	70,000 dalton peptide chain folded into five globular domains designated VH, Cμ1, Cμ2, Cμ3, and Cμ4
Fab	Proteolytic fragment consisting of L chain and Fd piece
Fd	Piece of μ chain containing VH and Cμ1 domain
F(ab')$_2$	Dimeric disulfide-linked proteolytic fragment consisting of two L chains and two Fd' pieces
Fd'	Piece of μ chain containing VH, Cμ1 and Cμ2 domains and hence having the cysteine residue that forms the inter μ chain disulfide bridge; few residues of the Cμ3 domain may also be present
Fab*	Proteolytic fragment consisting of L chain and Fd* piece
Fd*	Piece of a chain containing VH and Cμ1 domains and most of the Cμ2 domain but, lacking the cysteine residue that forms the inter μ chain disulfide bridge
(Fc)5μ	Disulfide-linked pentameric fragment containing five Fc monomers
Fc monomers	Dimer of two Fc pieces
Fc	Piece of μ chain containing the Cμ3 and Cμ4 domains and a few residues of the Cμ2 domain, including the cysteine residue that forms the inter μ chain disulfide bridge
Fc*	Piece of μ chain containing Cμ3 and Cμ4 domains and most or all of the Cμ2 domain
Fc'	Piece of μ chain containing the Cμ4 domain and most or all of the Cμ3 domain but none of the Cμ2 domain

3. Heat the equilibrated and sealed immobilized TPCK-trypsin column, the human IgM solution (1.0 mg/ml in Tris-Ca buffer), and 15 ml Tris-Ca buffer for 2 min at 60°C in the water bath.
4. Apply the human IgM solution (1.0 ml) to the column. After the sample has entered the top disk of the column, apply 0.2 ml Tris-Ca buffer to allow the sample to completely enter the gel bed.
5. Cap the bottom and top of the column, and incubate the column at 60°C for 30 min in the water bath.
6. After incubation, remove the top and bottom caps of the column, and elute the trypsin digestion products from the column using 4.0 ml Tris-Ca buffer.
7. Separate and concentrate the Fc5μ and Fab fragments from the digest using Centricon-30 and Centricon-100 microconcentrators (Amicon,

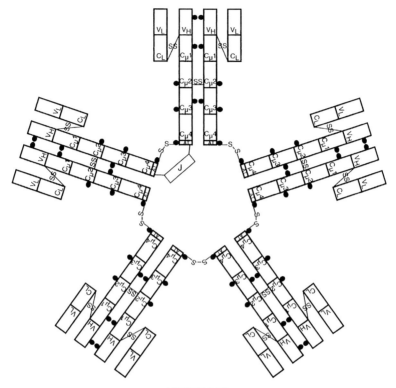

FIGURE 5.27

The pentameric IgM molecule contains 10 antigen binding sites in five Ig units bonded together by disulfide bridges and the J chain. Carbohydrate chains are present at numerous domain locations. •, Oligosacharides; AT=20-amino acid terminus of secretory IgM; J=Possible location of J chain.

Beverly, Massachusetts). Since Centricon-30 has a 30,000 MW cutoff, Fab fragments (45 kDa) will be retained during microconcentration. The Centricon-100 (MW cutoff of 100,000) will retain Fc5μ (350–450 kDa) during microconcentration. A diagram of this procedure is presented in Figure 5.30.

5.3.1.2 Immobilized Anhydrotrypsin

Anhydrotrypsin is a catalytically inactive derivative of trypsin in which the active-site residue Ser_{195} has been converted to a dehydroalanine residue.

Anhydrotrypsin selectively binds, under weakly acidic conditions, peptides that have arginine, lysine, and *S*-aminoethylcysteine at the C ter-

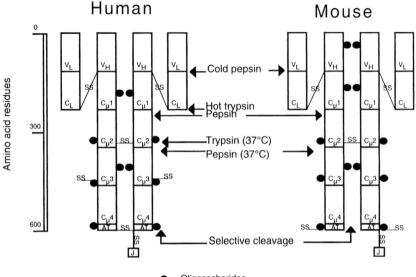

FIGURE 5.28

Cleavage sites for enzymatic digestion of human and mouse IgM. Note that only one Ig unit of the pentameric IgM molecule is illustrated. Depending on the temperature of the digest, different amino acid residues are exposed to the immobilized enzymes.

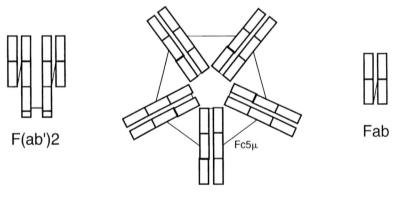

FIGURE 5.29

The digestion of human IgM with trypsin at 60°C yields F(ab')$_2$ (150 kDa), Fab (45 kDa), and Fc5μ (350–450 kDa) fragments.

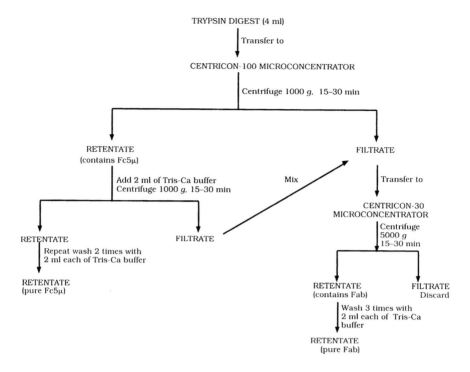

FIGURE 5.30

Schematic diagram of separation and concentration of the Fc5μ and Fab fragments generated by the digestion of human IgM with immobilized TPCK-trypsin.

minus. It does not bind peptides that have these amino acids at positions other than the C terminus. The affinity of these peptides for anhydrotrypsin is much greater than their affinity for trypsin.

Anhydrotrypsin has been shown to exhibit a much higher affinity for many product-type ligands (i.e., ligands that are similar to products made as a result of trypsin digestion), all of which contain an arginine residue with a free carboxyl group, than for their cognate substrate-type ligands (i.e., ligands that are similar to substrates that can be digested by trypsin) with the carboxyl group blocked. For example, the dissociation constant (K_d) for benzoyl-L-arginine is 0.2 mM and the K_d for benzoyl-L-arginine amide is 4.6 mM at pH 8.2 and 25°C. Native trypsin cannot discriminate between these two groups of ligands (i.e., K_d = 5.9 and 3.3 mM, respectively, for benzoyl arginine and its amide under the same conditions). The affinity of anhydrotrypsin for benzoyl-L-arginine is stronger at slightly acidic pH than at slightly alkaline pH, such as 8.0, which is optimum for the catalytic activity of native trypsin. The pK_a value of the ionized form of anhydrotrypsin responsible for the interaction with benzoyl-L-arginine is estimated to be 7.6.

Immobilized anhydrotrypsin is a versatile biospecific affinity support for the fractionation of tryptic digests of polypeptides and for the isolation of naturally occurring peptides that correspond to the products generated by the action of trypsin-like proteases. A protocol is given here for the isolation of peptides generated by tryptic digestion of ribonuclease, using immobilized anhydrotrypsin.

Isolation of Tryptic Peptides of Ribonuclease Using Immobilized Anhydrotrypsin

DIGESTION OF RIBONUCLEASE WITH TPCK-TRYPSIN

1. Carboxymethylate ribonuclease A (25 mg) according to the method described by Crestfield *et al.* (1963).
2. Dialyze the carboxymethylated protein against 0.1 M ammonium bicarbonate buffer, pH 8.0, and lyophilize.
3. Dissolve the carboxymethylated ribonuclease in 1.0 ml 0.1 M ammonium bicarbonate buffer, pH 8.0, and digest with TPCK-trypsin at 37°C for 12 hr with an enzyme/substrate ratio of 1:100 (w/w).

CHROMATOGRAPHY OF TRYPTIC PEPTIDES OF RIBONUCLEASE ON IMMOBILIZED ANHYDROTRYPSIN COLUMN

1. Pack immobilized anhydrotrypsin (1.0 ml gel; prepared as described by Ishi *et al.*, 1983) in a disposable polypropylene column (for column packing, see Section 4.2.1.1, Chapter 4).
2. Equilibrate the column by washing with 20 ml 50 mM sodium acetate buffer, pH 5.0, containing 20 mM $CaCl_2$ (binding buffer).
3. Mix the tryptic digest of ribonuclease (prepared as just described, 25 µl; 20 nmol) with 0.5 ml binding buffer and apply to the immobilized anhydrotrypsin column.
4. Wash the column with 20 ml binding buffer.
5. Elute the tryptic peptides bound to the column with 0.1 M formic acid.
6. Measure the absorbance of all fractions at 280 nm. Peptides in all fractions also can be detected by performing a Micro BCA protein assay test (Pierce).

A typical chromatographic profile of the tryptic peptides of ribonuclease on immobilized anhydrotrypsin is shown in Figure 5.31.

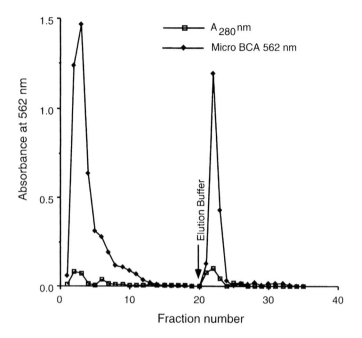

FIGURE 5.31

Affinity chromatography of the tryptic peptides of ribonuclease on an immobilized anhydrotrypsin column (1.0 ml gel). For experimental details, see the text.

5.3.1.3 Immobilized Chymotrypsin

Chymotrypsin preferentially catalyzes the hydrolysis of peptide bonds involving L-isomers of tyrosine, phenylalanine, and tryptophan. It also readily acts on amides and esters of susceptible amino acids.

Digestion of Oxidized Insulin B Chain Using Immobilized Chymotrypsin

PROTOCOL

1. Immobilize chymotrypsin to periodate-activated Sepharose CL-6B (see Section 3.3.1, Chapter 3) by the same protocol used for trypsin.
2. Dissolve oxidized insulin B chain (1.0 mg) in 1.0 ml 0.1 M ammonium bicarbonate, pH 8.0, and mix with the immobilized chymotrypsin (0.25 ml gel).
3. Mix the gel suspension for 18 hr at 37°C.

4. Centrifuge the digest suspension at 2000 rpm for 10 min. Extract the gel pellet twice with 0.5 ml 0.1 M ammonium bicarbonate, pH 8.0. Combine the three supernatants and lyophilize.
5. Dissolve the residue in 1 ml 1.0 M acetic acid. Analyze a 0.1-ml sample of the digest on an HPLC using a Waters DeltaPak C_{18} reverse-phase column (15 cm × 3.9 mm) with a 5–47.8% gradient of acetonitrile in water, both containing 0.1% TFA, over 15 min at a flow rate of 1.0 ml/min. Results of HPLC analysis are shown in Figure 5.32.

Digestion of Oxidized Ribonuclease A Using Immobilized Chymotrypsin

The protocol for the digestion of oxidized ribonuclease A is the same as that given for the digestion of oxidized insulin B chain. Results of HPLC analysis of the digest are shown in Figure 5.33.

5.3.1.4 Immobilized Anhydrochymotrypsin

Anhydrochymotrypsin is a catalytically inert derivative of chymotrypsin in which the active-site residue Ser_{195} is converted to a dehydroalanine residue. Although anhydrochymotrypsin is catalytically inactive, it has a high affinity for product-type peptides of α-chymotrypsin. Anhydrochymotrypsin specifically binds peptides that have L-tryptophan, L-tyrosine and L-phenylalanine at their C termini. It has no affinity for peptides that have other amino acids at their C termini. In addition, anhydrochymotrypsin will not bind peptides containing these three amino acids at positions other than the C terminus.

Immobilized anhydrochymotrypsin is a useful biospecific affinity support for the fractionation of chymotryptic digests of polypeptides and for the isolation of naturally occurring peptides that correspond to the prod-

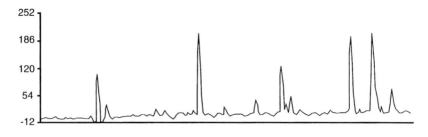

FIGURE 5.32

HPLC analysis and peptide maps of oxidized insulin B chain digested with immobilized chymotrypsin. For experimental details, see the text.

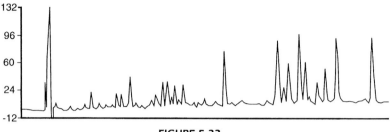

FIGURE 5.33

HPLC analysis and peptide maps of performic acid oxidized ribonuclease digested with immobilized chymotrypsin. For experimental details, see the text.

ucts generated by the action of chymotrypsin-like proteases. Immobilized anhydrochymotrypsin is also useful for the purification of chymotrypsin inhibitors.

Isolation of Chymotryptic peptides of Neurotensin Using Immobilized Anhydrochymotrypsin

DIGESTION OF NEUROTENSIN WITH TLCK-α-CHYMOTRYPSIN

1. Digest neurotensin with TLCK–α-chymotrypsin for 6 hr at 37°C in 0.1 M sodium bicarbonate, pH 8.1 (with an enzyme:substrate ratio of 1:100).
2. Add diisopropylfluorophosphate (Caution: highly toxic; use appropriate protective equipment) to a final concentration of 1 mM to stop the reaction. Adjust the pH of the sample to 5.0 with dilute acetic acid.

CHROMATOGRAPHY OF CHYMOTRYPTIC PEPTIDES OF NEUROTENSIN ON IMMOBILIZED ANHYDROCHYMOTRYPSIN

1. Pack immobilized anhydrochymotrypsin (5 ml; Pierce) in a disposable polypropylene column (for column packing, see Section 4.2.1.1, Chapter 4).
2. Equilibrate the column by washing with 50 ml 50 mM sodium acetate, pH 5.0, containing 20 mM CaCl$_2$ (binding buffer).
3. Apply the chymotryptic digest of neurotensin (100 nmol; see preceding protocol) to the immobilized anhydrochymotrypsin column.
4. Wash the column with 20 ml binding buffer.
5. Elute the bound chymotryptic fragments with a linear pH gradient of 20 ml of 0.1 M sodium formate, pH 4.5, and 20 ml of 0.1 M formic acid, pH 2.5, followed by elution with 20 ml of 0.1 M formic acid.

A typical chromatographic profile of chymotryptic peptides of neurotensin eluted from an immobilized anhydrochymotrypsin column is shown in Figure 5.34.

5.3.1.5 Immobilized Pepsin

Pepsin is an acidic endoprotease and is the principal proteolytic enzyme of vertebrate gastric juice. It hydrolyzes proteins and peptides favorably adjacent to aromatic and dicarboxylic L-amino acid residues, preferentially phenylalanine and leucine, but not at valine, alanine, or glycine. Pepsin also has an esterase activity.

Preparation of F(ab')$_2$ from IgG Using Immobilized Pepsin

Since the pioneering work by Porter on digestion of rabbit IgG by papain, proteolytic enzymes have been used widely in the determination of immunoglobulin structure and in the investigation of relationships between structure and function. Proteolytic studies have been applied in a number of areas, providing evidence for the division of immunoglobulins into sub-

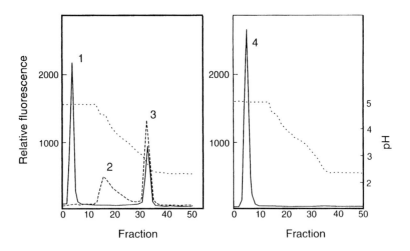

FIGURE 5.34

Affinity chromatography of a chymotrypsin digest of neurotensin on an immobilized anhydrochymotrypsin column (5.0 ml gel). For experimental details, see the text. (1: Ile–Leu; 2: Pyr–Leu–Tyr; 3: Glu–Asn–Lys–Pro–Arg–Arg–Pro–Tyr; 4: intact neurotensin: Pyr–Leu–-Tyr–Glu–Asn–Lys–Pro–Arg–Arg–Pro–Tyr–Ile–Leu). Fluorescamine, solid line; Tyrosine, dashed line; pH, dotted line.

5.3 Catalysis and Modification

classes and supporting the domain concept of immunoglobulin structure and function. Immunoglobulin fragments have been used to investigate the functions of isolated immunoglobulin domains and to raise domain-specific antisera. In addition, limited proteolysis has been used to isolate fragments from lymphocyte membrane immunoglobulins for immunochemical characterization and for the preparation of antisera.

Most specific enzymatic cleavages observed with IgG molecules occur in the unfolded regions of the peptide chains between domains; the hinge region is particularly susceptible to several enzymes. Cleavage at this point produces F(ab')$_2$ or Fab fragments and the Fc piece that may remain intact or be further degraded (Lamoyi and Nisonoff, 1983; Lamoyi, 1986).

The antigen-binding ability of the immunoglobulin molecule resides in the F(ab')$_2$ or Fab fragment; many studies of the properties of antibodies have involved the isolation of these fragments. In immunohistochemical studies, the small monovalent Fab fragments penetrate tissues better than intact immunoglobulin G and avoid the nonspecific binding by Fc receptors on tissues cells that normally present major problems in localization studies. For cancer therapy purposes, Fab and F(ab')$_2$ fragments, which lack the Fc region responsible for most effector functions, are believed to be preferable to intact antibody and have a higher diffusion capacity in the extravascular space because of their lower molecular weight. Therefore, good methods for the routine preparation of Fab and F(ab')$_2$ fragments are needed.

Pepsin normally cleaves at the C-terminal side chain of the inter-heavy-chain disulfide bonds and produces a bivalent antigen binding fragment, F(ab')$_2$, with a molecular mass of about 105,000 daltons (Fig. 5.35). The Fc portion undergoes extensive degradation.

PROTOCOL

1. Wash 0.25 ml (settled gel) of immobilized pepsin with 4×1 ml 20 mM sodium acetate buffer, pH 4.5 (digestion buffer), and suspend in 1.0 ml digestion buffer. (For preparation of immobilized pepsin, see Section 3.3.2, Chapter 3.)
2. Dissolve 10 mg human IgG solution in 1.0 ml digestion buffer and add to the immobilized pepsin gel suspension.
3. Mix the gel suspension in a shaker at 37°C for 2 hr. The gel should be in suspension during mixing, but mix gently to avoid frothing and denaturating the protein.
4. After 2 hr, add 10 mM Tris-HCl buffer, pH 7.5 (3.0 ml), to the gel suspension, mix, and centrifuge at 2000 g for 5 min.

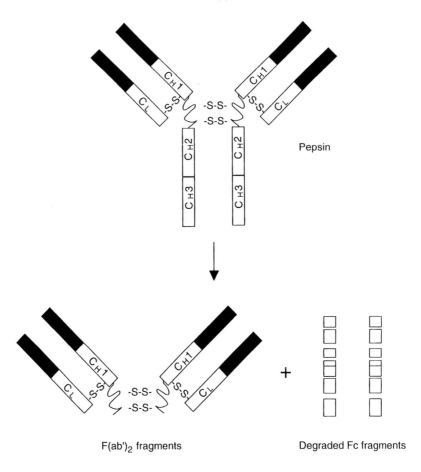

FIGURE 5.35

Pepsin cleavage of IgG occurs at the C-terminal side chain of the inter-heavy chain disulfide bonds, resulting in a bivalent antigen binding fragment, F(ab')$_2$, with a molecular mass of about 105,000 daltons.

---

5. Apply the supernatant liquid to an immobilized protein A column (5 ml gel) that was equilibrated previously by washing with 20 ml 10 mM Tris-HCl buffer, pH 7.5.
6. After the sample has entered the gel bed, wash the column with 15 ml 10 mM Tris-HCl buffer, pH 7.5, while 5.0-ml fractions are collected. The protein eluted is F(ab')$_2$.
7. Elute Fc and undigested IgG bound to the immobilized protein A column with 0.1 M glycine-HCl buffer, pH 2.8. The protein A gel may be reused.

A typical affinity chromatographic separation of F(ab')$_2$ from Fc and undigested IgG using an immobilized protein A column is shown in Figure 5.36.

Preparation of IgM Fragments Using Immobilized Pepsin

The proteolytic fragmentation of IgM with pepsin proceeds differently from that of IgG (described earlier). These changes are related to the differences in structure between the two molecules. In the pentameric IgM immunoglobulin structure, the μ chains are folded into multiple globular domains, whereas the hinge region typical in Ig molecules is replaced by the Cμ2 domain. IgM lacks the proline-rich amino acid sequence normally found in IgG molecules in this region; this makes the IgM molecule more susceptible to enzymatic cleavage.

Pepsin normally cleaves IgM at the site between domains Cμ2 and Cμ3 (see Fig. 5.28, Section 5.3.1.1). Using immobilized pepsin, it is possible to produce F(ab')$_2$, Fab, and Fv fragments from both human and mouse IgM (Rudders *et al.*, 1980; Beale, 1987). In addition, there is frequently cleav-

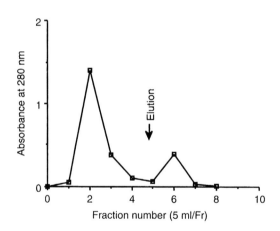

FIGURE 5.36

Affinity chromatography of an immobilized pepsin digest of human IgG separated on an immobilized protein A column (5.0 ml gel). Following the digestion of human IgG with immobilized pepsin, the digest (2.0 ml) is applied to the immobilized protein A column. The column is then washed with 15 ml 10 mM Tris–HCl buffer, pH 7.5, to recover the (Fab')$_2$ fragments. Bound Fc and undigested IgG are eluted with 0.1 M glycine–HCl buffer, pH 2.8. Fractions of 5.0 ml are collected. Chromatography is performed at room temperature.

age between domains to give Fc' and other fragments. Often, there is extensive degradation of the Fc portion.

In our own laboratiess, digestion of mouse IgM with immobilized pepsin produced F(ab')$_2$ in the early stages and Fab fragments during longer digestions, but no Fc5μ. Digestion of human IgM produced Fab, F(ab')$_2$, and Fab*, but no Fc5μ. These results indicate that the Fc portion is easily degraded into smaller peptides.

Using human monoclonal IgM, an immobilized pepsin digestion at 37°C was found to produce F(ab')$_2$ and Fv fragments. When this digestion was extended to 24 hrs, the immunoglobulin was degraded almost completely into small peptides and Fv fragments. The yield of Fv was about 27% using this procedure.

Thus, the dominant fragments produced by enzymatic digestion with immobilized pepsin (as shown in Fig. 5.37) are F(ab')$_2$, Fab, and Fv. Other portions of the molecule are extensively degraded. IgM F(ab')$_2$ fragments have a molecular mass ranging from 140 to 170 kDa, while Fab fragments are about 45 kDa.

PROTOCOL

1. Prepare immobilized pepsin according to the protocol described in Section 3.3.2, Chapter 3. Pack a 2-ml minicolumn with the immobilized enzyme according to the protocol in Section 4.2.1.1, Chapter 4.
2. Equilibrate the immobilized pepsin column with 6 ml digestion buffer (0.1 M sodium acetate, 0.15 M NaCl, pH 4.5).
3. Prepare an IgM solution at a concentration of 1 mg/ml in digestion buffer.
4. Preincubate separately for 5 min at 37°C the immobilized pepsin column, 1 ml IgM solution from Step 3, and about 15 ml digestion buffer.

FIGURE 5.37

Pepsin digestion of human and mouse IgM produces F(ab')$_2$, Fab, and Fv fragments.

5. Add 1 ml IgM solution to the immobilized pepsin column. Allow the solution to enter until the flow stops. Add a 0.2-ml portion of digestion buffer to the top of the column to move the IgM solution past the top disk and fully into the gel bed.
6. Incubate at 37°C for 1.5 hr.
7. Place the column in a clean 16 × 120 mm test tube. Add 0.2 ml digestion buffer to the column to elute some of the buffer from the bottom column tip. Discard the eluent.
8. Transfer to a clean tube, and apply 2 × 1 ml digestion buffer to collect the main fragment peak eluting from the column. Save the eluent.
9. Again transfer to a clean tube, and add 2 × 1 ml digestion buffer to elute any remaining protein fragments. Save the eluent.
10. Separate and concentrate the fragments produced from the digest (the eluents from Step 8 and 9) using Centricon-30 microconcentrators (Amicon, Beverly, Massachusetts). Since the Centricon-30 has a 30,000 MW cutoff membrane, the major fragments of IgM digestion [F(ab')$_2$ and Fab] will be retained during microconcentration, whereas the Fv fragment (MW 25,000) and smaller peptides will pass through.

5.3.1.6 Immobilized Papain

Papain is a sulfhydryl protease from *Carica papaya* latex. Papain has wide specificity and degrades most protein substrates. Preferential cleavage is at Arg-X, Lys-X, and Phe-X. Papain also has an esterase activity.

Since the pioneering work by Porter on papain digestion of rabbit IgG, papain has been used widely for the generation of Fab and Fc fragments of IgG (Coulter and Harris, 1983). When IgG is digested with papain in the presence of a reducing agent, one or more peptide bonds in the hinge region can be split, producing two types of fragments, Fab and Fc. The Fc fragment (so called because it readily crystallizes) is the heavy domain portion beneath the cleavage point (Fig. 5.38). The two univalent Fab fragments produced are capable of binding antigen and contain the VH and CH_1 domains of the heavy chain and the complete light chain. Fab fragments do not exhibit antigen precipitating activities. Therefore, it is advantageous to use Fab fragments in procedures in which it is desirable to bind antigen to antibody in solution without cross-linking or precipitation.

When Fc fragments are of interest, papain is the enzyme of choice since it yields an Fc fragment with an approximate molecular mass of 50,000 daltons. Papain-produced Fc fragments have been used to determine the IgG subclass specificity of Fc receptors induced in cells by infection with human cytomegalovirus.

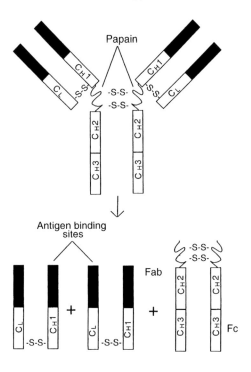

FIGURE 5.38

IgG digestion with papain in the presence of a reducing agent cleaves one or more peptide bonds in the hinge region, producing two types of fragments, Fab and Fc. The Fc fragment remains essentially intact and can be removed from the Fab by affinity chromatography on immobilized protein A.

Crystalline papain is often used for the digestion of IgG, however, it is subject to autodigestion. Some researchers use mercuripapain, which is less subject to autodigestion than crystalline papain. However, both enzymes require iodoacetamide to terminate digestion. Immobilized papain is not subject to autodigestion and does not require iodoacetamide for termination. Simply removing the digest from contact with immobilized papain gel by centrifugation or elution will stop the digestion. This makes digestion with immobilized papain a gentle and controllable reaction. Immobilized papain can be used for the efficient generation of Fab and Fc from IgG, avoiding contamination of final preparations with enzyme.

Preparation of Fab and Fc from IgG Using Immobilized Papain

PROTOCOL

1. Wash immobilized papain (0.5 ml settled gel; for preparation of immobilized papain, see Section 3.3.3, Chapter 3) with 4×2 ml 20 mM NaH$_2$PO$_4$, 20 mM cysteine-HCl, 10 mM EDTA-tetra sodium, pH 6.2 (digestion buffer), and suspend in 1.0 ml digestion buffer.
2. Dissolve 10 mg human IgG solution in 1.0 ml digestion buffer and add to the immobilized papain gel suspension.
3. Mix the gel suspension in a shaker at 37°C for 5 hr. Maintain the gel in suspension during mixing.
4. After 5 hr, add 10 mM Tris-HCl buffer, pH 7.5 (3.0 ml), to the gel suspension, mix, and centrifuge at 2000 g for 5 min.
5. Apply the supernatant liquid to an immobilized protein A column (5 ml gel; for preparation, see Section 3.4.4.1, Chapter 3) that was equilibrated previously by washing with 20 ml 10 mM Tris-HCl buffer, pH 7.5.
6. After the sample has entered the gel bed, wash the column with 15 ml 10 mM Tris-HCl buffer, pH 7.5, while 5.0-ml fractions are collected. The protein eluted is Fab.
7. Elute Fc and undigested IgG bound to the immobilized protein A column with 0.1 M glycine-HCl buffer, pH 2.8.

A typical affinity chromatographic separation of Fab from Fc and undigested IgG using an immobilized protein A column is shown in Figure 5.39.

5.3.1.7 Immobilized Bromelain

The proteolytic enzyme found in the juice of pineapple stem is called stem bromelain and the enzyme in the fruit is called fruit bromelain. Bromelain belongs to the same family of sulfhydryl proteases as papain. The enzyme generates fragments of IgG similar to those generated by papain (Milenic *et al.*, 1989). The bromelain-generated fragments of monoclonal antibodies (Mab) typically retain 100% of their immunoreactivity. Differences were observed between the bromelain- and papain-generated fragments when compared *in vivo*. Fragmentation of Mab B72.3 with bromelain has yielded a superior bivalent fragment for radioimmunolocalization.

Immobilized bromelain is prepared by the same method as that described for the preparation of immobilized papain (Section 3.3.3, Chapter 3).

Fragmentation of IgG with immobilized bromelain can be achieved by the same protocol described for immobilized papain.

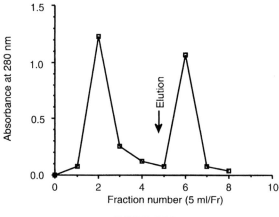

FIGURE 5.39

Affinity chromatography of an immobilized papain digest of human IgG on an immobilized protein A column (5.0 ml gel). Following the digestion of human IgG with immobilized papain, the digest (2.0 ml) is applied to an immobilized protein A column. The column is then washed with 15 ml 10 mM Tris–HCl buffer, pH 7.5, to recover the Fab fragments. The bound Fc and any undigested IgG is eluted with 0.1 M glycine–HCl buffer, pH 2.8. Fractions of 5.0 ml are collected. Chromatography is performed at room temperature.

5.3.1.8 Immobilized Carboxypeptidase Y

Carboxypeptidase Y (CPY) is an exopeptidase found in yeast. CPY has a broad specificity and catalyzes the release of L-amino acids, including proline, from the C terminus of a polypeptide chain. CPY also has esterase activity. Immobilized CPY has been used successfully in total hydrolysis, sequencing, and synthesis of peptides.

PROTOCOL

1. Carry out the deblocking reaction at pH 8.5 with the aid of a pH-stat. Dissolve Z-Asn-Phe-OEt (20 mg) in 160 ml of 25% methanol.
2. Start the reaction by addition of immobilized CPY (Pierce; 2.0 ml immobilized CPY, 25 ATEE units, washed with water and 25% methanol, and suction-dried to a moist cake) to the peptide ester solution.
3. When base consumption ceases, filter the reaction mixture.
4. Acidify the filtrate (pH 2–3), and concentrate the solution with a rotary evaporator at 40°C until precipitation occurs.
5. The crystallization of the peptide derivative is accomplished with ethanol/water. Add water to a 20% ethanolic solution of the peptide until the solution becomes turbid. Cool the solution to complete crystallization.

If desired, the immobilized CPY can be washed with 20% ethanol and stored in 50% glycerol containing 0.02% sodium azide at −20°C.

5.3.1.9 Immobilized *Staphylococcus aureus* V8 Protease

Staphylococcus aureus V8 protease (SAP) exhibits a high degree of specificity for the glutamate residues of peptides and proteins. Because of the narrow range of substrate specificity, SAP has become an enzyme of choice for peptide mapping and sequencing studies, much like trypsin. Its application in limited proteolysis studies of proteins is also well documented. The high substrate specificity of SAP makes it a potentially useful reagent for peptide bond synthesis as well. Recently the proteosynthetic activity of SAP has been demonstrated.

When a protease is used in a cleavage or semisynthetic reaction, it is generally necessary to inactivate the enzyme after the reaction and before attempting to isolate products. Since SAP is a very stable enzyme and no specific inhibitors of SAP are available, it is difficult to inactivate the enzyme. Thus the successful application of SAP in cleavage and semisynthetic reactions will be facilitated greatly by the development of a method for the rapid removal of the protease after the completion of cleavage or condensation reactions. Immobilized SAP has been used for condensation of α-globin fragments as well as for the cleavage of α-globins and performic acid-oxidized RNase A. Even after immobilization, the overall specificity of SAP for glutamate residues is preserved.

PROTOCOL FOR THE DIGESTION OF PERFORMIC ACID-OXIDIZED RNASE A USING IMMOBILIZED SAP

1. Dissolve performic acid-oxidized RNase A (1.0 mg; Sigma) in 1.0 ml 20 mM potassium phosphate buffer, pH 8.0, and mix with immobilized SAP (0.2 ml gel; see Section 3.3.4, Chapter 3, for the preparation of immobilized SAP).
2. Mix the gel suspension for 20 hr at 37°C.
3. Centrifuge the reaction mixture at 2000 rpm for 10 min and subject an aliquot of the supernatant liquid (0.1 ml) to reverse-phase HPLC on a Whatman Partisil ODS-3 column (4.6 × 250 mm) using a linear 5–50% gradient of acetonitrile containing 0.1% TFA. Maintain a flow rate of 1.0 ml/min and monitor the absorbance of the effluent at 210 nm.

For comparison, the digestion of performic acid-oxidized RNase A (1 mg/ml in 20 mM potassium phosphate buffer, pH 8.0) with soluble SAP can be carried out also using a (soluble) enzyme: substrate ratio of 1:250 (w/w).

The SAP peptide maps of the digest of performic acid-oxidized RNase A, generated using the soluble and the immobilized enzyme, are virtually identical, except for minor quantitative differences. Figure 5.40 shows the HPLC separation of an immobilized SAP digest of this protein. Digestions of performic acid-oxidized RNase also can be carried out with the soluble and immobilized enzymes at pH 4.0, the second pH optimum of SAP. The HPLC patterns are essentially identical to those obtained at pH 8.0. The maps of performic acid-oxidized RNase A are closely similar to those of RCM-RNase A reported by McWherter *et al.* (1984). The hydrolysis of RCM-RNase A by SAP during limited digestion occurs predominantly on the carboxyl side of glutamate residues. Based on the amino acid sequence of RNase A, six glutamoyl peptides are expected; five of the expected peptides chromatograph as distinct peaks during reverse-phase HPLC, and the sixth glutamoyl peptide emerges with the solvent front. The remarkable similarity between the peptide maps of oxidized RNase A digested with the soluble and immobilized enzyme clearly demonstrates that the overall specificity of SAP remains unperturbed, even after the immobilization process.

5.3.2 Immobilized Disulfide Reductants

Immobilized reducing agents have significant advantages over soluble reducing agents such as β-mercaptoethanol, cysteine, dithiothreitol (DTT), and dithioerythritol (DTE). Insoluble reducing agents can be used readily to reduce biological disulfides of all types (Gorecki and Patchornik, 1973, 1975). In this section, our studies on the use of immobilized lipoamide and

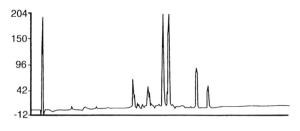

FIGURE 5.40

HPLC analysis and peptide maps of performic acid-oxidized ribonuclease A digested with immobilized *Staphylococcus aureus* V8 protease. For experimental details, see the text.

immobilized *N*-acetylhomocysteine in reducing peptides and proteins are described. (See Section 3.8, Chapter 3, for the preparation of these supports.) Various manufacturers provide such immobilized reductants (Pierce, Calbiochem, BioRad).

5.3.2.1 Reduction of Peptides Using Immobilized Reductants

GENERAL PROTOCOL

NOTE: For optimal reduction of peptides, the following steps should be performed at room temperature.

1. Pack an immobilized reductant gel (2 ml settled gel; for preparation of these types of supports, see Section 3.8, Chapter 3) in a disposable polypropylene column (for column packing, see Section 4.2.1.1, Chapter 4) and wash with 5 ml of 0.1 M sodium phosphate buffer, pH 8.0, containing 1 mM EDTA (equilibration buffer 1).
2. Reduce the sulfhydryl-containing column and activate by applying 10 ml fresh 10 mM DTT (15.4 mg DTT dissolved in 10 ml equilibration buffer 1).
3. Wash the column with 20 ml equilibration buffer 1 to remove free DTT.
4. Apply 1.0 ml peptide solution (dissolved in equilibration buffer 1) to the column.
5. After the sample has completely entered the gel bed, wash the column with 9 ml equilibration buffer 1, while 1.0-ml fractions are collected.
6. Monitor the elution of reduced peptide from the column by measuring the absorbance at 280 nm (if peptide absorbs at this wavelength) as well as by performing an Ellman's assay (Section 4.1.9, Chapter 4) for sulfhydryl groups using a small aliquot (10–20 µl) of each collected fraction.
7. Regenerate the sulfhydryl-containing support by following Steps 2 and 3. We have regenerated and reused such columns at least 5 times without any significant decrease in the reductive capacity.
8. Store the column in 0.02% sodium azide at 4°C.

5.3.2.2 Reduction of Proteins Using Immobilized Reductants

GENERAL PROTOCOL

NOTE: For optimal reduction of proteins, the following steps must be performed at room temperature.

1. Pack an immobilized reductant gel (2 ml settled gel; for preparation of these supports, see Section 3.8, Chapter 3) in a disposable polypropylene column (for column packing, see Section 4.2.1.1, Chapter 4) and wash with 5 ml 0.1 M sodium phosphate buffer, pH 8.0, containing 1 mM EDTA (equilibration buffer 1).
2. Reduce the sulfhydryl-containing column and activate by applying 10 ml fresh 10 mM DTT (15.4 mg DTT dissolved in 10 ml equilibration buffer 1).
3. Wash the column with 10 ml equilibration buffer 1 and 10 ml 0.1 M sodium phosphate buffer, pH 8.0, containing 1 mM EDTA and 6 M guanidine (equilibration buffer 2) to remove free DTT.
4. Apply 1.0 ml protein solution (dissolved in equilibration buffer 2) to the column.
5. After the sample has completely entered the gel bed, incubate the column at room temperature for 1 hr.
6. Wash the column with 9 ml equilibration buffer 2 while 2-ml fractions are collected.
7. Monitor elution of reduced protein from the column by measuring the absorbance at 280 nm as well as by performing an Ellman's assay for sulfhydryl groups (Section 4.1.9, Chapter 4) using a small aliquot (50–100 µl) of each collected fraction.
8. Regenerate the sulfhydryl-containing column by following Steps 2 and 3. We have regenerated and reused such columns at least 5 times without any significant decrease in the reductive capacity.
9. Store the column in 0.02% sodium azide at 4°C.

5.3.2.3 Reduction of Oxidized Glutathione Using Immobilized Reductants

Immobilized reductants are very efficient in reducing oxidized glutathione (Fig. 5.41). When 2, 4, and 8 µmol oxidized glutathione are applied to three individual immobilized lipoic acid columns, the recovery of reduced glutathione is very good (95%) as determined by the liberation of sulfhydryl groups using an Ellman's assay. The reductive capacity of each column (2 ml gel), when fully reduced with DTT, is about 30-40 µmol sulfyhydryl groups.

As evident from Figure 5.41, the amount of sulfhydryl groups released is proportional to the amount of oxidized glutathione applied to the sulfhydryl-containing column. The reduction of oxidized glutathione is very rapid. When 2 µmol oxidized glutathione are applied and incubated in the column for 0 hr and 1 hr, the chromatographic elution profile and the

5.3 Catalysis and Modification

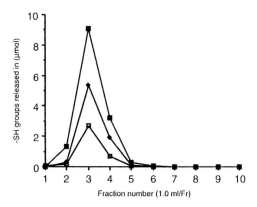

FIGURE 5.41
Three sulfhydryl agarose columns (2.0 ml gel in each column) are activated with a DTT solution and washed with equilibration buffer 1 as described in the text (Section 5.3.2.1). Then, 2 (□), 4 (◆), and 8 (■) μmol of oxidized glutathione are applied to the three columns. Each column is then eluted with 9 ml equilibration buffer 1 while 1-ml fractions are collected. The –SH content in each fraction is determined by performing an Ellman's assay using 20 μl of each fraction.

amount of sulfhydryl groups liberated at both time intervals are essentially same (3.4 μmol at 0 hr and 3.3 μmol at 1 hr).

5.3.2.4 Reduction of Insulin to Insulin A and B Chains using Immobilized Reductants

Immobilized reductants such as lipoic acid and *N*-acetylhomocysteine are very effective in reducing insulin to insulin A and B chains (Fig. 5.42). The recovery of insulin A and B chains is good (100%) as determined by measuring the absorbance at 280 nm. Approximately 0.91 μmol sulfhydryl groups (theoretical yield is 1.1 μmol) are liberated when 1.0 mg insulin is passed through the sulfhydryl-containing column.

We have observed that if insulin is applied and incubated for 15 min in the column, the recovery of insulin A and B chains is very poor (57%). Most likely, insulin A and B chains react with sulfhydryl groups of the sulfhydryl-containing column to form disulfides. Fractions eluted from the column can be analyzed by Tricine-PAGE under nonreducing conditions. Two bands corresponding to insulin A and B chains are typically observed.

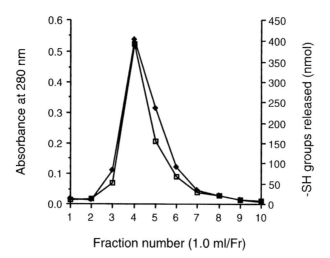

FIGURE 5.42
Bovine insulin (1 mg in 1 ml buffer) is applied to a freshly activated sulfhydryl agarose column. The column is then eluted with 9 ml equilibration buffer 1 (see text) while 1-ml fractions are collected. The absorbance at 280 nm of all fractions is measured (□). The –SH content of all fractions is determined by Ellman's assay procedure using 100 μl of each fraction (♦).

5.3.2.5 Reduction of Lysozyme and Ribonuclease A Using Immobilized Reductants

Immobilized reductants also can be used to reduce disulfide bonds in proteins. Results obtained using lysozyme and ribonuclease A as model proteins are shown in Figures 5.43 and 5.44. Approximately 2.4 μmol sulfhydryl groups are released when 5 mg lysozyme is reduced by sulfhydryl agarose, whereas approximately 1.8 μmol sulfhydryl groups are liberated on reduction of 5 mg ribonuclease A. Recovery of proteins is excellent.

In the absence of guanidine (6 M), lysozyme and ribonuclease are not reduced. These results show that, for a disulfide-reducing column to effectively reduce globular proteins, denaturing agents such as guanidine hydrochloride or urea must be used to unfold the three-dimensional structure so the disulfide bonds are accessible to the immobilized sulfhydryl groups.

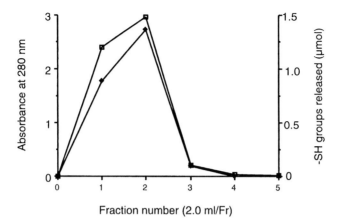

FIGURE 5.43

Lysozyme (5 mg dissolved in 1 ml equilibration buffer 2, see text) is applied to a freshly activated sulfhydryl agarose column and incubated in the column for 1 hr. The column is then eluted with 9 ml equilibration buffer 2 while 2-ml fractions are collected. The absorbance of all fractions is measured at 280 nm (□). The –SH content of all fractions is determined by the Ellman's assay procedure using 100 μl of each fraction (♦).

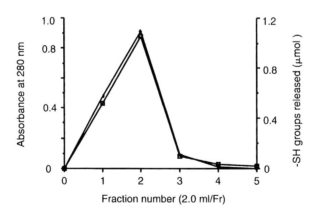

FIGURE 5.44

Ribonuclease A (5 mg dissolved in 1 ml of equilibration buffer 2; see text) is applied to a freshly activated sulfhydryl agarose column and incubated for 1 hr. The column is then eluted with 9 ml equilibration buffer 2 while 2-ml fractions are collected. The absorbance of all fractions is measured at 280 nm (□). The –SH content of all fractions is determined by the Ellman's assay procedure using 100 μl of each fraction (♦).

5.4 ANALYTICAL APPLICATIONS OF AFFINITY LIGANDS

Immobilized ligands may be used in many different formats to separate and measure specific analytes. A chromatographic separation of a substance that interacts with an affinity ligand can be the first step in quantifying its concentration in solution. Minicolumns of specific affinity gels can be used to separate the bound from the unbound fraction and to determine simply the percentage of the bound component or its concentration in the original solution. Systems involving high performance liquid affinity chromatography (HPLAC) can be constructed to speed this process and completely automate the measurement. Traditional immunoassays can be designed around immobilized affinity ligands, using such diverse formats as plates, beads, gels, and plastic devices. Antibodies, antigens, or biospecific affinity ligands may be used as the solid-phase adsorbant in such assays. Newer and faster flow-through immunoassay systems also can be designed using minicolumns of gels, beads, or custom devices that hold ligands immobilized on membranes. In addition, the electronic detection of affinity interactions has facilitated the development of powerful affinity electrode and biosensor systems that not only measure analyte concentrations, but also determine the quantitative characteristics of biospecific interactions.

The use of immobilized affinity ligands in analytical systems is being revolutionized. The specific interaction between a ligand and a target molecule is providing a powerful new tool for the development of a broad range of assay systems capable of measuring virtually any substance. The following sections discuss some of the most significant examples of how this technology is being used to solve analytical problems.

5.4.1 Use of Immobilized Aminophenyl Boronic Acid to Quantify Glycated Hemoglobin in Red Cell Hemolysates

Summary and Explanation of the Test

Diabetes mellitus is a clinical condition resulting from an absolute or relative deficiency of insulin. This deficiency manifests itself as, among other things, an inability to control the blood glucose level. The aim of insulin therapy in diabetes is to maintain a constant normal or near normal level of glucose in the blood. One of the most difficult tasks in managing diabetic patients is accurately assessing the degree of blood glucose control. Even patients with mild disease may show large fluctuations in blood glucose level. However, the measurement of glycated hemoglobin (GlyHb) has been accepted as a reliable indicator of the time-averaged mean blood glucose level in the diabetic patient.

5.4 Analytical Applications of Affinity Ligands

Glycated hemoglobins are modified by the attachment of glucose or other carbohydrate to one or more of the amine groups found in the hemoglobin molecule. Hemoglobin forms a stable derivative with glucose in a two-stage nonenzymatic process (Allen *et al.*, 1981). Glycation occurs gradually throughout the biological life span of the red blood cell (approximately 120 days). Within each red cell, the amount of GlyHb formed depends on the average concentration of glucose in the blood during the life of that red blood cell. Therefore, the measurement of GlyHb represents the time-averaged level of glucose in the blood.

A number of methods are currently in use for measuring GlyHb, some of which are often tedious and unreliable. Some procedures also can be very sensitive to small changes in temperature or pH. To avoid difficulties inherent in many currently used methodologies, an affinity chromatography method has been developed to provide a specific interaction with glycated hemoglobin (Mallia *et al.*, 1981; Klenk *et al.*, 1982).

Affinity chromatography has been found to have several advantages over other methods for the determination of GlyHb.

1. The method detects all glycated hemoglobins, not just hemoglobin molecules that are charge modified (HbA1).
2. The affinity method is unaffected by the presence of abnormal hemoglobins. Hemoglobin variants such as F, S, Wayne, and C do not interfere since the specificity of this method is for the *cis*-diols on glucose-modified hemoglobin.
3. This method is unaffected by moderate fluctuations in the assay temperature or the pH of the buffers.
4. Further, the affinity method is unaffected by the presence of "labile" glycated hemoglobins in the sample. Labile glycated hemoglobins are the unstable Schiff base intermediates (the aldimines) that form rapidly as the first step in the two-step nonenzymatic formation of glycated hemoglobin.
5. The affinity method is unaffected by the presence of carbamylated hemoglobin. Carbamylated hemoglobin occurs in uremic patients and has been shown to interfere in the ion-exchange method.
6. No special patient preparation is necessary. Reliable GlyHb results can be obtained from randomly drawn blood samples. The GlyHb result is unaffected by recent meals or physical exercise.

Clinical Significance

GlyHb testing provides unique information. GlyHb assay values (expressed as the percentage of total hemoglobin that is glycated) are useful in monitoring the effectiveness of diet/insulin or oral agent therapy in the normalization of serum glucose concentrations. The assay is useful to assess and track the degree of chronic glucose control in diabetics. GlyHb values in or near the normal range can be used to identify insulin-dependent diabetics who are at increased risk from hypoglycemia. Another major use of GlyHb testing is as an educational and motivational tool toward better glycemic control for all diabetic patients.

Interpretation of the Results

Elevated levels of GlyHb are observed in uncontrolled diabetics. Diabetics who control their disease may have GlyHb levels approaching the upper end of the normal range. The normal range for this test using an immobilized boronate affinity method (Pierce) is 3.8–6.0% glycated hemoglobin. Levels higher than these indicate poor glycemic control. At this time, there is little agreement on the GlyHb levels that should constitute "good" control of the disease. The best approach is to monitor the diabetic patient 2–4 times a year and to compare the most recent GlyHb level to the previous levels. Because the glucose modification of hemoglobin occurs slowly and depends on the circulating level of blood glucose, the GlyHb result represents a time-averaged blood glucose level. There is a time lag of 3–4 weeks before changes in the mean blood glucose level are reflected in the GlyHb result. Using the percentage GlyHb value, an assessment of long term control can be ascertained in the diabetic individual. Values lower than the normal range are of uncertain clinical significance.

Principle of the Procedure

Studies have delineated the chemical steps involved in the posttranslational nonenzymatic glycation of hemoglobin in red blood cells (Bunn *et al.*, 1978). As shown in Figure 5.45, the aldehyde group of glucose forms a reversible Schiff base linkage with the N-terminus of the α- and β- chains and certain lysine residues of the hemoglobin molecule. This labile adduct (aldimine) then undergoes an Amadori rearrangement to the stable ketoamine, commonly referred to as GlyHb. The coplanar *cis*-diol groups of glycated hemoglobins can then interact with immobilized aminophenyl boronic acid to form a reversible five-membered ring complex. This complex can be dissociated with sorbitol.

GlyHb can be separated and quantified using this immobilized affinity ligand. In practice, the sample containing a mixture of glycated and nonglycated hemoglobin variants is applied to the prepacked columns. Glycated hemoglobins will be retained by the column and nonglycated entities will pass through unretained. After appropriate washing, the bound glycated he-

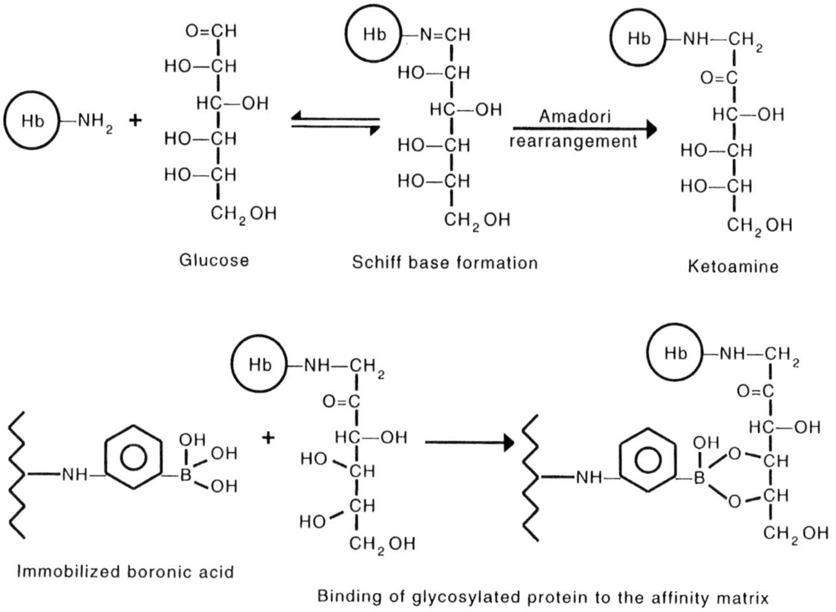

FIGURE 5.45

Glucose modifies hemoglobin nonenzymatically to produce an intermediate labile Schiff base that can undergo an Amadori rearrangement to form a stable ketoamine structure. Immobilized boronic acid can specifically bind the adjacent diols on the sugar modification, producing a method of separation and measurement of the glycated species.

moglobins are eluted with a sorbitol-containing buffer. Relative ratios of glycated and nonglycated fractions are quantified by measuring the absorbance of the hemoglobin at 414 nm.

PROTOCOL FOR THE MEASUREMENT OF GLYHB USING IMMOBILIZED AMINOPHENYL BORONIC ACID

a. Preparation of Packed Cell Hemolysate

Draw venous blood aseptically in standard blood collection tubes containing EDTA or fluoride as the anticoagulant (purple or grey tops). Mix all specimens well, prior to preparation of hemolysate, by this procedure.

1. Centrifuge the anticoagulated whole blood at 1000 g for 10 min at room temperature to separate plasma from red blood cells (RBCs).
2. Remove the plasma and white blood cells (buffy coat) by aspiration using a glass or plastic transfer (Pasteur) pipette.
3. Wash the packed cells by adding 9 parts saline (0.9% sodium chloride) to 1 part RBCs. Mix by inversion, centrifuge at 1000 g for min, and again aspirate the top layer.
4. Repeat the saline wash twice.
5. Add an equal volume of 0.9% NaCl saline to the washed packed red blood cells and mix gently.
6. Prepare a packed cell hemolysate by mixing 1 part washed RBCs with 10 parts deionized water. For example, add 50 µl washed saline-suspended RBCs to 0.5 ml water in a 13 × 100 mm test tube. Gently swirl or invert to mix the hemolysate and let stand for 5 min.

The clear packed cell hemolysate may be used immediately or it may be stored frozen at $-20°C$ for up to one month or at -70 to $-90°C$ for up to 6 months, pending analysis.

b. Affinity Separation of GlyHb on Immobilized Aminophenyl Boronic Acid

1. Pack a minicolumn containing 0.5 ml immobilized aminophenyl boronic acid (GlycoGel; Pierce) using the procedure outlined in Section 4.2.1.1, Chapter 4.
2. Equilibrate the column by adding at least 2.0 ml of 0.2 M ammonium acetate, 50 mM $MgCl_2$, pH 8.0 (binding buffer).
3. Suspend the equilibrated column in a 16 mm × 125 mm test tube that has been marked NB (not bound or nonglycated hemoglobin fraction) and add 50 µl hemolysate just prepared to the top disk of the column.
4. Allow the sample to pass through the top disk and follow with 0.5 ml binding buffer to insure complete transfer of the sample through the disk and into the gel. The column flow will automatically stop when the liquid level reaches the top disk.
5. Add 5.0 ml binding buffer to the column. When the NB fraction has been collected, the total volume will be 5.55 ml (0.05 ml sample, 0.5 ml chase, and 5.0 ml wash). Mix the test tube contents well. The final buffer/NB fraction is stable for at least 2 hr at room temperature.
6. Transfer the column to a clean 16 mm × 125 mm test tube that has been marked B (bound or glycated hemoglobin fraction). Add 3.0 ml 50 mM Tris, 0.1 M sorbitol, pH 8.5 (elution buffer) and collect the entire frac-

tion. Mix the test tube contents well. The final buffer/B fraction is stable for at least 2 hr at room temperature.

7. Set a spectrophotometer to 414 nm and zero the absorbance with deionized water. Read and record the absorbance of the NB fraction and the B fraction.

c. Calculation of the Percentage of Glycated Hemoglobin

Using the following equation, caluclate the percentage of GlyHb.

$$\% \text{ GlyHb} = \frac{(100)(3.0)(A_{414} \text{ bound fraction})}{(3.0)(A_{414} \text{ bound fraction}) + (5.55)(A_{414} \text{ not-bound fraction})}$$

In this equation, 3.0 represents the volume (ml) of the bound glycated fraction and 5.55 represents the volume of the not-bound nonglycated fraction.

A typical elution profile for the separation of glycated from nonglycated hemoglobin using immobilized aminophenyl boronic acid is shown in Figure 5.46.

To maintain reproducibility in this separation and, thus, precision in the measurement of GlyHb, the manufacture of the affinity gel must be controlled finely. The level of ligand immobilized on the support directly con-

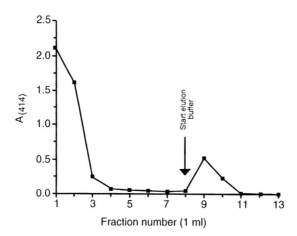

FIGURE 5.46

The elution profile for the affinity separation of glycated hemoglobin (eluted, bound fraction) from nonglycated hemoglobin (first peak, not bound) on a 0.5-ml column of immobilized boronic acid.

tributes to the efficiency of GlyHb binding and, ultimately, to the measured percentage. In addition, the column packing process must insure that the same amount of gel is used in each column so column-to-column or run-to-run precision is acceptable. All these parameters are important considerations for the successful use of affinity chromatography in any analytical test.

The measurement of glycated hemoglobin was one of the first clinical tests developed using a separation step based on affinity chromatography. The power and specificity of immobilized ligand systems, whether designed around a chromatographic approach or as a batch separation procedure, can be applied to the measurement of almost any diagnostically important parameter.

5.4.2 Use of HPLAC for Automated Affinity Assays

High performance liquid affinity chromatography is a technique that exploits the speed and separation efficiency of HPLC and couples with it the specificity of affinity chromatography (Nilsson and Mosbach, 1981). HPLC systems usually make use of small beaded matrices with tight pore size distribution characteristics. The uniformity and small diameter of the beads provides optimal separations and extremely high peak resolution capability. Common support materials usually consist of silica derivatives or polymeric matrices that can withstand the high pressures and high flow rates required in these separations without physical collapse. Analytical HPLC systems now provide complete control over the entire chromatography process, from sample application, buffer changes, flow rate, and back pressure control to integrated peak detection with computer-assisted data handling.

When immobilized affinity ligands are incorporated into such systems, the speed and resolution abilities inherent in the HPLC environment provide the potential for rapid separation and analysis of specific target analytes (Larsson et al., 1983). The control and reproducibility provided by computerized HPLC systems makes such affinity separations extraordinarily precise, so the purification technique can be extended easily to sensitive analytical analysis.

Silica is probably the most popular matrix for HPLC separations because of its excellent dimensional stability under high pressures and flow rates. It can also withstand solvent changes with no changes in bed volume and no degradation of the porous structure of the beads. To prepare affinity supports from silica based matrices, the chemical derivatization process usually involves using a functional silane derivative that can be modified subsequently to provide the necessary coupling groups for ligand immobilization. The silane coating also serves to eliminate the nonspecific binding character of the silica support. A common choice for this coating step is the use of the epoxy compound glycidoxypropyl silane (see Section 1.2.3, Chapter 1).

The coated silica support then contains terminal epoxides that can be used directly for the immobilization of ligands containing amine, hydroxyl, or sulfhydryl groups. The epoxide also may be opened to provide two adjacent hydroxyls for further derivatization. Oxidation of these hydroxyls with sodium periodate yields a formyl functionality suitable for coupling amine-containing ligands by reductive amination. The hydroxyls also may be activated by a number of chemistries suitable for hydroxylic matrices. Some of the more common methods of activation used with silica supports include the reactive sulfonate chemistries using the tresyl or tosyl activation procedures.

One of the main drawbacks of using silica is its instability in alkaline pH environments (pH >8.0). To be safe from hydrolysis and decomposition of the matrix, it is recommended that the working pH not exceed 7.0. Compared with other chromatographic support materials, this limitation places severe restrictions on the coupling chemistry and on the operation of the resulting column. For this reason, some commercially available polymeric supports may prove to be more practical for the preparation and use of HPLAC matrices. In particular, the small particle size varieties of HEMA and TSK supports, both prepared from methacrylate-based monomers, may be better choices than silica in an affinity environment. These polymeric supports may be derivatized easily with no constraints on pH or the solvents used during coupling or chromatography.

Other types of affinity chromatography support materials also can be used in conjunction with an HPLC instrument. The use of larger bead sizes with such matrices as Toyopearl, HEMA, Trisacryl, and even the soft gel supports such as cross-linked agarose work well if care is taken to control back pressure and flow rate. The only deficiency with larger beads comes in the loss of resolution when multiple peaks are being separated on elution. However, if only a single bound fraction is being analyzed, the use of larger beads will make the handling and immobilization steps much easier than the use of very small particles that cannot be readily filtered or manipulated in bulk.

Applications of Analytical HPLAC

Virtually any ligand that can be immobilized to a chromatography support can be used also in an HPLAC format. Dozens of examples can be found in the literature ranging from low-molecular-weight biospecific ligands, dyes, and chelators to various immunospecific antibodies and other protein molecules. (For a review, see Larsson *et al.*, 1983.) The true power of HPLAC, however, is not merely in its purification potential, but in the precision and resolution of the resulting separations. In this regard, HPLAC is a very powerful tool for automating analytical determinations.

For instance, an immobilized affinity ligand can be used in an HPLAC setup to measure the concentration of a target molecule in a sample solution. A specific example of this approach is the determination of mouse IgG concentration using immobilized protein A (Hammen *et al.*, 1988). A sample containing immunoglobulin is injected into the affinity column and washed free of any unbound protein. The bound IgG fraction is then eluted with acetic acid. The quantity of bound immunoglobulin (as analyzed by the integrated area under the peak) is proportional to the initial concentration of IgG in the sample (Fig. 5.47). Using a small 1-ml column of the affinity matrix, a complete separation including regeneration takes only 10 min. Calibrating the instrument with several standards of known IgG

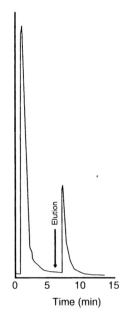

FIGURE 5.47
The separation of mouse IgG on an HPLAC support consisting of a 5 cm × 5 mm column of immobilized protein A.

concentration allows rapid determination for unknown samples. Since column performance is reproducible over dozens of cycles, such a system creates one of the quickest methods for measuring mouse IgG concentrations in complex solutions. This procedure can be applied to measuring immunoglobulin concentrations in serum, cell culture broth, or ascites fluid with similar results.

HPLAC also can provide affinity separations of multiple binding components in a mixture through a process called differential elution. In this technique, a complex mixture containing more than one substance capable of binding an affinity support is applied to the column and allowed to bind. The noninteracting substances are washed off the column, and an elution buffer is applied that can selectively elute the bound components based on their relative affinity for the immobilized ligand. Usually such elution buffers are rather mild compared with those that may be chosen to elute a bound substance in a traditional affinity purification scheme. The result is that peaks are separated during the elution step, from the weakest interacting substance to the strongest interacting one. The trick is to formulate the elution buffer so good resolution is obtained without severe peak broadening or tailing caused by insufficient elution buffer strength to break the affinity interactions quickly.

In some cases, the binding buffer itself can serve as the differential elution buffer. By choosing a binding buffer that provides only a weak affinity interaction, so interacting substances are retarded and not tightly bound as they pass through, several components in a complex mixture may be separated into peaks.

Using this technique, immobilized concanavalin A can be used to separate molecules with carbohydrate components that interact to varying degrees with the lectin. Figure 5.48 shows the separation of three glycosides in less than 30 min using only a single buffer.

Another way to separate several interacting substances in an HPLAC system is to use the relatively new technique called weak affinity chromatography (Ohlson *et al.,* 1989). In the weak affinity approach, the affinity ligand is chosen purposely to have a low affinity binding constant for its complementary target molecule(s). For instance, a particular monoclonal antibody might be selected as a ligand because of its weak binding character toward the antigen. A properly constructed weak affinity chromatography support will have powerful separation capabilities for substances with binding site similarities. In addition, a relatively weak affinity ligand will provide milder elution conditions under which to purify labile substances.

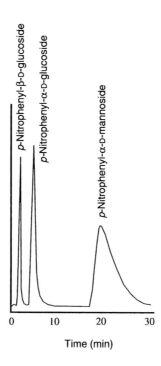

FIGURE 5.48

The differential separation of three glycosides using an HPLAC support consisting of a 10 cm × 5 mm column of immobilized concanavalin A. The isocratic separation was performed at 1 ml/min using the buffer 0.025 M Tris, 0.25 M NaCl, 0.5 mM CaCl$_2$, 0.5 mM MgCl$_2$, pH 6.8.

5.4.3 Immunoassay Techniques Using Immobilized Affinity Ligands

An immunoassay involves the detection or measurement of an analyte using the specific interaction between an antibody molecule and an antigen. [For an excellent text, see Tijssen (1985).] Typically, either the antibody or the antigen is tagged with a detectable component that can be used to identify or quantify the analyte in solution. In most cases, the analyte is either the antibody or the antigen molecule, and the immune complex that forms as the result of their interaction creates the foundation for the assay. The detection system may consist of a radiolabel, a fluorescent tag, or an enzyme conjugate.

The immobilization methods described throughout this book, especially in Section 3.4 (Chapter 3), are all useful in producing viable immunoassay systems. However, our intent in this discussion is not to review immunoassay techniques in general. Many other books give excellent insight on how to design and develop assays based on antibody–antigen interactions. Rather, this section is dedicated to some new immunoassay concepts in which immobilized affinity ligands are used in unique flow-through systems to measure analytes of interest. The following examples illustrate how affinity systems based on immunoassay principles are being used currently to solve analytical problems.

5.4.3.1 Enzyme Immunoassays Using Flow-Through Affinity Chromatography Columns (Immunography)

Enzyme linked immunoassay (EIA) systems were first developed as alternatives to radioimmunoassays (RIA). In an EIA, the detection is done using an enzymatic label in place of a radioisotope, eliminating problems with safety, expense, and short shelf life inherent in an RIA. The concept behind both assays is identical; only the detection reagents differ.

There are two basic types of EIAs, homogeneous and heterogeneous. The homogeneous enzyme immunoassay uses no solid-phase adsorbant and commonly is used to quantify low molecular weight molecules such as digoxin, amphetamines, and other drugs. In this assay system, an enzyme-labeled antigen (e.g., a drug) is added to a sample containing an unknown concentration of the same antigen. An antibody specific to the antigen is then added and the antibody and antigen bind in a way that inhibits the enzymatic activity of the label. Comparison with the response of known concentrations of standards produces a quantifiable result.

Heterogeneous EIAs involve the separation of the unbound antibody or antigen in a liquid phase from a solid phase to which the antigen–antibody complex is bound. These systems are commonly called enzyme-linked immunosorbent assays (ELISA), since an immobilized ligand attached to an insoluble matrix is used in the separation step. The solid phase can be agarose, cellulose, silicone-rubber, glass, or plastic in various configurations, for example, polystyrene beads or microtiter plates. The assay systems described in this section are exclusively heterogeneous.

A major type of heterogeneous assay is the sandwich assay. In this case, a specific antibody is immobilized on the solid phase. A sample containing the antigen is then added and allowed to bind. In the direct ELISA detection method, after sufficient incubation time and washing, a labeled primary antibody (with specificity to the antigen) is added to the solid phase and allowed to incubate. The unbound antibody label is washed away and the remaining solid phase contains the antigen "sandwiched" between the antibody adsorbed to the solid phase and the labeled antibody. Enzymatic substrate is then added; the resulting colorimetric reaction is proportional to the amount of antigen in the test solution.

In the indirect ELISA sandwich assay, an unlabeled primary antibody is first bound to the captured antigen. After incubation and washing, a labeled secondary antibody is added and given time to bind. The secondary antibody is from a different species than the primary antibody and is raised against the immunoglobulin class of the species from which the primary antibody originated. After a final wash, enzymatic substrate is added and the resulting colorimetric reaction is proportional to the amount of antigen present. The indirect method usually has increased detectability over the direct method. To increase sensitivity further, additional bridging steps can be incorporated in which another unconjugated antibody can be added after the primary antibody and before the labeled antibody.

All ELISA sandwich assays make use of an immobilized ligand, usually an antigen or antibody, to specifically interact with the analyte of interest. The choice of solid phase to be employed in the ELISA ultimately governs the development of the test protocol. When wells or tubes function as the solid phase, assays are performed by adding the reactants directly to the vessels. Separation is accomplished after each incubation step by removing the solution through aspiration and subsequently washing the surfaces with buffer. The use of other supports such as beads, sticks, or disks still requires the use of a reaction vessel, but in these cases the adsorbant is usually physically removed from the surrounding solution to be effectively washed.

However varied these assay techniques may seem, most involve one common parameter: the batchwise loading of reactants onto an insoluble support. Those supports that consist of macroscopic nonporous polymers are rather limited in such batch methods because of the characteristically small surface area available for immobilization and interaction with the surrounding reaction medium. Diffusion characteristics at the surface of these supports limit the speed at which an assay can be performed. In some cases, overnight incubations have become necessary because of the slow reaction kinetics at the surface of the adsorbant. Although the use of porous supports such as agarose has circumvented the problem of surface area and capacity somewhat, the common batchwise loading technique still results in problems because of the difficulties associated with separating and washing polydisperse supports with small particle size. Tedious centrifugation steps are often required to wash the support free of reactants.

To overcome the problems of batch manipulation and slow diffusion rates at the surface of an adsorbant, we have combined the efficiency of affinity chromatography with the analytical sensitivity of enzyme-linked immunoassays. We call this combination immunography. Designed around a minicolumn to hold the solid phase during an assay, the immunography approach creates a dynamic flowing immunoassay system with high capacity and rapid flow characteristics. The result is a significant decrease in incubation times and a dramatic increase in the overall speed of an assay.

The reactions involved in the immunography approach are not unlike those seen in typical heterogeneous enzyme assays, except that they are accomplished using a chromatography system. As illustrated in Figure 5.49, a sample is applied to an affinity column to which a ligand is attached that has specificity for the analyte of interest. The agarose column quickly binds the analyte and removes it from solution. Next, an antibody–enzyme conjugate is applied that has specificity for the bound analyte. The conjugate complexes with the available antigenic sites and the excess is washed off with buffer. Finally, a substrate solution is added that will develop in response to the bound antibody–enzyme conjugate. After an incubation

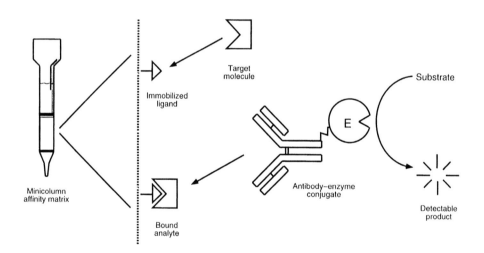

FIGURE 5.49

The "immunography" approach to an immunoassay involves the use of a minicolumn packed with the appropriate affinity matrix. The affinity ligand may consist of a biospecific molecule directed against the desired target analyte, or it may be an immunoaffinity matrix using an antibody. The flow-through characteristics of the affinity column provide enhanced reaction rates and thus decrease the time of an assay. All the manipulations of ELISA techniques, adding specific antibody or antibody–enzyme conjugate, are done chromatographically. The substrate solution is added to the column last and after a brief incubation period, the substrate color is eluted and quantified.

period, the developed color is eluted from the column and its absorbance determined. Various indirect and bridging techniques can be built from this basic protocol to increase sensitivity and detectability of the analyte.

Figure 5.50 shows the design of the minicolumn format. This type of minicolumn is common to small-scale purification methods (see Chapter 4). Here we use the same format, only in an immunoassay procedure. The gel bed is packed in the space between two porous disks. These disks hold the gel in place and regulate the flow of buffer through the matrix. When a solution is applied to the column and flows through to the level of the top disk, the force of capillary action in the disk automatically stops the flow. A significant result of this behavior is that the amount of solution applied to the column will always be the same as that eluted, making quantitative separations possible.

Using this configuration, the gel itself becomes part of a dynamic system, serving not only as immunoadsorbent, but also as reaction chamber for the binding of antibody–enzyme conjugate and incubation chamber for substrate color development.

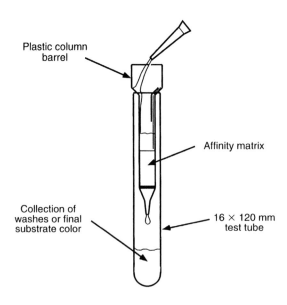

FIGURE 5.50

The minicolumns used for flow-through ELISA assays consist of small plastic devices that can conveniently hold the 0.5–1.0 ml of affinity gel required. They come with top and bottom caps that seal the columns for storage, and top and bottom disks to hold and bracket the gel bed.

5.4 Analytical Applications of Affinity Ligands

Such affinity-based systems can use an antigen, an antibody, or a biospecific affinity molecule that can capture the analyte of interest as the immobilized ligand. A low molecular weight affinity ligand could be used instead of an immobilized antibody and result in significant cost savings in assay development. This type of affinity ligand need not have absolute specificity for the analyte, as long as the specific interaction between the antibody–enzyme conjugate and the bound analyte is preserved without nonspecific binding.

We have developed two examples of affinity-based ELISA involving the measurement of human C-reactive protein (CRP) and human fibronectin in serum or plasma. The following protocols describe the preparation and use of the CRP immunography assay system. The fibronectin assay is performed identically, but immobilized gelatin is used as the affinity chromatography matrix.

Measurement of Human CRP Using Immunography

Following tissue injury, inflammation, infection, and malignant transformation, an increase in the number of so-called acute phase proteins occurs in plasma. One of the first of these proteins to be identified is C-reactive protein, so named because it precipitates pneumococcal somatic C polysaccharide. Clinical measurement of this protein is useful in monitoring the activity of many disease states that cause inflammatory or necrosing disorders. Postoperative CRP monitoring is an effective and sensitive test for the development of sepsis.

To determine human CRP using an affinity chromatography mediated ELISA, immobilized phosphoethanolamine (PEA) is used as a specific ligand for the analyte. CRP will bind with good affinity to the immobilized PEA; subsequent use of a specific antibody–enzyme conjugate directed against CRP will quantify its concentration in serum or plasma.

The materials required for this procedure are listed here.

1. *Immobilized phosphoethanolamine* PEA is immobilized on Sepharose Cl-6B using the periodate oxidation/reductive amination coupling procedure (Section 2.2.1.4, Chapter 2). PEA is reacted at a level of 4 mg per ml gel. The affinity matrix is then packed into minicolumns (Section 4.2.1.1, Chapter 4) at a volume of 0.5 ml per column.

2. *Assay/wash buffer* The assay/wash buffer is used for column equilibration, sample loading, antibody–enzyme conjugate application, and all washing steps. The buffer is 0.02 M Tris, 0.15 M NaCl, 0.01 M $CaCl_2$, 0.1% BSA, 0.02% sodium azide, pH 8.0.

3. *Substrate buffer*   The substrate buffer is used to dilute the enzyme substrate and to wash through the final substrate color development for collection in test tubes. The buffer consists of 1 M diethanolamine, 1 mM MgCl$_2$, pH 9.8.

4. *Anti-human CRP–/ /–alkaline phosphatase conjugate*   This antibody–enzyme conjugate is used to detect and measure bound human CRP. A 1 mg/ml solution of this conjugate is typically diluted 1000–2000-fold for use in the assay. The conjugate may be prepared by glutaraldehyde cross-linking according to standard procedures.

5. *CRP standards*   For comparison to unknowns, a CRP standard solution (Scripps Laboratories, La Jolla, California) is prepared in the assay/wash buffer at a concentration of 50 µg/ml. The solution then may be diluted serially to generate a linear standard curve for the calculation of unknown CRP concentrations.

6. *Enzyme substrate solution*   The alkaline phosphatase substrate, *p*-nitrophenyl phosphate, is dissolved at a concentration of 2 mg/ml in substrate buffer. Prepare fresh.

PROTOCOL

The following protocol makes use of the immobilized PEA minicolumns suspended in 16 × 120 mm test tubes. Each addition of sample or wash should be done by adding the required amount directly to the top disk of the packed column. This method of addition is especially important when applying the sample, conjugate, and substrate solutions. Try to avoid extraneous drops along the inner sides of the column, since carryover into subsequent steps may cause anomalous results.

NOTE: All eluents off the columns are discarded until the collection of the final substrate color for absorbance measurements.

1. Equilibrate the columns with 2 ml assay/wash buffer.
2. Apply 100 µl undiluted (clarify, if necessary, by centrifugation) human plasma or serum to each column for the unknown test samples. Apply 100 µl appropriately diluted CRP standards to their correspondingly labeled columns. Use at least four serially diluted standards (starting at 50 µg/ml; stock solution) to accurately measure the unknowns. Make at least one column control by adding only assay/wash buffer in place of sample.
3. Follow the sample additions with 0.5 ml assay/wash buffer to rinse the samples into the affinity gels and partially clean the sides of the column. Next, wash with an additional 5 ml assay/wash buffer.

4. Apply 500 µl antibody–enzyme conjugate diluted at least 1:1000 with assay/wash buffer.
5. Incubate 10 min.
6. Wash with 0.5 ml, 2 ml, and then 5 ml assay/wash buffer to completely remove any unbound conjugate.
7. Transfer the columns to fresh test tubes and add 500 µl substrate solution to each column.
8. Incubate for 30 min at room temperature to develop substrate color.
9. Elute developed color with 2 ml substrate buffer and collect the entire eluent.
10. Measure the absorbance of each fraction at 405 nm.

Plot the concentration of the standards against their measured absorbance values. The test should be linear in approximately the range of 0.5–200 µg of CRP per ml plasma or serum. The clinically useful range for this parameter is from a normal value of about 3 µg/ml to an acute-phase sepsis-related value over 50 µg/ml. Figure 5.51 shows the standard curve resulting from this assay system.

The use of such affinity chromatography separations in EIA systems also can be done using immobilized antibodies or antigens. The flow-through nature of the column format provides rapid interaction times and enough

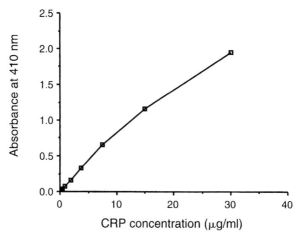

FIGURE 5.51

A standard curve generated by assaying various known concentrations of human C reactive protein (CRP) using an immunography ELISA system consisting of immobilized phosphoethanolamine.

capacity to effect concentration to measure even highly dilute analytes. A similar approach using an automated system involving a centrifugal analyzer has been reported in the literature (Freytag et al., 1984; Leflar et al., 1984).

5.4.3.2 Enzyme Immunoassays Using Flow-Through Affinity Membrane Systems

Similar to the flow-through affinity chromatography based ELISA just described is the technology referred to as enzyme-linked immunofiltration assay (ELIFA) using flow-through membranes (Valkirs and Barton, 1985; Jsselmuiden et al., 1989). All immunography ELISA principles described previously apply to ELIFA, except that ELIFA takes advantage of filtering solutions through a membrane instead of a chromatography support to bind the components of an assay. This facilitates "immunoconcentration" on the surface of the membrane and serves to bind high levels of ligand compared with those that can be immobilized on a plastic surface such as a microtiter plate. This system (in principle) not only overrides the effects of limiting diffusion at the surface of nonporous materials, but allows the measurement of the analyte molecules even if their concentrations are low in solution.

Limiting diffusion is a phenomenon common to reactions that occur at a liquid–solid interface, for example, that observed with ELISA on the surface of a plastic microtiter plate. Since the reactants are depleted in a zone close to the surface, reaction progress is dependent on the diffusion of new reactants from the bulk solution to the surface. One method for overriding the effects of limiting diffusion is to pull unreacted molecules in the solution to the reaction surface by using a porous solid phase such as a membrane.

ELIFA can be used for dot-blot type assays in which the final colored product is precipitated on the surface of a membrane or, with the appropriate device to collect the soluble substrate color at the end of an assay, in more traditional EIA procedures. Although the dot blots are acceptable for qualitative analysis, a reflecting densitometer is required for quantitative measurements. Using ELIFA in flow-through EIA, a soluble substrate solution can be measured using a spectrophotometer or a microplate reader.

In the diagnostic industry, ELIFA technology has resulted in home testing devices in which samples and reagents are filtered through a small membrane to perform an assay. Usually these devices pull the solutions through the membrane by capillary action created by a porous material placed at the back of the membrane. The home pregnancy test devices manufactured by a number of companies are perhaps the best examples of this technology. A single membrane about 1 cm in diameter is held in a plastic device and backed by adsorbant material. A sample is placed on the membrane, and is drawn through as the adsorbant material soaks up the solution. As it contacts the ligand on the membrane surface (e.g., an antibody against β-HCG), the antigen binds. Next, a drop or two of a specific antibody–enzyme conjugate is added and also is drawn through the membrane, specifically binding the antigen that was previously bound. A final addition of substrate solution results in the deposit of a visible insoluble colored product on the membrane if the antigen was of sufficient concentration in the original sample. The entire test is made possible by the increase in interaction efficiencies created by filtering the solutions past the membrane surface.

The filtration of solutions past a membrane surface decreases a standard 1 hr incubation step at 37°C with plastic ELISA plates to about 5 min at room temperature. This is because the replenishment of reactants at the reaction surface eliminates the limiting diffusion effect, a phenomenon associated with surface-bound reactions. Pierce Chemical developed an apparatus for this immunofiltration approach called the Easy-Titer ELIFA System. The basic principles of immunoassays run on the device are virtually the same as in traditional ELISA systems; the only difference is that the solutions are pulled through a membrane using a mild vacuum. In the final step of the assay, a colorless substrate solution is drawn past the membrane-bound enzyme immunoconjugate, and the soluble colored solution produced as a result of contacting the immobilized enzyme is quantitatively transferred into a standard 96-well ELISA plate. Measurements for the assay can then be made using a microplate reader.

The Pierce apparatus sandwiches a membrane between a 96-well sample application plate and a vacuum chamber (see Fig. 5.52). The device is designed and optimized for: (1) uniform and slow flow rates past the membrane, (2) uniform "pull-through" times for solutions in all wells across the 96-well top, and (3) the precise and quantitative transfer of colored substrate product into a standard microtiter plate for analysis in an automated microtiter plate reader. Most binding steps of the assay can be carried out without an intervening wash step since unbound ligand or receptor molecules pass through the membrane and into a waste chamber. However, after binding the final component in the assay, that is, the enzyme conjugate, a wash step is performed to completely remove any unbound conjugate in the membrane and transfer cannula. After this step, the apparatus is opened between the membrane support plate and the vacuum chamber (the membrane remains sandwiched at this point) and the microtiter plate is put in place. The substrate is then loaded into the sample application plate wells and color develops as it is drawn past the surface-bound enzyme and is transferred into the corresponding well in the microtiter plate below.

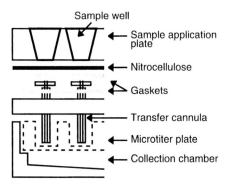

FIGURE 5.52
A cutaway schematic of the Pierce Easy-Titer ELIFA device. The top plate contains 96 sample wells which correlate to the 96 wells of a microplate. Each sample well is open to a membrane that is sandwiched immediately underneath. The membrane has an affinity ligand immobilized on its surface (either covalently or passively adsorbed) that has specificity for the analyte of interest. Special transfer cannulas are channels connecting the sample wells with the underlying microtiter plate wells. Gaskets prevent sample and reagent cross-talk between wells. The collection chamber is used (without a microplate inserted) for any washes or other reagent additions that are not to be retained. At the end of an immunoassay procedure, the final addition of substrate solution is slowly pulled through the affinity membrane, color is developed by contact with bound antibody–enzyme conjugate, and the solution is collected in the microplate below. After disassembly, the microplate contents can be measured using a standard microplate reader.

The membrane used in an ELIFA assay can be nonactivated for passive adsorption of ligand on the surface. Thus, nitrocellulose works extremely well, having high capacity for noncovalently binding a wide variety of protein molecules. Activated membranes with specific covalent coupling chemistries also can be used, especially to couple small molecules that do not adhere well to nitrocellulose. See the section on activated membranes in Chapter 1 for the types and chemistries commercially available.

Depending on the antibody–antigen system used, ELIFA has the potential to be more sensitive than traditional ELISA. Assay sensitivity can be increased by drawing ligand and receptor solutions more slowly past the membrane surface or by increasing the reactant concentrations. The apparatus also allows for immunoconcentration by drawing large volumes of dilute solutions through the membrane.

The steps of the simple assay listed here illustrate the accelerated reaction kinetics that can be achieved as limiting diffusion of traditional ELISA

5.4 Analytical Applications of Affinity Ligands

is overridden. Similar accelerated reaction kinetics can be obtained in dot-blot assays for antibody detection when solutions are slowly filtered through a membrane.

PROTOCOL FOR ELISA USING THE EASY-TITER ELIFA SYSTEM

The protocol described here is a simplified version of the detailed protocol found in the instruction booklet accompanying the product (Pierce Chemical).

NOTE: The vacuum relief valve must be opened and closed after each step to release the vacuum.

1. Assemble the Easy-Titer unit with the nitrocellulose membrane or other suitable membrane in place.
   1. Clamp collection chamber and membrane support plate together.
   2. Align top gasket over the cannulas.
   3. Wet a piece of nitrocellulose and place over the top gasket covering all 96 wells.
   4. Position the sample application plate using the guide pins attached to the membrane support plate.
   5. Put the four thumb screws into place and tighten until the plates hit the position stops. (Lightly tighten two of the thumb screws at diagonal corners. Then alternate to the other two thumb screws and tighten them to a slightly greater degree. Continue this alternating and tightening action.)
2. Pipette 200 µl antigen-containing sample into the appropriate wells. Pull samples through the membrane in 5 min by adjusting the pump rate.
3. Block with 200 µl of 3% BSA solution by pulling the solution through in 5 min.
4. Add 200 µl primary antibody and pull through in 5 min.
5. Add 200 µl enzyme-conjugated secondary antibody and pull through in 5 min.
6. Wash any excess enzyme-conjugated secondary antibody out of the cannulas by adding 200 µl buffered solution to each sample well and pulling through in 30 sec. Repeat for a total of 3 washes.
7. Open the vacuum relief valve and keep open. Place a microtiter plate into the collection chamber. Reassemble unit and close the vacuum relief valve.
8. Add 200 µl substrate solution to the wells and pull through into the microtiter plate in 5 min.
9. Read collected substrate solutions in an automatic ELISA plate reader.

If the product of the enzyme/substrate reaction is too dark in color, the pump flow rate should be increased; if it is too light, the flow rate should be decreased. The process of adding the substrate solution can be repeated to obtain the desired color intensity of the product. Concentrations of the antigen, antibody, or enzyme conjugates should be similar to those used in ELISAs.

The total time required for an EIA by flow-through membrane technology with the Easy-Titer unit is 35–40 min. The protocol described here uses passively immobilized ligands on a nitrocellulose membrane, but covalent coupling chemistry using preactivated membranes of various compositions also works well.

5.4.4 Affinity Electrodes and Biosensors

Principles of Biosensor Construction

Advances in the immobilization of affinity ligands and innovations in the emerging field of bioelectronics have combined to produce revolutionary new detection devices. Affinity electrodes and biosensors are based on the specific interaction between receptors, enzymes, or antibody molecules and their specific target analytes (Guilbault, 1984; Scheller *et al.*, 1985). The application of this technology to measurement systems has created novel analytical detection devices for such diverse fields as diagnostics, therapeutics, process control, waste and environmental monitoring, computer technology, and the kinetic analysis of the interaction of various biological substances.

The basic operation of a biosensor is illustrated in Figure 5.53 (Taylor, 1991). Common to every device is a support material to which an immobilized affinity ligand is attached. This ligand may be an enzyme that is designed to monitor the presence of its specific substrate in solution. It may be an antibody that can measure its complementary antigen, or an antigen to detect specific antibodies. The affinity ligand may also be any biospecific molecule that interacts with a particular receptor protein (or *vice versa*). It can even be an immobilized intact living organism (cellular) that can act on specific substances in the solutions with which it comes in contact.

To detect the interaction of these affinity pairs, a functional biosensor needs an electronic transducer that senses the subtle chemical changes that take place between the immobilized ligand and the specific analyte. The detection process may involve the monitoring of electrical effects such as potentiometric changes, amperometric fluctuations, or capacitance differences; optical effects such as light absorption, scattering, or refractive in-

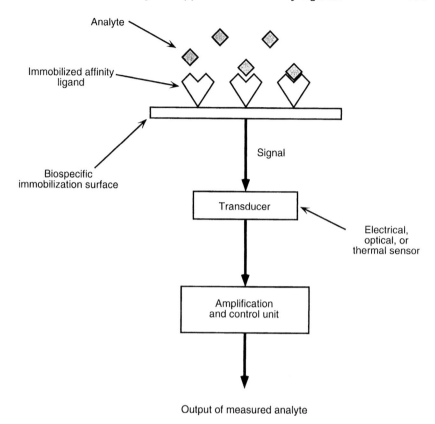

FIGURE 5.53

Basic components of a biosensor device. A biospecific immobilization surface contains a covalently attached affinity ligand with specificity for the target analyte of interest. When sample is added, the affinity interaction creates a signal that can be monitored by a sensing transducer. The transducer can use changes in electrical, thermal, or optical parameters to detect the binding of specific target molecules. The amplification and control unit is often a computerized device able to translate the transducer-sensed information into calculated measurements.

---

dex; changes in density or mass; acoustical effects such as changes in amplitude, frequency, or phase of a sound wave; or thermal differences using sensitive calorimeters. The electronic detector then sends its signals to an amplification device that also may process and compute the concentration of the analyte in solution. The output and control of these instruments may be as simple as reading a needle gauge on a device such as a pH meter or as complex as sophisticated computerized instruments with programmable interfaces.

The principles of biosensor operation have been employed for decades with oscilloscopes designed to monitor slight changes in electrical phenomena. For instance, an olfactory organ such as the antenna of a butterfly can be placed between the input leads of an oscilloscope and used to detect the interaction of olfactory receptors with various volatile substances. In this case, the initial amplifiers of the receptor–ligand interaction are the olfactory nerves that generate action potentials along their length in response to the binding of specific substances. Important qualitative information can be obtained in such a system, but is of little quantitative use.

In modern biosensor design, a synthetic receptor–ligand surface is constructed that has specificity for a single substance. Since the surface is monospecific and the response varies in proportion to the quantity of ligand in the sample solution, quantitative analytical measurements are possible.

The first such biosensor was constructed in 1962 (Clark), and used an immobilized glucose oxidase enzyme associated with a Clark pO_2 electrode to detect and measure glucose in solution. This simple system eventually became the basis for a functional glucose analyzer. Hundreds of biosensors have been constructed since that time, detecting every conceivable interaction from biological systems to chemically based affinity pairs.

Immobilization Technology in Biosensor Design

Techniques for attaching affinity ligands to the biospecific immobilization surface of a biosensor are not unlike the techniques and chemistries developed for coupling to other matrices (Gaber and Farmer, 1984; Beissinger *et al.,* 1986). Both noncovalent adsorption and entrapment procedures, as well as chemically created covalent linkages, are used to immobilize biosensor ligands. Regardless of the coupling method, the goal is to form a local concentration of the affinity ligand across the biospecific surface. Correct orientation and retention of activity are important in this process, especially for ligands containing active sites that must interact with specific analytes after immobilization.

The biospecific immobilization surface may consist of inorganic, synthetic, or natural gelatinous polymers, or even films of biological molecules, sometimes cross-linked on a supporting membrane. Common matrices are glass, gelatin, polyacrylamide, cross-linked proteins, collagen, membranes consisting of cellulose derivatives, nylon, polyvinylchloride, polyvinylalcohol, films of lipid bilayers or liposomes, and various composites created from combinations of these supports.

5.4 Analytical Applications of Affinity Ligands

In general, entrapment or adsorption procedures do not yield stable affinity systems for biosensor design. Such sensors may work for brief periods in the laboratory, but the weak bonds created by noncovalent attachment usually cause severe leakage of the biomolecule off the surface and degradation of performance with use. Entrapment, however, does provide a viable immobilization means when attaching cellular ligands to the surface, since the cells are typically surrounded by a polymerized or gelatinous membrane and are unable to break free.

For small ligands, as well as for proteins and enzymes, covalent coupling provides the best approach to biosensor fabrication. The optimal covalent immobilization techniques involve the use of activation chemistries identical to those outlined throughout this book. The available functionalities on the biospecific surface must be activated to allow the coupling of the appropriate chemical groups on the target ligand. The chemistries most often used for this purpose include cyanogen bromide, carbonyldiimidazole (CDI), water-soluble carbodiimides (e.g., EDC), and reductive amination (Thomas and Broun, 1976; Yamamoto *et al.,* 1980; Beissinger *et al.,* 1986; Matsuoka *et al.,* 1987). Cross-linking agents such as glutaraldehyde and other bifunctional compounds that have two reactive portions are also used extensively (Katz, 1990). Sometimes cross-linkers are used to chemically stabilize a noncovalently adsorbed ligand, but they are even more versatile as true covalent immobilizing agents.

Particularly for enzyme immobilization, glutaraldehyde provides a rapid, simple, and economical coupling technique that preserves enzymatic activity in most instances. Many of the reaction schemes using glutaraldehyde cross-linking are illustrated as the formation of Schiff bases that may or may not be reduced subsequent to reaction with ligand. Although this cross-linking mechanism is certainly plausible, a much more stable and likely reaction involves vinyl addition to a glutaraldehyde polymer (see Fig. 5.54). In this case, secondary amine bonds are formed without reduction and the multipoint attachment creates a network of ligand on the surface of the biosensor support. This reaction mechanism for glutaraldehyde is certainly more likely to result in the observed stability of the cross-linked products, unlike bonds established solely from Schiff base interactions.

Another promising immobilization technology for use in affinity biosensors is the modification of gold foil electrodes with dithiobis-succinimidyl propionate (DSP; Pierce) (Katz, 1990). The structure of the bifunctional reagent is shown in Figure 2.4 (Chapter 2). DSP has, in addition to its active NHS groups on either end, a disulfide linkage in the middle that will rapidly chemisorb to gold surfaces. The resulting stability exceeds even that offered by covalent silane bonds with glass. NHS-activated gold electrodes

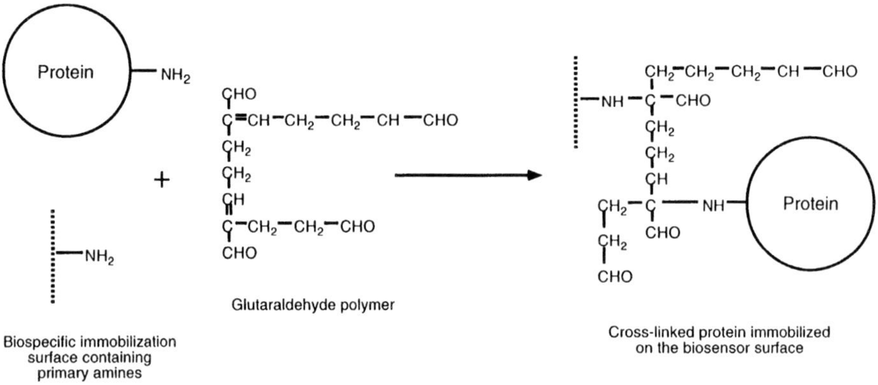

414   5. Selected Applications

FIGURE 5.54

Glutaraldehyde cross-linking is a common method of affinity ligand immobilization for biosensor devices. The majority of protocols does not make use of the aldehyde residues in a reductive amination procedure. Instead, these protocols couple through a process of vinyl addition to the double bonds of a glutaraldehyde polymer.

---

subsequently coupled with enzymes or redox-sensitive ligands no doubt will find widespread use in certain electrochemical measurements.

The activation of gold foil surfaces with DSP is quick and efficient.

1. Gold foil is incubated for 10 min at room temperature in 0.01 M isopropanolic DSP (400 mgs per 100 ml) solution after which the foil is thoroughly rinsed with fresh isopropanol and ethanol.
2. An affinity ligand containing a primary amine functionality is coupled via the NHS groups according to the descriptions in Chapter 2.

Successful immobilization technology in biosensor design should yield a stable product with a high retention in affinity binding activity. The sensor must be able to withstand widely varying storage conditions without loss of sensitivity. The operational temperature fluctuations that a biosensor may undergo range from 4° to 37°C. At least 1 year stability of the immobilized affinity ligand is necessary for commercial viability.

Most biosensor or affinity electrode designs reported in the literature have had trouble meeting these minimum criteria necessary to form commercial products. Thousands of reports exist on various types of biosensors, but few real products have resulted from those efforts. The major problems include matrix fouling and instability, size of the instrument, and the slow speed of biosensor response.

5.4 Analytical Applications of Affinity Ligands 415

The Pharmacia BIAcore system may be a breakthrough in biosensor utility. The instrument is the first general purpose and customizable biosensor device commercially available. The system allows real-time measurement of such biospecific parameters as binding site recognition, affinity constants, reaction specificity, kinetics, multiple binding, and cooperativity of an interaction. The system uses replaceable sensor chips to which ligands can be immobilized for analysis.

As a transducer or detector of the biospecific interaction, the BIAcore system uses a phenomenon called surface plasmon resonance (SPR). SPR makes use of laser light scanning of the immobilization surface to register the changes in mass occurring as a result of receptor–ligand interactions. The immobilization surface of the sensor chip is constructed of a glass support overlaid with a gold film. The metal is coated covalently with a layer of carboxylated dextran that creates a biocompatible matrix with low non-specific binding character. The desired ligand is immobilized on the carboxylated dextran in a two-stage reaction using the water-soluble carbodiimide EDC and N-hydroxysuccinimide (NHS) (Fig. 5.55). This procedure

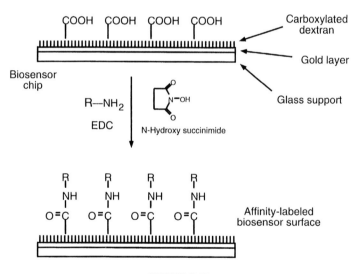

FIGURE 5.55

The Pharmacia BIAcore device makes use of replaceable biosensor chips containing the immobilization surface. The matrix consists of a glass plate overlaid with a gold surface. Carboxylated dextran has been layered on this surface to create functionalities for coupling ligands. The recommended procedure uses the water-soluble carbodiimide EDC to form amide linkages with amine-containing affinity ligands.

is similar to the EDC/Sulfo-NHS method described in Section 2.2.1.6, Chapter 2.

The SPR effect occurs between the incoming photons of a wedge of plane-polarized laser light and the electrons of the gold underlying surface of the sensor chip. At a certain angle of incidence, the energy of the light is completely transferred to the electrons and no light is reflected. This is called the resonance angle or angle of nonreflectance. As ligand is coupled to the carboxylated dextran coating this angle changes slightly in response to the added mass in the vicinity of the metal surface. When interacting molecules bind to this ligand, further changes occur in the resonance angle and these changes can be correlated to the quantitative characteristics of the biospecific interaction. The result is a powerful analytical system built on the basic technology of immobilized affinity ligands.

Of all the systems currently in operation using immobilized affinity ligands, the biosensors perhaps hold the greatest promise. It is not too difficult to imagine a future in which almost any biological substance can be measured quickly and analyzed through biosensor affinity interactions. Biosensor devices will contain preformulated matrices with dozens of different affinity ligands immobilized on chips that can measure a battery of biochemically or diagnostically important parameters in seconds. All of this will be made possible by the technology that allows successful immobilization of a host of affinity ligands on solid surfaces.

APPENDIX

Table of Suppliers

Supplier	Affinity matrices	Activated supports	Activation reagents	Immobilized ligands	Prepacked columns	Columns and accessories	Affinity purification kits
Aldrich P.O. Box 355 Milwaukee, WI 53201			✔				
Alltech 2051 Waukegan Rd. Deerfield, IL 60015	✔	✔		✔	✔	✔	
American Qualex 14620 E. Firestone Blvd. La Mirada, CA 90638		✔					
Amicon 24 Cherry Hill Drive Danvers, MA 01923	✔	✔		✔	✔	✔	✔
Beckmann 2500 Harbor Blvd. Fullerton, CA 92634		✔		✔	✔	✔	✔
Bio-Probe 14272 Franklin Ave. Tustin, CA 92680		✔	✔	✔	✔		✔
Bio-Rad 3300 Regatta Blvd. Richmond, CA 94804	✔	✔	✔	✔	✔	✔	✔
Bodman P.O. Box 2221 Aston, PA 19014	✔	✔		✔		✔	
Boehringer Mannheim P.O. Box 50414 Indianapolis, IN 46250		✔	✔	✔	✔		✔
Calbiochem P.O. Box 12087 San Diego, CA 92112		✔	✔	✔	✔		✔

(Continues)

Appendix

Table of Suppliers (continued)

Supplier	Affinity matrices	Activated supports	Activation reagents	Immobilized ligands	Prepacked columns	Columns and accessories	Affinity purification kits
Chromatochem 2837 Fort Missoula Rd. Missoula, MT 59801		✔		✔	✔	✔	
Collaborative Res. Inc. 2 Oak Park Bedford, MA 01730				✔			
E-Y Lab 107 N. Amphlett Blvd. San Mateo, CA 94401					✔		✔
Gelman Scientific 600 South Wagner Rd. Ann Arbor, MI 48106				✔			
Genex Corp. 16020 Industrial Dr. Gaithersburg, MD 20877				✔	✔		✔
I.B.F. Biotech. 7151 Columbia Gateway Dr. Columbia, MD 21046	✔	✔		✔	✔	✔	
ICN P.O. Box 5023 Costa Mesa, CA 92626	✔	✔	✔	✔	✔		✔
Immunicon 1310 Masons Mill II Huntington Valley, PA 19006				✔			
J.T. Baker 222 Red School Lane Phillipsburg, NJ 08865	✔	✔	✔		✔	✔	
Multiple Peptide Systems 3550 General Atomics Court San Diego, CA 92121				✔			
Nalge Co. 75 Panorama Creek Dr. Rochester, NY 14602		✔		✔			
Perseptive Biosystems University Park at MIT 38 Sidney St. Cambridge, MA 02139	✔			✔	✔	✔	
Perstorp Biolytica S-22370 Lund, Sweden				✔	✔	✔	✔
Pharmacia LKB 800 Centennial Ave. Piscataway, NJ 08854	✔	✔	✔	✔	✔	✔	✔
Pierce P.O. Box 117 Rockford, IL 61105	✔	✔	✔	✔	✔	✔	✔

Table of Suppliers (continued)

Supplier	Affinity matrices	Activated supports	Activation reagents	Immobilized ligands	Prepacked columns	Columns and accessories	Affinity purification kits
Polyscience 400 Valley Road Warrington, PA 18976	✔	✔	✔	✔	✔	✔	✔
Repligen 1 Kendal Square, Bldg. 700 Cambridge, MA 02139				✔			
Schleicher & Schuell 10 Optical Ave. Keene, NH 03431		✔		✔	✔		✔
Sepracor 33 Locke Drive Marlborough, MA 01752				✔	✔	✔	
Sigma P.O. Box 14508 St. Louis, MO 63178	✔	✔	✔	✔	✔	✔	✔
Sterogene 140 E. Santa Clara St. #15 Arcadia, CA 91006		✔		✔			
Tosohaas Rohm & Hass Bldg. Independence Mall West Philadelphia, PA 01757	✔	✔		✔	✔	✔	
Waters 34 Maple St. Milford, MA 01757	✔	✔		✔	✔	✔	
Zymed 458 Carlton Court So. San Francisco, CA 94080				✔	✔		✔

REFERENCES

Page numbers listed are not inclusive and indicate only the first page of the journal article used.

Ackers, G. K. (1964). *Biochemistry* **3**, 723.
Adams, R., Bachman, W. E., Fieser, L. F., Johnson, J. R., and Snyder, H. R. (1942). *In* "Organic Reactions," Vol. 1., p. 303. Wiley Press.
Afting, E. G., and Becker, M. I. (1981). *Biochem. J.* **197**, 519.
Akerstrom, B., and Bjorck, L. (1986). *J. Biol. Chem.* **261**, 10240.
Alberts, B. M., Amodio, F. J., Jenkins, M., Gutmann, E. D., and Ferris, F. J. (1968). *Cold Spring Harbor Symp. Quant. Biol.* **33**, 289.
Alhanaty, E., Bashan, N., Moses, S., and Shaltiel, S. (1979). *Eur. J. Biochem.* **101**, 283.
Allen, D. W., Schroeder, W. B., and Balog, J. (1981). *J. Amer. Chem. Soc.* **80**, 1628.
Anderson, G. W., Zimmerman, J. E., and Callahan, F. M. (1964). *Am. Soc.* **86**, 1839.
Andersson, L. (1988). "ISI Atlas of Science: Biochemistry/1988." ISI, .
Andersson, L., and Porath, J. (1986). *Anal. Biochem.* **154**, 250.
Arnold, F. H. (1991). *Bio/Technology* **9**, 151.
Aucouturier, P., Pineau, N., Brugier, J-C., Mihaesco, E., Duarte, F., Skvaril, F., and Preud'homme, J-L. (1989). *J. Clin. Lab. Analysis* **3**, 244.
Axen, R., Porath, J., and Ernback, S. (1967). *Nature (London)* **214**, 1302
Barrett, M. J. (1977). U. S. Patent No. 4,001,583.
Bartling, G. J., Brown, H. D., and Chattopadhyay, S. K. (1973). *Nature (London)* **243**, 342.
Baues, R. J., and Gray, G. R. (1977). *J. Biol. Chem.* **252**, 57.
Bayer, E. A., Ben-Hur, H., and Wilchek, M. (1987). *Anal. Biochem.* **161**, 123.
Beale, D. (1987). *Exp. Comp. Immunol.* **11**, 287.
Beale, D., and Van Dort, T. (1982). *Comp. Biochem. Physiol.* **71B**, 475.
Beaty, N. B., and Lane, M. D. (1982). *J. Biol. Chem.* **257**, 924.
Beissinger, R. L., Farmer, M. C., and Gossage, J. L. (1986). *Trans. Amer. Soc. Artif. Inter. Organs* **32**, 58.
Belew, M., Juntti, N., Larsson, A., and Porath, J. (1987). *J. Immunol. Meth.* **102**, 173.

Bhatia, S. K., Shriver-Lake, L. C., Prior, K. J., Georger, J. H., Calvert, J. M., Bredehorst, R., and Ligler, F. S. (1989). *Anal. Biochem.* **178**, 408.
Biagioni, S., Sisto, R., Ferraro, A., Caiafa, P., and Turano, C. (1978). *Anal. Biochem.* **89**, 616.
Bjorck, L., and Kronvall, G. (1984). *J. Immunol.* **133**, 969.
Bolton, E. T., and McCarthy, B. J. (1962). *Proc. Natl. Acad. Sci. U.S.A.* **48**, 1390.
Borch, R. F., Bernstein, M. D., and Durst, H. D. (1971). *J. Am. Chem. Soc.* **93**, 2897.
Bradford, M. M. (1976). *Anal. Biochem.* **72**, 248.
Britten, R. J. (1963). *Science* **142**, 963.
Brocklehurst, K. Carlsson, J., Kierstan, M. P. J., and Crook, E. M. (1974). *Meth. Enzymol.* **34**, 531.
Brown, B. A., Drozynski, C. A., Dearborn, C. B., Handjian, R. A., Liberatore, F. A., Tulip, T. H., Tolman, G. L., and Haber, S. B. (1988). *Anal. Biochem.* **172**, 22.
Bunn, H. F., Gabbay, K. H., and Gallop, P. M. (1978). *Science* **200**, 21.
Caron, M. G., Srinivasan, Y., Pitha, J., Kociolek, K., and Lefkowitz, R. (1979). *J. Biol. Chem.* **254**, 2923.
Cayley, P. J., Dunn, S. M. J and King, R. W. (1981). *Biochemistry* **20**, 874.
Clark, L. C., and Lyons, C. (1962). *Ann. N. Y. Acad. Sci.* **148**, 133.
Cleland, W. W. (1964). *Biochemistry* **3**, 480.
Clonis, Y. D., Atkinson, A., Bruton, C. J., and Lowe, C. R. (1987). "Reactive Dyes in Protein and Enzyme Technology." Stockton Press, New York.
Coleman, P. L., Walker, M. M., Milbrath, D. S., Stauffer, D. M., Rasmussen, J. K., Krepski, L. R., and Heilmann, S. M. (1990). *J. Chrom.* **512**, 345.
Collier, R., and Kohllaw, G. (1971). *Anal. Biochem.* **42**, 48.
Coulter, A., and Harris, R. (1983). *J. Immunol. Meth.* **59**, 199.
Crestfield, A. M., Moore, S., and Stein, W. H. (1963). *J. Biol. Chem.* **238**, 622.
Cuatrecasas, P. (1970). *J. Biol. Chem.* **245**, 3059.
Cuatrecasas, P., Wilcheck, M., and Anfinson, C. B. (1968). *Proc. Natl. Acad. Sci. U.S.A.* **61**, 636.
Dey, P. M. (1984). *Eur. J. Biochem.* **140**, 385.
Dey, P. M., Naik, S., and Pridham, J. B. (1982). *FEBS Lett.* **150**, 233.
Domen, P. L., Nevens, J. R., Mallia, A. K., Hermanson, G. T., and Klenk, D. C. (1990). *J. Chrom.* **510**, 293.
Douglas, J. T., Wu, Q. X., Agustin, G. P., and Madarang, M. G. (1988). *Leprosy Rev.* **59**, 37.
Dubois, M., Gilles, K. A., Hamilton, J. K., Rebers, P. A., and Smith, F. (1956). *Anal. Chem* **28**, 350.
Dulaney, J. T. (1979). *Mol. Cell. Biochem.* **21**, 43.
Duncan, R. E., and Gilham, P. T. (1975). *Anal. Biochem.* **66**, 532.
Dzau, V. J., Slater, E. E., and Haber, E. (1979). *Biochemistry* **18**, 5224.
Eldjarn, L., and Jellum, E. (1963). *Acta Chem. Scand.* **17**, 2610--2621.
Eliasson, M. et al. (1988). *J. Biol. Chem.* **263**, 4323.
Ellman, G. L. (1958). *Arch. Biochem. Biophys.* **74**, 443.
Engvall, E., Jonsson, K., and Perlmann, P. (1971). *Biochim. Biophys. Acta* **251**, 427.
Ey, P. L., Prowse, S. J., and Jenkin, C. R. (1978). *Immunochem.* **15**, 429.
Farooqui, A. A. (1980). *J. Chrom.* **184**, 335.

Faulmann, E. L., Duvall, J. L., and Boyle, M. D. P. (1991). *Bio/Techniques* **10**, 748.
Finaly, T. H., Troll, V., Levy, M., Johnson, A. J., and Hodgins, L. T. (1978). *Anal. Biochem.* **87**, 77.
Fleet, G. W. J., Porter, R. R., and Knowles, J. R. (1969). *Nature (London)* **224**, 511.
Fornstedt, N., and Porath, J. (1975). *FEBS Lett.* **57**, 187.
Freytag, J. W., Lau, H. P., and Wadsley, J. J. (1984). *Clin. Chem.* **30**, 1494.
Gaber, B. P., and Farmer, M. C. (1984). *Prog. Clin. Biol. Res.* **165**, 179.
Gilham, P. T. (1971). *Meth. Enzymol.* **21**, 191.
Glatz, J. F. C., and Veerkamp, J. H. (1983). *J. Biochem. Biophys. Met.* **8**, 57.
Goding, J. W. (1986). *In* "Monoclonal Antibodies: Principles and Practice," pp. 6–. Academic Press, Orlando, Florida.
Gorecki, M., and Patchornik, A. (1973). *Biochim. Biophys. Acta* **303**, 36.
Gorecki, M., and Patchornik, A. (1975). U. S. Patent No. 3,914,205.
Grabarek, Z., and Gergely, J. (1990). *Anal. Biochem.* **185**, 131.
Gravel, R. A., Lam, K. F., Mahuram, D., and Kronis, A. (1980). *Arch. Biochem. Biophys.* **201**, 669.
Gray, G. R. (1974). *Arch. Biochem. Biophys.* **163**, 426.
Green, A. A. (1931). *J. Biol. Chem.* **93**, 495.
Green, N. M. (1965). *Biochem. J.* **94**, 230.
Green, N. M., and Toms, E. J. (1973). *Biochem. J.* **133**, 687.
Guilbault, G. G. (1984). "Analytical Uses of Immobilized Enzymes." Marcel Dekker, New York.
Guire, P. E. (1978a). *Enzyme Eng.* **3**, 63.
Guire, P. E. (1978b). U. S. Department of Commerce, National Technical Information Service. PB 287 343.
Guire, P. E. (1988). U. S. Patent No. 4,722,906
Guire, P. E. (1990a). U. S. Patent No. 4,973,493.
Guire, P. E. (1990b). U. S. Patent No. 4,979,959
Guire, P. E., and Chudzik, S. J. (1989). U. S. Patent No. 4,826,759.
Hammen, R. F., Pang, D., Remington, K., Thompson, H., Judd, R. C., and Szuba, J. (1988). *BioChromatography* **3**, 54.
Harlow, E., and Lane, D. (1988). *In* "Antibodies: A Laboratory Manual," pp. 510–. Cold Spring Harbor Laboratory Press, Cold Spring Harbor, New York.
Harlow, E., and Lane, D. (1988b). *In* "Antibodies: A Laboratory Manual," pp. 312–. Cold Spring Harbor Laboratory Press, Cold Spring Harbor, New York.
Harris, R. G., Rowe, J. J. M., Stewart, P. S., and Williams, D. C. (1973). *FEBS Lett.* **29**, 189.
Hauri, H-P., and Bucher, K. (1986). *Anal. Biochem.* **159**, 386.
Hayes, C. E., and Goldstein, I. J. (1974). *J. Biol. Chem.* **249**, 1904.
Hearn, M. T. W. (1987). *Meth. Enzymol.* **135**, 102.
Hecht, S. (1977). *Tetrahedron* **33**, 1671.
Helseth, D. L. Jr., and Veis, A. (1984). *Proc. Natl. Acad. Sci. U.S.A.* **81**, 3302.
Henrikson, K., Allen, S. H. G., and Maloy, W. L. (1979). *Anal. Biochem.* **94**, 366.
Hewlett, G. (1990). *Bio/Technology* **8**, 565.
Hixson, H. F., and Nishikawa, A. H. (1973). *Arch. Biochem. Biophys.* **154**, 501.
Hjerten, S., and Mosbach, K. (1962). *Anal. Biochem.* **3**, 109.

Hoare, D. G., and Koshland, D. E. (1966). *J. Amer. Chem. Soc.* **88**, 2057.
Hoare, D. G., and Koshland, D. E. (1967). *J. Biol. Chem.* **242**, 2447.
Hoffman, W. L., and O'Shannessy, D. J. (1988). *J. Immunol. Met.* **112**, 113.
Hofmann, K., and Axelrod, A. E. (1950). *J. Biol. Chem.* **187**, 29.
Hofmann, K., Melville, D. B., and duVigneaud, V. (1941). *J. Biol. Chem.* **141**, 207.
Holleman, W., and Weiss, L. J. (1978). *J. Biol. Chem.* **251**, 1691.
Holloway, P. W. (1973). *Anal. Biochem.* **53**, 304.
Honda, T., Fujita, A., Tsubakhihara, Y., and Morihara, K. (1986). *J. Chrom.* **376**, 385.
Hoyer, L. W., and Shainoff, J. R. (1980). *Blood* **55**, 1056.
Hutchens, T. W., and Porath, J. (1986). *Anal. Biochem.* **159**, 217.
Hutchens, T. W., and Porath J. O. (1987a) Clin. Chem. **33**, 1502.
Hutchens, T. W., and Porath, J. (1987b) Biochemistry **26**, 7199.
Imagawa, M., Yoshitake, S., Hemaguchi, Y., Ishikawa, E., Niitsu, Y., Urushizaki, I., Kanazawa, R., Tachibana, S., Nakazawa, N., and Ogawa, H. (1982). *J. Appl. Biochem.* **4**, 41.
Inman, J. K., and Dintzis, H. M. (1969). *Biochemistry* **8**, 4074.
Ishii, S., Yokosawa, H., Kumazaki, T., and Nakamura, I. (1983). *Meth. Enzymol.* **91**, 378.
Issekutz, A. C. (1983). *J. Immunol. Met.* **61**, 275.
Ito, N., Abe, I., Noguchi, K., Kazama, M., Shimura, K., and Kasai, K.-I. (1987). *Biochem. Internat.* **15**, 311.
Jany, K. D., Keil, W., Meyer, H., and Kiltz, H. H. (1976). *Biochim. Biophys. Acta* **453**, 62.
Jayabaskaran, C., Davison, P. F., and Paulus, H. (1987). *Prep. Biochem.* **17**, 121.
Jellum, E. (1964). *Acta Chem. Scand.* **18**, 1887–1895.
Ji, T. H. (1979). *Biochim. Biophys. Acta* **559**, 39.
Ji, T. H., and Ji, I. (1982). *Anal. Biochem.* **121**, 286.
Johnson, B. J. B. (1981). *Biochemistry* **20**, 6103.
Johnson, D. A., Gautsch, J. W., Sportsman, J. R., and Elder, J. H. (1984). *Gene Anal. Tech.* **1**, 3.
Jsselmuiden, O. E., Herbrink, P., Meddens, M. J. M., Tank, B., Stolz, E., and Van Eijk, R. V. W. J. (1989). *J. Immunol. Meth.* **119**, 35.
Kabayashi, H., Kusakabe, I., and Murakami, K. (1982). *Anal. Biochem.* **122**, 308.
Kagedal, L. (1989). *In* "Protein Purification: Principles, High Resolution Methods and Applications" (J-C. Janson, J-C. and L. Ryden eds.), pp. 227–. VCH Publishers, New York, New York. pp 227.
Kamicker, B. J., Schwartz, B. A., Olson, R. M., Drinkwitz, D. C., and Gray, G. R. (1977). *Arch. Biochem. Biophys.* **183**, 393.
Katz, E. Y. (1990). *J. Electroanal. Chem.* **291**, 257.
Kawasaki, T., Kawasaki, N., and Yamashina, I. (1989). *Meth. Enzymol.* **179**, 310.
Klenk, D. C., Hermanson, G. T., Krohn, R. I., Fujimoto, E. K., Mallia, A. K., Smith, P. K., England, J. D., Wiedmeyer, H., Little, R. R., and Goldstein, D. E. (1982). *Clin. Chem.* **28**, 2088.
Kohn, J., and Wilchek, M. (1982). *Enzyme Microb. Technol.* **4**, 161

References

Kozutsumi, Y., Kawasaki, T., and Yamashina, I. (1980). *Biochem. Biophys. Res. Commun.* **95,** 658.
Kunitz, M. (1945). *Science* **101,** 668.
Kunitz, M. (1947). *J. Gen. Physiol.* **30,** 311.
Lamoyi, E. (1986). *Meth. Enzymol.* **121,** 652.
Lamoyi, E., and Nisonoff, A. (1983). *J. Immunol. Meth.* **56,** 235.
Larsson, P. O., Glad, M., Hansson, L., Mansson, M. O., Ohlson, S., and Mosbach, K. (1983). *In* "Advances in Chromatography" (Giddings *et al.,l* eds.) Vol. 21, pp. 41–. Marcel Dekker, New York.
Lee, R. T., and Lee, Y. C. (1979). *Carbohydrate Res.* **77,** 149.
Leflar, C. C., Freytag, J. W., Powell, L. M., Strahan, J. C., Wadsley, J. J., Tyler, C. A., and Miller, W. K. (1984). *Clin. Chem.* **30,** 1809.
Lehtonen, O.-P., and Viljanen, M. K. (1980). *J. Immunol. Meth.* **34,** 61.
Liener, I. E., Sharon, N., and Goldstein, I. J. (1986). "The Lectins. Proerties, Functions, and Applications in Biology and Medicine." Academic Press, San Diego.
Lihme, A., and Heegaard, P. M. H. (1990). *Anal. Biochem.* **192,** 64.
Lindberg, U., and Persson, T. (1971). *Eur. J. Biochem.* **54,** 411.
Lindmark, R., Thoren-Tolling, K., and Sjoquist, J. (1983). *J. Immunol. Meth.* **62,** 1.
Lis, H., and Sharon, N. (1984). *Biol. Carbohydr.* **2,** 1.
Litman, R. M. (1968). *J. Biol. Chem.* **243,** 6222.
Little, J. N., Waters, J. L., Bombaugh, K. J., and Pauplis, W. J. (1971). *In* "Gel Permeation Chromatography" (K. H. Altgelt L. Segal eds.), p. 205. Dekker, New York.
Lonnerdal, B., and Keen, C. L. (1982). *J. Appl. Biochem.* **4,** 203.
Lowe, C. R., and Dean, P. D. G. (1974a). *In* "Affinity Chromatography," pp. 222. John Wiley & Sons, New York.
Lowe, C. R., and Dean, P. D. G. (1974b). *In* "Affinity Chromatography," pp. 33. John Wiley & Sons, New York.
Lowe, C. R., Harvey, M. J., Craven, D. B., and Dean, P. D. G. (1973). *Biochem. J.* **133,** 499.
Lowe, C. R., Harvey, M. J., and Dean, P. D. G. (1974). *Eur. J. Biochem.* **41,** 347.
McCutchan, T. F., Gilham, P. T., and Soll, D. (1975). *Nucleic Acids Res.* **2,** 853.
Majumder, H. K., Maitra, U., and Rosenberg, M. (1979). *Proc. Natl. Acad. Sci. U.S.A.* **76,** 5110.
Male, K. B., Nguyen, A. L., and Luong, J. H. T. (1990). *Biotechnol. Bioeng.* **35,** 87.
Mallia, A. K., Hermanson, G. T., Krohn, R. I., Fujimoto, E. K., and Smith, P. K. (1981). *Anal. Lett.* **14,** 649.
Manil, L., Motte, P., Permas, P., Troalen, F., Bohuon, C., and Bellet, D. (1986). *J. Immunol. Meth.* **90,** 25.
March, S. C., Parikh, I., and Cuatrecasas, P. (1974). *Anal. Biochem.* **60,** 149.
Mares-Gula, M., and Shaw, E. (1965). *J. Biol. Chem.* **240,** 1579.
Markwardt, F., Landmann, H., and Walsmann, P. (1968). *Eur. J. Biochem.* **6,** 502.
Matsumae, H., Minobe, S., Kindan, K., Watanabe, T., Sato, T., and Tosa, T. (1990). *Biotech. Appl. Biochem.* **12,** 129.

Matsumoto, I., Kitagaki, H., Akai, Y., Ito, Y., and Seno, N. (1981). *Anal. Biochem.* **116,** 103.
Matsuoka, H., Tanioka, S., and Karube, I. (1987). *Anal. Lett.* **20,** 63.
Means, G. E., and Feeney, R. E. (1968). *Biochemistry* **7,** 2192.
Merkle, R. K., and Cummings, R. D. (1987). *Meth. Enzymol.* **138,** 232.
Moore, E. C., Peterson, D., Yang, L. Y., Yeung, C. Y., and Neff, N. F. (1974). *Biochemistry* **13,** 2904.
Moss, L. G., Moore, J. P., and Chan, L. (1981). *J. Biol. Chem.* **256,** 12655.
Nethery, A., Raison, R. L., and Eastbrook-Smith, S. B. (1990). *J. Immunol. Meth.* **126,** 57.
Neurath, A. R., and Strick, N. (1981). *J. Virol. Meth.* **3,** 155.
Nevens, J. R., Mallia, A. K., Wendt, M. W., and Smith, P. K. (1992). *J. Chrom.,* in press.
Ngo, T. T. (1986). *BioTechnology* **4,** 134
Ngo, T. T., Yam, C. F., Lenhoff, H. M., and Ivy, J. (1981). *J. Biol. Chem.* **256,** 11313.
Nicolson, G. L., Blaustein, J., and Etzler, M. E. (1974). *Biochemistry* **13,** 196.
Nilsson, K., and Mosbach, K. (1980). *Eur. J. Biochem.* **112,** 397
Nilsson, K., and Mosbach, K. (1981). *Biochem. Biophys. Res. Commun.* **102,** 449.
Nilsson, K., and Mosbach, K. (1987). *Meth. Enzymol.* **135,** 65.
Nopper, B., Kohen, F., and Wilchek, M. (1989). *Anal. Biochem.* **180,** 66.
O'Carra, P., and Barry, S. (1972). *FEBS Letters* **21,** 281.
O'Carra, P., Barry, S., and Griffin, T. (1973). *Biochem. Soc. Trans.* **1,** 289.
Ohlson, S. (1988). European Patent No. 0 290 406.
Ohlson, S., et al. (1989). *Trends Biotech.* **7,** 179.
O'Shannessy, D. J., and Quarles, R. H. (1985). *J. Applied Biochem.* **7,** 347.
O'Shannessy, D. J., and Wilchek, M. (1990). *Anal. Biochem.* **191,** 1.
O'Shannessy, D. J., Dobersen, M. J., and Quarles, R. H. (1984). *Immunol. Lett.* **8,** 273.
Pace, B., and Pace, N. R. (1980). *Anal. Biochem.* **107,** 128.
Palmer, J. L., and Nissonoff, A. (1963). *J. Biol. Chem.* **238,** 2393.
Parikh, I., Sica, V., Nola, E., Puca, G. A., and Cuatrecasas, P. (1974). *Meth. Enzymol.* **34,** 670.
Paul, R., and Anderson, G. W. (1962). *J. Org. Chem.* **27,** 2094.
Plaut, A. G., and Tomasi, T. B., Jr. (1970). *Proc. Natl. Acad. Sci. U.S.A.* **65,** 318.
Poe, M., Breeze, A. S., Wu, J. K., Short, C. R., and Hoogsteen, K. (1979). *J. Biol. Chem.* **254,** 1799.
Poonian, M. S., Schlabach, A., and Weissbach, A. (1971). *Biochemistry* **10,** 424.
Porath, J. (1974). *Meth. Enzymol.,* **34,** 13.
Porath, J. (1987). U. S. Patent No. 4,696,980.
Porath, J., and Axen, R. (1974). *Meth. Enzymol.* **34,** 41.
Porath, J., and Belew, M. (1983). In "Affinity Chromatography and Biological Recognition" (I. M. Chaiken, M. Wilchek, and I. Parikh, eds.) pp. 173. Academic Press, San Diego.
Porath, J., and Olin, B. (1983). *Biochemistry* **22,** 1621.
Porath, J., Carlsson, J., Olsson, I., and Belfrage, G. (1975). *Nature London* **258,** 598.
Porath, J., Olin, B., and Granstrand, B. (1983). *Arch. Biochem. Biophys.* **225,** 543.
Porath, J., Maisano, F., and Belew, M. (1985). *FEBS Lett.* **185,** 306.

Potuzak, H., and Dean, P. D. G. (1978). *FEBS Lett.* **88**, 161.
Potuzak, H., and Wintersberger, U. (1976). *FEBS Lett.* **63**, 167.
Rao, Y. S., and Filler, R. (1986). *In* "Heterocyclic Compounds" (I. J. Turchi, ed.), Vol. 45, pp. 361– . Wiley, New York.
Regnier, F. E., and Noel, R. (1976). *J. Chromatogr. Sci.* **14**, 316.
Renthal, R. P., Jordan, S., Steinman, A., and Stryer, L. (1976). *In* "Concanavalin A Tool" (H. Bittiger and H. Schnebli, eds.), pp. 429-434. Wiley, New York.
Riddles, P. W., Blakely, R. L., and Zermer, B. (1979). *Anal. Biochem.* **94**, 75.
Roitt, I. (1977). *In* "Essential Immunology," p. 21. Blackwell Scientific, London.
Rosenberg, M. (1974). *Nucleic Acids Res.* **1**, 653.
Rosenberg, M., and Gilham, P. T. (1971). *Biochim. Biophys. Acta* **246**, 337.
Rosenberg, M., Wiebers, J. L., and Gilham, P. T. (1972). *Biochemistry* **11**, 3623.
Rousseaux, J., *et al.* (1983). *J. Immunol. Meth.* **64**, 141.
Rudders, R. A., Andersen, J., and Fried, R. (1980). *J. Immunol.* **124**, 2347.
Sahni, G., Mallia, A. K., and Acharya, A. S. (1991). *Anal. Biochem.* **193**, 178.
Sakakabara, S., and Inukai, N. (1965). *Bull. Chem. Soc. Japan* **38**, 1979.
Scheller, F., Schubert, F., Renneberg, R., Muller, H. G., Janchen, M., and Weise, H. (1985). *Biosensors* **1**, 135.
Schmer, G. (1972). *Hoppe-Seyler's Z. Physiol. Chem.* **353**, 810.
Schneider, C., Newman, R. A., Sutherland, D. R., Asser, U., and Greaves, M. (1982). *J. Biol. Chem.* **257**, 10766.
Schott, H., Rudloff, E., Schmidt, P., Roychoudhury, R., and Kossel, H. (1973). *Biochemistry* **12**, 932.
Schrock, A. K., and Schuster, G. B. (1984). *J. Am. Chem. Soc.* **106**, 5228.
Schwartz, B. A., and Gray, G. R. (1977). *Arch. Biochem. Biophys.* **181**, 542.
Seppala, I., Sarvas, H., Peterfy, F., and Makela, O. (1981). *Scand. J. Immunol.* **14**, 335.
Shainoff, J. R. (1980). *Biochem. Biophys. Res. Commun.* **95**, 690.
Shainoff, J. R., and Dardik, B. N. (1981). *J. Immunol. Meth.* **42**, 229.
Shepard, E. G., De Beer, F. C., von Holt, C., and Hapgood, J. P. (1988). *Anal. Biochem.* **168**, 306.
Sica, V., Nola, E., Parikh, I., Puca, G. A., and Cuatrecasas, P. (1973a). *Nature (New Biol.)* **244**, 36.
Sica, V., Nola, E., Puca, G. A., Parikh, I., and Cuatrecasas, P. (1973b). *Fed. Proc.* **32**, 1297.
Singhal, R. P., Bajaj, R. K., Buess, C. M., Smoll, D. B., and Vakharia, V. N. (1980). *Anal. Biochem.* **109**, 1.
Smith, P. K., Krohn, R. I., Hermanson, G. T., Mallia, A. K., Gartner, F. H., Provenzano, M. D., Fujimoto, E. K., Goeke, N. M., Olson, B. J., and Klenk, D. C. (1985). *Anal. Biochem.* **150**, 76.
Sorensen, K., and Brodbeck, U. (1986). *J. Immunol. Meth.* **95**, 291.
Sorensen, P., Farber, N. M., and Krystal, G. (1986). *J. Biol. Chem.* **261**, 9094.
Staros, J. V., Wright, R. W., and Swingle, D. M. (1986). *Anal. Biochem.* **156**, 220.
Steers, E., Cuatrecasas, P., and Pollard, H. (1971). *J. Biol. Chem.* **246**, 196.
Stell, J. G. P., Warne, A. J., and Lee-Woolley, C. (1989). *J. Chrom.* **475**, 363.
Stewart, J. M., and Young, J. D. (1984). *In* "Solid Phase Peptide Synthesis," 2d Ed. pp. 20 and 33. Pierce Chemical Co., Rockford, Illinois.

Stich, T. M. (1990). *Anal. Biochem.* **191**, 343.
Sulkowski, E. (1985). *Trends Biotech.* **3**, 1.
Sundberg, L., and Porath, J. (1974). *J. Chrom.* **90**, 87.
Taylor, R. F. (1991). *In* "Protein Immobilization: Fundamentals and Applications," pp. 263. Marcel Dekker, New York.
Tenner, A. J., Lesavre, P. H., and Cooper, N. R. (1981). *J. Immunol.* **127**, 648.
Ternyck, T., and Avrameas, S. (1972). *FEBS Lett.* **23**, 24.
Then, R. L. (1979). *Anal. Biochem.* **100**, 122.
Thomas, D., and Brown, G. (1976). *Meth. Enzymol.* **44**, 901.
Tijssen, P. (1985). *In* "Practice and Theory of Enzyme Immunoassays," pp. 297. Elsevier, New York.
Towbin, H., Stauhelin, T., and Gordon, J. (1979). *Proc. Natl. Acad. Sci. U.S.A.* **76**, 4350.
Traut, R. R., Casiano, C., and Zecherle, N. (1989). *In* "Protein Function, A Practical Approach" (T. E. Creighton, ed.), pp. 101– . IRL Press.
Travis, J., Bowen, J., Tewksbury, D., Johnson, D., and Pannell, R. (1976). *Biochem. J.* **157**, 301.
Uy, R., and Wold, F. (1977). *Anal. Biochem.* **81**, 96.
Valkirs, G. E., and Barton, R. (1985). *Clin. Chem.* **31**, 1427.
Vance, D. E., Weinstein, D. B., and Steinberg, D. (1984). *Biochim. Biophys. Acta* **792**, 39.
Villarejo, M. R., and Zabin, I. (1973). *Nature (New Biol.)* **242**, 50.
Vogt, R. F., Phillips, D. L., Henderson, L. O., Whitfield, W., and Spierto, F. W. (1987). *J. Immunol. Meth.* **101**, 43.
Vretblad, P. (1976). *Biochim. Biophys. Acta* **434**, 169.
Vunakis, H. V., and Langone, J. (1980). *Meth. Enzymol.* **70**, 127.
Watson, D. H., Harvey, M. J., and Dean, P. J. (1978). *Biochem. J.* **173**, 591.
Weetall, H. H. (1976). *Meth. Enzymol.* **44**, 134.
Weinstein, D. B. (1979). *Circulation (Suppl. II)* **60**, 204.
Weissbach, A., and Poonian, M. (1974). *Meth. Enzymol.* **34**, 463.
Weith, H. L., Wiebers, J. L., and Gilham, P. T. (1970). *Biochemistry* **9**, 4396.
Weston, P. D., and Avrameas, S. (1971). *Biochem. Biophys. Res. Commun.* **45**, 1574.
Whiteley, J. M., Henderson, G. B., Russell, A., Singh, P., and Zevely, E. M. (1977). *Anal. Biochem.* **79**, 42.
Wilchek, M., and Bayer, E. A. (1990). *Meth. Enzymol.* **184**, .
Wilchek, M., and Miron, T. (1985). *Appl. Biochem. Biotech.* **11**, 191.
Worthington Enzyme Manual (1988). Worthington Biochemical Corporation, Freehold, New Jersey 07728
Yamamoto, N., Nagasawa, Y., Sluto, S., Tsulomura, H., Sawai, M., and Okumura, H. (1980). *Clin. Chem.* **26**, 1569.
Yoshitake, S., Imagawa, M., Ishikawa, E., Niitsu, Y., Urushizaki, I., Nishiura, M., Kanazawa, R., Kurosaki, H., Tachibana, S., Nakazawa, N., and Ogawa, H. (1982). *J. Biochem.* **92**, 1413.
Zimmerman, D., and Van Regenmortel, M. H. V. (1989). *Arch. Virol.* **106**, 15.

Books on Immobilized Affinity Ligand Technology

Carr, P. W., and Bowers, L. D. (1980). "Immobilized Enzymes in Analytical and Clinical Chemistry, Fundamentals and Applications." John Wiley & Sons, New York.

Chaiken, I. M., Wilchek, M., and Parikh, I. (1983)."Affinity Chromatography and Biological Recognition." Academic Press, New York.

Chibata, I. (1978). "Immobilized Enzymes: Research and Development." John Wiley & Sons, New York.

Dean, P. D. G., Johnson, W. S., and Middle, F. A. (1985). "Affinity Chromatography: A Practical Approach." IRL Press, Oxford.

Dunlap, R. B. (1973). "Immobilized Biochemicals and Affinity Chromatography." Plenum Press, New York.

Egly, J-M. (1979). "Affinity Chromatography and Molecular Interactions." INSERM, Paris.

Gribnau, T. C. J., Visser, J., and Nivard, R. J. F. (1982). "Affinity Chromatography and Related Techniques: Theoretical Aspects/Industrial and Biomedical Applications." Elsevier, New York.

Gutcho, S. J. (1974). "Immobilized Enzymes: Preparation and Engineering Techniques." Noyes Data Corporation, Park Ridge, New Jersey.

Hutchens, T. W. (1989). "Protein Recognition of Immobilized Ligands." Alan R. Liss, New York.

Jakoby, W. B., and Wilchek, M. (1974). "Methods in Enzymology," Vol. 34. Academic Press, New York.

Janson, J-C., and Ryden, L. (1989). "Protein Purification. Principles, High Resolution Methods, and Applications." VCH Publishers, New York.

Lowe, C. R., and Dean, P. D. G. (1974). "Affinity Chromatography." John Wiley & Sons, New York.

Mohr, P., and Pommerening, K. (1985). "Affinity Chromatography: Practical and Theoretical Aspects." Marcel Dekker, New York.

Mosbach, K. (1976). "Methods in Enzymology," Vol. 44. Academic Press, New York.
Mosbach, K. (1987). "Methods in Enzymology," Vol. 135. Academic Press, New York.
Mosbach, K. (1987). "Methods in Enzymology," Vol. 136. Academic Press, New York.
Mosbach, K. (1987). "Methods in Enzymology," Vol. 137. Academic Press, New York.

Schott, H. (1984). "Affinity Chromatography, Template Chromatography of Nucleic Acids and Proteins." Marcel Dekker, New York.

Scouten, W. H. (1981). "Affinity Chromatography: Bioselective Adsorption on Inert Matrices." John Wiley & Sons, New York.

Sundaram, P. V., and Eckstein, F. (1978). "Theory and Practice in Affinity Techniques." Academic Press, New York.

Taylor, R.F. (1991). "Protein Immobilization. Fundamentals and Applications." Marcel Dekker, New York.

Turkova, J. (1978). "Affinity Chromatography." Elsevier, New York.

Wingard, L. B., Jr., Katchalski-Katzir, E., and Goldstein, L. (1976). "Immobilized Enzyme Principles." Academic Press, New York.

INDEX

Acetic anhydride, 90, 117, 139, 156, 158, 160
  use to cap excess amine terminal
    spacers, 158
Acetonitrile, 135
Acetophenone, 127
Acetylation, 117
Acetylenes, 127
N-Acetyl-galactosamine, 324
N-Acetyl-D-glucosamine, 323
  immobilized, 322
N-Acetyl-DL-homocysteine thiolactone, 279
  immobilization of, 277
N-Acetyl-homocysteine, 279
  immobilized, 277, 383, 385
N-Acetyl-galactosamine
  coupling by epoxy, 164
  immobilized, 325
N-Acetyl-glucosamine, 164, 250, 326, 337
  coupling by epoxy, 161
  immobilized, 322
Acid protease, 172, 205
Ackers, G. K., 310
Acrylamide, 16, 19
N-Acryloyl-2-amino-2-hydroxymethyl-1,
    3-propane diol, 19
N-Alkyl carbamate linkage, 64
Activated carbon, 105
Activated charcoal, 351
Activation
  criteria to consider, 52
  general considerations, 51
  methods, 51–136
Activation level, 281
  assay by ninhydrin, 282
  measurement of, 281
Active groups
  blocking excess, 156
Active hydrogen, 120, 127, 128, 131, 177, 178,
    186, 190, 193, 194, 211
Activity
  antibody binding, 334
  determination of, 313
  effect of ligand density, 313
  measurement of, 315
Acute phase proteins, 403
Acylation, 361
O-Acylisourea, 81
Adams, R., 127
Adipic dihydrazide, 43, 110, 112, 287, 288
β-Adrenergic receptors, 187, 188
Adsorption, 413
  of antibody to polystyrene, 232.
  See also passive adsorption
Affi-Gel 10, 57
Affi-Gel 15, 57
Affinity electrodes, 388, 410, 414
Affinity purification, 317
Affinity techniques, 294
Afting, E.G., 172
Agar
  entrapment of DNA in, 257
Agarose, 1, 2, 4, 5, 6, 25, 54, 57, 75, 85, 90,
    142, 154, 161, 166, 214, 223, 281, 331,
    395, 399, 400, 401. See also Sepharose
  boronate derivative, 273
  cross-linked, 7
  freeze-dried, 8
  handling, 8

431

Agarose *(continued)*
  immobilization methods, 8
  mechanical stability, 9
  non-cross-linked, 8, 27
  periodate oxidation of, 69
  pH compatibility, 9
  pore structure, 7
  solvent compatibility, 9
  stability, 8
  structure and properties, 6
Akerstrom, B., 246
Alanine, 150, 372
β-Alanine, 150-152
  immobilization protocol, 151
Alberts, B. M., 256
Albumin, 174, 195, 246, 248, 275, 351
  coupling via CNBr, 196
  purification, 342
  removal of fatty acids from, 351
Aldehyde, 33, 49, 69, 76, 110, 115, 127, 197, 200, 204, 206, 207, 208, 209, 254, 255, 262
  reduction of, 70
Aldehydo-RNA, 270
Aldimine, 390
Alhanaty, E., 358
Alkaline phosphatase, 404
Alkaline phosphatase-avidin, 319
Alkyl chloroformates, 90
Alkylation, 115
Allen, D. W., 389
Allyl dextran, 23
Allyl glycidyl ether, 37
Alprenolol, 187, 188
Alumina, 11
  coating with silane, 14
Aluminum, 180
Amadori rearrangement, 390
Amethopterin, 186
Amide groups
  activation with glutaraldehyde, 77
Aminated epoxides, 153
Amine reactive chemistries, 53
Amines, 395
Amino acids
  use as spacers, 150
Aminoalkylated, 127
4-Aminobenzoic hydrazide
  immobilized, 262

p-Aminobenzamidine, 167, 169, 318, 317, 355, 359
p-Aminobenzylalkyl group, 120
6-Aminocaproic acid, 147, 167, 352
  modification with ethylene diamine, 147, 148
  use as a spacer, 146
S-2-Aminoethylcysteine, 361, 366
Aminoethyl polyacrylamide, 116
Aminoethyl-cellulose, 272
4-Amino-3-hydroxy-6-methylheptanoic acid, 172
Aminophenyl boronic acid
  immobilized, 388, 390, 392
  ligand loading, 393
3-Aminopropionic acid. *See* β-Alanine
Aminopropyl triethoxysilane, 14
Ammonia, 127
Ammonium sulfate, 183, 309, 317, 322, 324, 326, 327, 344
AMP, 273
Amperometric, 410
Amphetamines, 399
Anderson, G. W., 57, 64, 343
Andersson, L., 179, 180
Androsterone, 190
Angle of nonreflectance, 416
Anhydrides, 153
Anhydrochymotrypsin
  immobilized, 370, 371
3-Anhydrogalactose, 6
Anhydrotrypsin, 203
  immobilized, 365, 368
Aniline, 97, 130
Anthraquinone, 174
Anti-a$_2$-macroglobulin
  coupling via protein A/DMP, 224
  immobilized, 329
Anti-antibody, 363
α$_1$-Antichymotrypsin, 174
Anti-HSA
  immobilized, 329
Anti-human CRP-//-alkaline phosphatase
  conjugate, 404
Anti-human serum albumin, 223, 228
Antibiotic, 354
Antibody, 42, 49, 62, 95, 112, 195, 206, 210, 225, 232, 235, 239, 333, 334, 377, 388, 396, 398, 403, 405, 407, 410. *See also* Immunoglobulin

anti-human serum albumin, 223
  coupling through carbohydrates, 230
  coupling through sulfhydryls, 230
  coupling to glutaraldehyde treated microplates, 240
  coupling to hydrazide polystyrene balls, 236, 237
  coupling to microplates, 240, 241
  coupling to microplates using photochemical reagents, 241
  coupling to polystyrene, 231, 233
  coupling to polystyrene plates using vinyl monomers, 242
  coupling to polystyrene using Sulfo-SANPAH, 241
  fragments, 204
  fragmentation by pepsin, 218, 372
  fragmentation by papain, 218, 377
  IgG structure, 218
  immobilization, 217, 331
  immobilized, 42, 329
  labeled, 400
  monoclonal, 218
  oxidation by periodate, 230
  oxidized, 238
  passive adsorption to polystyrene, 231
  pH denaturation, 308
  primary, 400
  purification, 210, 317
  secondary, 400
  stucture, 218
Antibody-enzyme conjugate, 401, 402, 403, 404, 405, 407, 409
Anticoagulant, 391
Antigen, 49, 210, 218, 224, 226, 228, 233, 239, 244, 251, 333, 334, 377, 388, 397, 398, 403, 405, 407, 409, 410
  immobilization, 210
  labeled, 399
  purification, 317, 329
Antigen binding, 362, 373
  activity, 112
  site, 218, 224, 226, 228, 244, 363
Antigen-antibody complex, 399
Antineoplastic agent, 186
Antisera, 210, 373
Antithrombin III, 174
α1-Antitrypsin, 174
Apolipoproteins, 352
Aprotinin, 356

  immobilized, 355, 356, 357
Arginine, 167, 206, 361, 364, 367
  trypsin cleavage at, 203
Arnold, F. H., 179
Artocarpus integrifolia, 254, 325, 326
Aryl azide, 132
Aryl functionalities, 43
Arylchlorosulfonyl groups, 43
Arylsulfonamides, 43
Ascites, 210, 331, 336, 337, 338, 341, 397
Aspartate, 150, 207, 211, 222
ATEE, 380
ATP, 174
Aucouturier, P., 254, 326
Autolytic products, 317, 319
Avidin, 44, 138, 169, 171, 172, 197, 201, 308, 320
  assay of activity, 288
  conjugates, 319
  coupling via CNBr, 199
  coupling via reductive amination, 200
  monomeric, 200
  purification, 319
Avrameas, S., 77
Axelrod, A.E., 170
Axen, R., 53, 104, 105
Azlactone, 4, 5, 28, 30, 90
  activation and coupling chemistry, 90
  formation of, 90
  ligand coupling protocol, 95
  reactivity, 28, 90
Azlactone beads, 28, 90
  chemical stability, 29
  handling, 30
  immobilization methods, 30
  mechanical stability, 28
  structure and properties, 28

Bacillus sp., 248
BAEE, 203, 205, 318
  measurement of papain activity using, 206
Bandieria simplicifolia, 158
Barrett, M. J., 239
Barry, S., 144
Bartling, G. J., 64
Barton, R., 406
Baues, R. J., 115, 116, 158, 160
Bayer, E. A., 110, 169, 198, 288, 319

BCA, 284, 368
  assay of hydrazides, 287
  reagent preparation, 284
Beale, D., 364, 375
Beaty, N. B., 200
Becker, M. I., 172
Beissinger, R. L., 412, 413
Belew, M., 179, 274, 340, 341
Benzamidine, 167, 204
α-N-Benzoyl-L-arginine, 203, 367
Benzylamine, 185
Bhatia, S. K., 100
Biagioni, S., 265
Bicinchoninic acid. See BCA
Bifunctional cross-linking reagents, 236
Binding buffer, 296, 297, 304, 312, 314, 319, 329, 334
  addition of blocking agents, 307
  addition of counterligand, 308
  degassing, 301
  effect of pH, 304
  effect of salt strength, 305
  for α_2-macroglobulin on metal chelate supports, 345
  for albumin on immobilized Cibacron Blue, 342
  for avidin on immobilized iminobiotin, 320
  for chymotryptic fragments on immobilized anhydrochymotrypsin, 371
  for glycated hemoglobin on immobilized boronic acid, 392
  for IgM purification on immobilized C1q, 336
  for immunoaffinity binding of HSA on anti-HSA, 329
  for immunoaffinity of α2-macroglobulin on anti-α_2-mac, 331
  for immunoglobulin binding proteins 332
  for isotype elution of mouse IgGs on protein A, 335
  for measurement of capacity, 314
  for phosphoproteins on iron chelate affinity support, 343
  for purification of IgM on immobilized MBP, 338
  for removal of trypsin using immobilized soybean trypsin inhibitor, 358
  for removing trypsin on immobilized aprotinin, 356
  for thiophilic adsorption, 341
  for trypsin on immobilized benzamidine, 319
  for tryptic fragments on immobilized anhydrotrypsin, 368
  inclusion of salt, 307
  optimization, 304
  presence of metal ions, 305
  recipes, 305
  sample equilibration in, 313
  use in differential elution techniques, 397
Binding capacity, 314
  flow rate dependence, 310
Binding site, 137
Bio-Beads SM-2, 346, 350
Bio-Gel P, 16
Biosensor, 388, 410, 413, 414, 416
  design, 412
  ligand immobilization, 412
Biospecific immobilization surface, 412
Biotin, 44, 138, 169, 170, 197, 200, 201, 308, 319
Biotin binding sites, 290
Biotin-p-nitrophenyl ester, 288, 289
Biotinylated protein, 169, 288
Bisoxirane, 107, 118, 153, 161, 163, 164, 188
Bjorck, L., 246
Blocking agents, 42, 139, 156, 233, 236, 244
  best choices of, 157
  common additives, 307
  for immunoassays, 307
  prevention of nonspecific binding to polystyrene, 232
Blood clotting factors, 174
Blood collection tubes, 391
Blood glucose, 388
Blood proteins
  fractionation of, 174
Blotting, 48
BLOTTO, 307
Bolton, E. T., 258
Borch, R. F., 69, 70, 115
Boronic acid
  complex formation with RNA, 272
  immobilized, 388, 390, 391
Bovine serum albumin, 307
Bradford, M. M., 290
Britten, R. J., 257
Brocklehurst, K., 104, 105
Brodbeck, U., 284

Bromelain, 379
Bromoacetic acid, 99, 181, 183
Bromoacetyl, 98, 214
  activation protocol, 98
Brown, G., 100, 413
BSA, 195, 233, 284, 307, 350, 354, 403, 409
  removal of endotoxin from, 355
Bucher, K., 233, 307
Buffy coat, 392
Bunn, H. F., 390
1,4-Butanediol diglycidyl ether, 118, 153, 161, 163, 181, 268

C-Reactive protein. *See* CRP
C-Terminal, 366, 370, 380
C1q, 250, 336
  immobilization, 336
  isolation, 336
Cab-O-Sil, 351, 352
Cadmium, 180
Calcium, 304, 309, 319, 327, 337, 356, 358, 366, 368, 371, 403
  presence in binding buffers, 306
Calorimeters, 411
cAMP, 273
Capacitance, 410
Capacity, 3, 4
  determination of, 313
  effect of flow rate, 314
  effect of ligand density, 313
  effect of sample, 313
  measurement by overload, 314
  total, 313
Capping agent. *See* Blocking agent
Carbodiimide, 49, 81, 82, 90, 147, 148, 167, 170, 213, 236, 263, 279, 413, 415. *See also* EDC, CMC, and DCC
Carbohydrate, 252, 397
  oxidation by sodium periodate, 110, 113
  reducing, 161
  reductive amination coupling, 115
Carbohydrate binding proteins. *See* lectins
Carbonates, 64
Carbonyl diimidazole, 52, 57, 64. *See also* CDI
  activation and coupling chemistry, 64
Carbonyl reactive chemistries, 110–118
Carboxamide, 18, 19
Carboxyl proteases, 172
Carboxylated dextran, 415, 416

Carboxymethyl cellulose, 2
Carboxypeptidase Y, 380
Carica papaya latex, 206, 377
Carnation Instant Milk, 233
Caron, M. G., 188
Carrier protein, 214
Casein, 233, 307
Castor bean lectin, 321, 324, 325
Castor beans
  extraction of, 323, 324
Catalysts
  immobilized, 360–386
Catecholamine hormones, 187
Cathepsin D, 172
Cayley, P. J., 186
CDI, 11, 21, 23, 33, 36, 52, 58, 64, 142, 184, 185, 282, 291, 413
  activation of carboxylate groups, 64
  activation of hydroxyl groups, 64
  coupling of DADPA, 143
  coupling of octylamine, 184
  coupling of peptides, 214
  hydrolysis rate, 65
  ligand coupling protocol, 68
  ligand leakage, 65
Cell culture supernatant, 210, 331, 397
Cellibiose, 161
Cells
  immobilization, 413
Cellulose, 9, 47, 54, 75, 257, 263, 265, 267, 268, 269, 399, 412
  coupling DNA to, 255, 257, 265, 268
  coupling of nucleotide homopolymers, 270
  coupling oligo(dT) to, 264
  handling, 11
  immobilization methods, 11
  structure and properties, 10
Cellulose acetate
  entrapment of DNA, 257
Chaotropic, 183, 197, 308, 334, 341
  Hofmeister series, 309
Chelation compounds, 179
Chelators, 396
  as elution agents, 309
Chimeric protein, 247
Chitin, 1
  immobilized, 326
Chloramine T, immobilized, 41
Chlorosulfonic acid, 43
Chlorotriazine, 176

Cholesterol
  removal of, 351
Chromatography. See also Columns
  column packing and use, 295
  large scale, 300
Chudzik, S. J., 50, 135
Chymotrypsin, 203, 358, 369, 370
  immobilized, 369
  removal of, 356
  chymotryptic digests
    fractionation of, 370
Cibacron Blue F3GA, 174, 175, 176
  immobilized, 342
Cibacron Brilliant Blue FBR-P, 175
Clark, L. C., 412
Clark pO_2 electrode, 412
Cleland, W. W., 275
Cleland's reagent, 276
Clonis, Y. D., 174, 176
CM-Cellulose, 267
  coupling of DNA to, 267
CMC, 263
CNBr, 21, 23, 27, 33, 36, 52, 53, 55, 65, 146, 166, 196, 199, 201, 245, 251, 252, 258, 336, 338, 354, 358, 413
  coupling of albumin, 196
  coupling of anti-human serum albumin, 223
  coupling of MBP, 250
  coupling of nucleic acids, 258
  coupling of poly(A), 259
  ligand coupling protocol, 56
  ligand leakage, 55
CoA dependent enzymes, 174
Coagulation factors, 166
Cobalt, 180
Coleman, P. L., 90
Collagen, 412
Collier, R., 308
Columns
  dimensions versus flow rate, 311
  height, 300
  packing, 301
    large, 300
      packing and use, 300
      use of, 302
    mini, 295
      packing and use, 294
Complement, 336, 337
Complement proteins, 174

Concanavalin A, 252, 254
  immobilization protocol, 253
  use in HPLAC, 397
Conjugates, 363
Contact time, 310
Contaminants
  list of those removed by affinity
    techniques, 345
  removal of, 345–360
Controlled pore glass, 11. See also CPG
  structure and properties, 12
Coomassie Blue G-250
  assay of coupled protein, 290
Coordination complexes, 179, 180, 181
Coordination sites, 179
Copper, 180, 284, 287
Coulter, A., 218, 377
Counter ligand, 201, 308, 309, 341
Covalent chromatography, 151
CPG, 11
  chemical stability, 12
  coating with silane, 14
  covalent coating, 12
  immobilization methods, 12
  mechanical stability, 12
  nonspecific binding, 14
CPY, 380
Crestfield, A. M., 368
Critical micelle concentration, 346
Cross-linking agent, 43, 50, 239, 241
  photoreactive, 132
CRP, 403, 404
  clinical range, 405
Cuatrecasas, P., 55, 104, 120, 138, 139, 153, 287, 310, 314
Cummings, R.D., 252, 253
Cyanate esters, 54
Cyanide, 75
Cyanogen bromide, 8, 53, 166. See also CNBr
  activation and coupling chemistry, 53
  coupling of heparin, 166
Cyanuric chloride, 95, 119, 266
  activation protocol, 97
  coupling of DNA, 265
  ligand coupling protocol, 97
Cyclization catalyst, 90
1-Cyclohexyl-3-(-2-morpholinethyl)
  carbodiimide metho-p-toluene-
  sulfonate. See CMC

Cysteine, 107, 151, 156, 179, 206, 211, 214,
    215, 230, 244, 275, 276, 379, 382
Cystine, 214
Cytomegalovirus, 377

D-Galactose, 6, 327
D-Glucosamine, 165
D-Glucuronic acid, 165
DADPA, 123, 130, 131, 143, 170, 177, 186,
    190, 194, 213, 277, 279, 282, 292
  immobilization protocol, 142
  neutralization of, 142
  succinylated, 167, 213
  succinylation of, 154
  use as a spacer, 141
DAH. See Diaminohexane
Dardik, B. N., 69
DCC, 264
Deamidation
  of albumin, 351
Dean, P. G. D., 81, 255, 313
Degassing, 301
Dehydroalanine, 365, 370
Dehydrogenases, 174
Delipidation, 351
Denaturation, 308
Denaturing agents, 386
Densitometer, 406
Density
  ligand, 313
Detergents, 346
  removal of, 347–350
  removal using Bio-Beads SM-2, 350
  removal using Extracti-Gel, 347
Detoxi-Gel, 311
Dextran, 23, 54
Dextran sulfate, 167
Dey, P. M., 116
Diabetes mellitus, 388
N,N'-Diallyltartradiamide, 19
Diaminodipropylamine, 99, 108, 112, 123,
    131, 168, 172, 191, 194. See also
    DADPA
1,6-Diaminohexane, 186
  use as a spacer, 144
Diazonium, 43, 120, 176
  activation and coupling protocol, 123
  alternative to, 128
  chemistry of activation and coupling, 120

coupling of DNA, 260
coupling of peptides, 216
disadvantages of, 120
peptide coupling, 216
Dichlorotriazine, 175
Dicyclohexylcarbodiimide. See DCC
Dicyclohexylurea, 265
Diethanolamine, 404
Diethyl ether, 324
Diffusion, 310, 400, 406, 408
1,5-Difluoro-2,4-dinitrobenzene, 43
Digoxin, 399
Dihydrofolate reductase, 186
Dihydrolipoamide
  immobilized, 277
N,N-Diisopropylethylamine, 97
Diisopropylfluorophosphate, 371
Dimethyl pimelimidate (DMP)
  use in antibody immobilization, 224
Dintzis, H. M., 120, 149, 160, 286
2,2'-Dipyridyl disulfide, 104, 105
Disaccharides, 116
  reducing, 116
  reductive amination coupling, 116
Disks, porous, 295, 296, 297, 299, 402, 404
N,N'-Disuccinimidyl carbonate (DSC,) 58
Disuccinimidyl succinate, 58
Disulfides, 59, 211, 214, 218, 226, 235, 358,
    360, 373
  of IgM, 363
  immobilized disulfide reductants, 151, 277,
    382, 383
  interaction with gold surfaces, 413
  reduction of, 214, 226, 275, 382, 385, 386
5,5'-Dithiobis(2-nitrobenzoic acid). See
    Ellman's reagent
Dithiobis-succininimidyl propionate. See DSP
Dithioerythritol. See DTE
Dithiothreitol. See DTT
Divinylsulfone, 21, 23, 27, 33, 36, 52, 88, 107,
    119, 164, 214, 274
  activation and coupling chemistry, 88
  coupling of mannose, 164
  ligand coupling protocol, 89
  preparation of thiophilic supports, 274
DMP, 226
  use in antibody immobilization, 224
DMSO, 135
DNA, 256, 257, 263, 267, 268, 269, 270
  coupling by esterification, 267

DNA *(continued)*
  coupling by irradiation, 257
  coupling to cellulose by adsorption, 255
  coupling to polyacrylamide, 258
  coupling via carbodiimide, 263
  coupling via CNBr, 258
  coupling via cyanuric chloride, 265
  coupling via diazonium, 259
  coupling via epoxy, 268
  immobilization by entrapment, 257
  immobilization protocols, 255
  single stranded
    coupling via diazonium, 260
Domains, 373, 375
Domen, P. L., 73, 112, 226, 229
Dot-blot, 406, 409
Douglas, J. T., 233, 307
Drugs, 127, 185
DSC, 59, 60, 63
DSP, 59, 60, 413
DTE, 275, 276, 382
DTT, 188, 214, 275, 276, 352, 382, 383, 384
Dubois, M., 161
Dulaney, J. T., 253
Duncan, R. E., 273
DVS. *See* Divinylsulfone
Dyes, 174, 176, 396
  immobilized, 173, 341
Dzau, V. J., 172

Easy-Titer ELIFA System, 407
EDA. *See* Ethylene diamine
EDC, 81, 99, 110, 112, 146, 148, 150, 153, 167, 169, 170, 173, 186, 190, 213, 236, 279, 413, 415
  activation and coupling chemistry, 81
  coupling of peptides, 213
  hydrolysis rate, 81
  ligand coupling protocol, 83
  use with sulfo-NHS, 82, 85, 416
EDC/Sulfo-NHS, 416
EDTA, 305, 309, 337, 352, 379, 383, 384, 391
Effector functions
  of immunoglobulins, 363, 373
EGTA, 309
EIA, 399, 405, 406, 410
  heterogeneous, 399
  homogeneous, 399
Eldjarn, L., 277

Eliasson, M., 247
ELIFA, 406, 408
ELISA, 41, 45, 399, 400, 406, 407, 408
  affinity-mediated, 403
  limiting diffusion, 406
  sandwich assay, 400
Ellman, G. L., 293
Ellman's reagent, 109, 215
  measurement of sulfhydryls, 383, 384
  measurement of immobilized sulfhydryls, 293
  preparation of, 294
Elution
  buffers, 315
    for immunoglobulin binding proteins, 333
    gentle, 333
    in immunoaffinity techniques, 333
    isotype, 334
  chaotropic agents, 308
  couterligands, 308
  differential, 309, 397
  gradient, 310
  in immunoaffinity chromatography, 308
  metal ion removal, 309
  methods of, 307
  pH shift, 308
  in weak affinity chromatography, 397
Emphase, 90
Endoprotease, 344, 359, 372
Endotoxin, 311, 345
  removal of, 352, 353, 355
Engvall, E., 233
Enterokinase, 317, 359
Entrapment, 413
Enzacryl, 18
Enzymatic digestion, 318
  of immunoglobulins, 363
Enzyme conjugate, 398
Enzyme-linked immunoassay. *See* EIA
Enzyme-linked immunosorbent assays. *See* ELISA
Enzyme-antibody conjugates, 253
Enzyme-linked immunofiltration assay. *See* ELIFA
Enzymes, 166, 195, 345, 372, 410, 413
  immobilization of, 202
  immobilized, 360, 376
    advantages, 202
    uses, 360

measurement of activity, 315
milk clotting, 172
Epichlorohydrin, 153, 181, 182
Epoxides, 14, 21, 23, 27, 33, 36, 37, 69, 70, 88, 107, 118, 153, 164, 181, 214, 269, 270, 395
  activation protocol, 118
  coupling of alprenolol, 188
  coupling of DNA, 268
  coupling of iminodiacetic acid, 180
  coupling of lactose, 161
  coupling of N-acetylgalactosamine, 164
  coupling of N-acetylglucosamine, 164
  coupling of nucleotide homopolymers, 270
  ligand coupling protocol, 119
Erythroagglutinin, 323
Esterase, 205, 377, 380
Esterification, 267
Esters, 127
17β-Estradiol, 128,188, 190
Estradiol-17-β-hemisuccinate, 190
Estrogen, 188
Ethanolamine, 56, 95, 119, 139, 156, 167, 196, 199, 201, 206, 207, 226, 251, 267, 270
Ethylene diamine, 18, 158, 160, 181, 183, 282
  modification of polyacrylamide, 149
  use as a spacer, 147
Ethylene dimethacrylate, 34
Ethylene glycol, 309
1-Ethyl-3-(3-dimethylaminopropyl) carbodiimide hydrochoride. *See* EDC
17α-Ethynylestradiol-3-methyl ether, 194
Ethynyl steroids, 194
Eupergit, 5, 37
  chemical stability, 37
  handling, 39
  immobilization methods, 37
  mechanical stability, 37
  structure and properties, 37
Exopeptidase, 380
Extracti-Gel D, 346, 348, 349
  detergent removal protocol, 348
  exclusion limit, 348
Ey, P. L., 245, 334

F(ab′)$_2$, 109, 205, 218, 235, 363, 366, 372, 373, 374, 375, 377
  coupling to TNB-thiol matrices, 109
  coupling through sulfhydryls, 230
  coupling via iodoacetyl, 226
  reduction with mercaptoethylamine, 226
Fab, 205, 206, 218, 362, 363, 373, 375, 377, 379
  interaction with protein G, 246
Fab*, 376
Factor Xa, 358
Farmer, M. C., 412
Farooqui, A. A., 166
Fatty acids
  removal from albumin, 351
  removal of, 351
Faulmann, E. L., 251
Fc, 205, 218, 224, 226, 246, 247, 251, 331, 336, 237, 245, 363, 373, 374, 376, 377, 379
  carbohydrate, 221
  degradation by pepsin, 218
  interaction with protein G, 246
  protein A interaction, 245
  receptors, 377
Fc′, 376
Fc5μ, 362, 364, 366, 376
Fe$_3$O$_4$, 25
Feeney, R. E., 115
Ferric chloride, 343
Ferric ions
  immobilized, 343
Ferritin-avidin, 319
α-Fetoprotein, 174
Fiberglass, 257
Fibronectin, 403
Filler, R., 90
Fines, 3, 34
Finaly, T. H., 97
Fleet, G. W. J., 132
Flow rate, 300, 310, 394, 395
  during column packing, 301
  effect on capacity, 314
  factors influencing, 311
  optimization, 310
Fluorescent tag, 398
Fluoride, 391
2-Fluoro-1-methylpyridinium toluene-4-sulfonate. *See* FMP
FMP, 80, 245
  activation and coupling chemistry, 79
  coupling of peptides, 214
  ligand coupling protocol, 80
Folic acid, 186

Formaldehyde, 44, 127, 128, 131, 177, 178, 191, 194, 216
Formyl groups, 33, 49, 69, 110, 133, 151, 160, 206, 222, 228, 236, 240, 254, 255, 360, 395. *See also* Aldehydes
Fornstedt, N., 164
Fraction collectors, 300
Fragmentation
 of IgM, 375
 of immunoglobulins, 362, 373
Free radical addition, 44, 188
Freytag, J. W., 406
Frontal analysis, 314
Fv, 205, 363, 375, 376, 377

Gaber, B. P. 412
Galactose, 324, 325, 326
β-D-Galactosidase, 208, 209
Gallium, 180
Gel filtration, 23, 26, 317, 325
Gelatin, 233, 307, 412
 immobilized, 403
Gilham, P. T., 264, 272, 273
Glass, 5, 11, 59, 119, 399, 412
 coating, 307
 nonspecific effects, 306
 stability, 12
Glatz, J. F. C., 351
α-Globin fragments, 381
Glucosamine, 166
Glucose, 10, 388, 390
 biosensor, 412
Glucose analyzer, 412
Glucose oxidase
 immobilized, 412
Glucose-modified hemoglobin. *See* GlyHb
Glutamic acid, 150, 207, 211, 222, 381
Glutamine, 206
Glutamoyl peptide, 382
Glutaraldehyde, 18, 27, 49, 77, 78, 236, 240, 404, 413
 activation of polystyrene microplates, 239
 modification of aminophenyl groups of polystyrene, 240
Glutaric anhydride, 153, 156
Glutathione, 277, 352, 384
 reduction of, 384
Glycated hemoglobin. *See* GlyHb

Glycation
 of hemoglobin, 389
Glycerol, 204, 292, 293
Glycidol, 21, 33, 36, 69, 70, 151, 246, 247
 oxidation by periodate, 76
Glycidoxypropyl silane, 395
γ-Glycidoxypropyl trimethoxysilane, 14
Glycidyl methacrylate, 31, 37
Glycine, 150, 156, 202, 206, 306, 308, 372, 379
Glyco Gel, 392
Glycolipids, 253
Glycopeptide, 366
Glycoproteins, 252
 coupling to a hydrazide matrix, 110
 oxidation by sodium periodate, 113
Glycosaminoglycans, 158
Glycosidases, 116, 202
Glycosides, 397
GlyHb, 388, 389, 390, 392
 elevated levels, 390
 measurement by affinity chromatography, 389
 percent, 393
Goding, J. W., 218
Gold, 415, 416
 activation with DSP, 414
 disulfide interaction with, 413
Gold foil electrodes, 59, 413
Gold surfaces, 59
Goldstein, I. J., 326
Gorecki, M., 277, 382
Gradient elution, 310
Graft copolymer, 44
Graft copolymerization, 242
 to polystyrene, 240
Gram-negative bacteria, 352, 355
Gravel, R. A., 200
Gravity flow, 311
Gray, G. R., 115, 116, 158, 160
Green, A. A., 197, 309, 321
Griffonia simplicifolia, 321
 purification of lectin, 326
Guanidine hydrochloride, 308, 309, 319, 343, 386
Guanine, 259
Guilbault, G. G., 410
Guire, P. E., 50, 135
Gutcho, 360

Index

H_2SO_4, 43
HABA, 321
Hammen, R. F., 396
Harlow, E., 210, 233, 307
Harris, R. G., 141, 218, 377
Hauri, H. -P., 233, 307
Hayes, C. E., 326
HbA_1, 389
β-HCG, 407
HDL, 352
Hearn, M. T. W., 64
Heavy chains, 218, 220, 226, 228, 235, 254, 325, 373, 377
Hecht, S., 273
Heegaard,P. M. H. 274, 340
Helseth, D. L., 172
HEMA, 34, 395
  chemical stability, 34
  handling, 36
  immobilization methods, 36
  mechanical stability, 34
  particle size, 36
  structure and properties, 34
Hemagglutinin, 253, 323
Hemoglobin, 14
  abnormal, 389
  carbamylated, 389
  glycation of, 389, 390
  measurement of pepsin activity, 205
Hemolysate, 392
Henrikson, K., 200
Heparin, 165
  coupling to agarose, 166
Hepatocytes, 352
Heterobifunctional cross-linking reagents, 100, 234
Heterogeneous immunoassays, 399
Hewlett, G., 356
Hexanediamine, 234
Hexosamine, 166
High performance liquid affinity chromatography. See HPLAC
Hinge region, 218, 226, 235, 373, 375
Histidine, 179, 206
  immobilized, 354, 355
Hixson, H.F., 167
Hjerten, S., 16
HNO_3, 43, 240
Hoffman, W.L., 110
Hofmann, K., 170

Hofmeister series, 93, 183, 309
Holleman, W., 167
Holloway, P. W., 350
Homocysteine, 151
  immobilized, 215
Homogeneous immunoassays, 399
Honda, T., 167
Hormone, 188, 253
Hormone receptors, 253
Horse serum, 307
Hoyer, L. W., 69
HPLAC, 310, 394
  analytical applications, 395
  in analytical determinations, 396
  differential elution, 397
  matrices, 395
  weak affinity chromatography, 397
HPLC, 33, 36, 40, 252, 313, 318, 339, 361, 362, 370, 381, 382, 394
  use with polymeric supports, 15
HSA, 195, 228, 328, 342
Hutchens, T.W., 274, 340
Hydrazide, 18, 43, 49, 77, 78, 110, 132, 140, 149, 237, 270, 287, 288, 331
  activation and coupling chemistry, 110
  assay of using BCA, 287
  coupling of antibody, 228, 230
  coupling of oxidized RNA, 272
  coupling to oxidized antibody, 229
  ligand coupling protocol, 113
  reaction with TNBS, 286
Hydrazine, 18, 43, 149
Hydrazone, 77, 110, 237, 270
Hydrolases, 174
Hydrophobic adsorption, 232
  on polystyrene, 42
Hydrophobic interactions, 305, 306, 308, 309, 340
  influence of buffer constituents, 184
Hydrophobic interaction chromatography, 183, 346, 353
Hydroxyazobenzoic acid. See HABA
2-Hydroxyethyl methacrylate, 34
N-Hydroxy succinimide ester. See NHS ester
N-Hydroxysulfosuccinimide. See sulfo-NHS
Hydroxyl reactive chemistries, 118–119
Hydroxyls, 395
Hypoglycemia, 390

IgA, 220, 246
  interaction with jacalin, 254, 325, 326
  interaction with protein B, 251
  structure, 221
IgD, 220, 246
  interaction with jacalin, 254, 326
IgE, 220
IgG, 91, 174, 220, 223, 230, 332, 335, 336, 341, 373, 375, 377
  coupling to polystyrene beads, 236
  fragmentation, 372
  fragmentation by papain, 218, 377, 379
  fragmentation with bromelain, 379
  fragments, 204, 205, 206, 379
  interaction with protein A, 245
  interaction with protein G, 246
  mouse
    measurement using immobilized protein A, 396
    subclass fractionation, 336
  purification by thiophilic adsorption, 340
  purification of, 331
  purification using immobilized protein A, 245
  subclasses, 334, 335, 341
IgM, 210, 220, 244, 246, 304, 336, 339, 362, 366, 376, 377
  binding to mannan binding protein (MBP), 250
  fragmentation by pepsin, 376
  fragmentation of, 362, 375
  fragments, 204, 205, 363
  human versus mouse, 363, 365
  purification of, 336, 337
  structure, 221
  trypsin digestion, 363
^{125}I labeled, 360
Imagawa, M., 100
Imidazole, 291
Imidazolyl carbamate, 64
Imidocarbonates, 54
2-Iminobiotin, 170
  immobilized, 319
3,3'-Iminobispropylamine. See DADPA
Iminodiacetic acid, 41, 118, 180, 181
  immobilized, 343, 344, 360
2-Iminothiolane, 107, 109
Imminium salt, 115
Immobilized ligands
  analytical applications, 388–416
  in biosensors, 410–416
  catalysis and modification applications, 360–387
  density of, 313
  immobilization of, 137–279
  immunoassay applications, 398–410
  purification applications, 317–345
  scavenging applications, 346–360
Immobilized protein
  assay of using BCA, 284
  assay using Coomassie Blue G-250, 290
  detection of using Coomassie Blue, 290
  measurement by difference, 291
Immune complex, 398
Immunoadsorbent, 402
Immunoaffinity, 15, 42, 109, 217, 220, 224, 244, 306, 311, 317, 328, 329
  elution conditions, 308
  gentle elution, 333
Immunoaffinity supports
  preparation of, 210
Immunoassays, 48, 169, 198, 209, 231, 233, 234, 251, 307, 388, 398
  blocking agents, 307
  ELIFA, 407
  flow-through affinity systems, 49, 398–410
  heterogeneous, 399
  using polystyrene balls, 41
Immunoconcentration, 406, 408
Immunoconjugates, 202, 407
Immunocytochemical staining, 169, 198
Immunogenic, 213
Immunogens, 345
Immunoglobulin, 41, 197, 214, 234, 247, 275, 304, 305, 331, 333, 341, 373, 396, 400. *See also* Antibody, IgA, IgE, IgD, IgG, and IgM
  adsorption on thiophilic matrices, 341
  binding proteins, 210, 247, 251, 304, 331
  immobilization of, 244
  binding to thiophilic supports, 274
  binding to immobilized peptides, 212
  classes of, 220
  coupling using protein A/DMP, 224
  enzymatic digestion of, 364
  F(ab')$_2$, 109, 205, 218, 235, 363, 372, 373, 374, 375, 377
    coupling to TNB-thiol matrices, 109
    coupling through sulfhydryls, 230
    coupling via iodoacetyl, 226

reduction with mercaptoethylamine, 226
Fab, 205, 206, 218, 362, 363, 364, 373, 375, 377, 379
  interaction with protein G, 246
Fab*, 376
Fc, 205, 218, 224, 226, 246, 247, 251, 331, 336, 237, 245, 363, 373, 374, 376, 377, 379
  binding protein affinities, 249
  carbohydrate, 221
  degredation by pepsin, 218
  interaction with protein G, 246
  protein A interaction, 245
  receptors, 377
Fc', 376
Fc5µ, 362, 363, 364, 376
Fv, 205, 363, 375, 376, 377
fragments, 202, 204, 205, 362, 372
  uses of, 362
glycosylation, 221
hinge region, 218, 226, 235, 373, 375
IgG structure, 218
immobilization methods, 218, 331
interaction with jacalin, 325
interaction with protein A, 245
interaction with protein A/G, 248
interaction with protein G, 246
isotype elution buffers, 334
J chain, 221
measurement of using HPLAC, 396
purification by Fc binding proteins, 244
purification by thiophilic interaction chromatography, 274
purification o,f 341
reduction of, 244, 226
site directed immobilization, 226
Immunography, 401, 403, 406
Immunohistochemical, 373
Immunoreactivity, 379
Immunotoxin, 324
Incubation
  of sample, 311
Indium, 180
Inhibitors, 381
  protease, 358, 359, 360, 371
Inman, J. K., 120, 149, 160, 286
Insulin, 277, 388, 390
  reduction of, 385

Insulin B chain, 361
  oxidized, 361, 369
Inukai, N, 57
Iodination, 41
Iodine, 105
Iodoacetamide, 188, 378
Iodoacetic acid, 99
Iodoacetyl, 98, 112, 140, 214, 230, 331
  activation protocol, 98
  coupling of reduced antibody, 228
  coupling of reduced antibody or fragments by, 226
  coupling to reduced antibody, 230
Iodobeads, 41
Ion exchange effects, 52
Ion exchange interactions, 306, 317, 347, 389
Iron, 180
Iron chelate chromatography, 343
Iron oxide, 25
Ishi, S., 368
Isocyanate, 43
Isopropanol, 59
Isothiocyanate, 43
Isotype elution buffers, 334
Isourea, 55, 146
Issekutz, A. C., 354
Ito, N., 167

J chain, 221
Jacalin, 252, 254, 321, 326
  coupling protocol, 255
  purification of, 325
  specificity, 254
Jack fruit seeds
  extraction of, 254, 325, 326
Jany, K. D., 167
Jayabaskaran, C., 107
Jellum, E., 277
Ji, T. H., 132
Johnson, B. J. B., 273
Johnson, D. A., 233, 307
Jsselmuiden, O. E., 406

Kabayashi, H., 172
Kagedal, L., 179
Kamicker, B. J., 115, 116
Katz, E. Y., 60, 413

Kawasaki, T., 250, 337
Keen, C. L., 179
Ketoamine, 390
Ketones, 69, 110, 115, 127
KI$_3$, 105
kinases, 174
Klenk, D. K., 389
Kohllaw, G., 308
Kohn, J., 54
Kozutsumi, Y., 338
Kronvall, G., 246
Kunitz, M., 358

Lactose, 161, 324
  coupling by epoxy, 161
  immobilized, 324
Lamoyi, E., 373
Lane, M. D., 200, 210, 233, 307
Langone, J. ,210
Larsson, A., 394, 396
Latex particles, 42
Leashes. *See* Spacers
Lectins, 116, 161, 163, 166, 252, 254, 312, 327, 328, 397
  common ones used in affinity systems, 252
  immobilization protocols, 252
  lectin affinity chromatography, 252
  melibiose bindin,g 158
  purification of, 321
  purification using polyacrylamide supports, 16
Lee, R. T., 115, 116
Leflar, C. C., 406
Lehtonen, O. -P., 233
Leucine, 205, 372
Liener, I. E., 321
Ligand. *See also* Immobilized ligands
  coupling of, 137–279
  measurement by difference analysis, 291
  density of, 4, 183, 281, 313
  measurement by direct absorbance scan, 291
  measurement of, 281
  flexibility and binding orientation, 139
  leaching, 52, 55, 65
  radiolabel analysis of, 75
  small, 137
Light chains, 218, 220, 377

Lihme, A., 274, 340
Lindberg, U., 258
Lindmark, R., 245
Linear flow rate, 310. *See also* Flow rate
Linker. *See* Spacer
Lipemic sera, 302
Lipid A
  interaction with polymyxin B, 354
Lipidex 1000, 351
Lipids, 302, 312, 345, 352, 412
  extraction from seeds, 322
  removal of, 351, 352
Lipoamide
  immobilized, 279, 382, 384, 385
Lipoic acid, 279
Lipopolysaccharide, 345, 354
  removal of, 353
Lipoproteins, 174, 343, 352
  removal of, 351
Liposomes, 412
Lis, H., 253
Litman, R. M., 257
Little, J. N., 310
Lonnerdal, B., 179
Lowe, C. R., 81, 138, 281, 310, 313
Lyotropic agent, 93
Lysine, 150, 151, 169, 198, 206, 211, 222, 361, 365, 390
  trypsin cleavage at, 203
Lysozyme, 277
  reduction of, 386

α_2-Macroglobulin, 174, 224, 344, 345
  carrier-fixed, 355
  immunoaffinity purification, 329
  purification of, 344
Magnesium, 309
Magnetic beads, 25
Magnogel AcA, 26
Majumder, H. K., 273
Male, K. B., 167
Maleic anhydride, 361
Maleimide, 214, 216, 234, 235, 244
  activation and coupling chemistry, 100
  activation protocol, 103
  ligand coupling protocol, 103
Mallia, A. K., 389
Maltose, 161
Manganese, 180

Manil, L., 210
Mannan
  immobilized, 338
Mannan-binding protein. *See* MBP
Mannich condensation reaction, 44, 126, 132, 292
  alternative to diazonium coupling, 123
  coupling chemistry, 127
  coupling of drugs, 186
  coupling of estradiol, 190
  coupling of ethynyl steroids, 194
  coupling of phenol red, 176
  coupling of THC, 193
  coupling of thymol blue, 177
  ligand coupling protocol, 131
  peptide coupling, 216
Mannose, 164, 250, 337
  coupling by divinylsulfone, 164
March, S. C., 56
Mares-Gula, M., 167
Marihuana, 193
Markwardt, F., 167
Matrix
  activation of, 51–136
  agarose, 6–9. *See also* Sepharose
  azlactone beads, 28–31
  capacity, 3
  cellulose, 9–11
  chemical stability, 2
  column packing, 295, 301
  controlled pore glass, 11–14
  definition, 1
  degassing, 301
  Eupergit, 37–39
  fines, 3
  flow characteristics, 3
  freeze dried, 102
  functional groups, 2
  general considerations, 1–6
  HEMA, 34–37
  ligand density, 4. *See also* Ligand, density of
  magnetic, 26
  measurement of gel volume, 22
  mechanical stability, 2
  membranes, 45–50
  nonporous, 15
  nonspecific effects, 4
  particle size distribution, 3
  polyacrylamide beads, 16–19
  polymeric, 15
  polystyrene, 41–45
  pore size, 4
  Poros, 39–41
  Sephacryl, 22–25
  silica, 11–14
  surface area, 4
  synthetic supports, 15
  Toyopearl, 31–34
  Trisacryl, 19–22
  Ultrogel AcA, 25–28
  user friendly, 5
Matsumae, H., 355
Matsumoto, I., 115, 116
Matsuoka H., 413
MBP, 250, 304, 337, 338, 339
  binding properties to IgM, 250
  coupling via CNBr, 251
  immobilized, 337
  purification of, 338
McCarthy, B. J., 258
McCutchan, T. F., 272
McWherter, 382
Means, G. E., 115
Melibiose, 160, 164, 326
  coupling to polyacrylamide, 158
  immobilized, 326, 328
Membranes, 45–50, 307, 406, 409, 412
  chemical composition of, 47
  immobilization chemistry, 49
  immobilization yields, 49
  in ELIFA, 406
  pore structure, 46
  structure and properties, 45
  use in ELIFA, 406
Mercaptans, 275
2-Mercaptoethanol, 156, 165, 274, 275, 276, 345, 382
2-Mercaptoethylamine, 156, 226, 230, 244
2-Mercaptopyridine, 105
Mercuripapain, 378
Mercury, 180
Merkle, R.K., 252, 253
Metal chelate affinity chromatography, 179, 309, 329, 344, 360
Metal chelators ,179
Metal ions, 309
  in binding buffers, 305

Metals, 345
　immobilized, 179
Methacrylamide, 37
Methacrylate, 34, 395
Methacrylate derivatives, 31
Methanesulfonic acid, 240
Methotrexate, 186
α-Methyl alanine, 90
Methyl a-D-mannopyranoside, 254
1-Methyl-2-pyridone, 79
N,N'-Methylene bisacrylamide, 16, 23, 28, 90
N,N'-Methylene bismethacrylamide, 37
$MgCl_2$, 392, 404
　presence in binding buffers, 306
Microtiter plates, 231, 307, 399, 406, 407, 409
　assay of coupled protein using BCA, 284
　derivatization using photoreactive cross-linkers, 135
Milenic, 379
Minicolumns, 109, 125, 289, 296, 300, 314, 320, 329, 331, 335, 343, 348, 354, 356, 358, 363, 368, 371, 376, 383, 384, 388, 392, 401, 403, 404
　packing and use, 295
　use in chromatography, 297
　use in coupling reactions, 297
　use in immunoassays, 402
Miron, T., 58, 60
Modification reagents
　immobilized, 360
Monoclonal antibodies, 218, 332, 379, 397
Monomers
　composition of acrylamide, 19
　composition of azlactone beads, 28, 90
　composition of Eupergit, 37
　composition of HEMA, 34
　composition of Sephacryl, 22
　composition of Toyopearl, 31
　composition of Trisacryl, 19
　cross-linking, 19
　functional, 15
Monomeric avidin, 199
　preparation of, 200
Moore, S., 273
Mosbach, K., 16, 85, 360, 394
Moss, L. G., 268
Mouse IgG . See also Immunoglobulin, IgG
　measurement using immobilized protein A HPLAC, 397

Mouse IgG1, 341. See also Immunoglobulin, IgG
　purification of, 340
Mucopolysaccharide, 165

$Na_2S_2O_4$, 240
NAD^+, 174, 176
$NADP^+$, 174, 176
$NaIO_4$. See Periodate
$NaNO_2$, 126
Naphthalene, 174
NaSCN, 183, 342
Nethery, A., 336
Neurath, A. R., 240
Neurotensin, 371
Nevens, J., 244, 250, 337
Ngo, T. T., 79, 133
NHS-Activated gold electrodes, 59
NHS Ester, 52, 57, 65, 100, 132, 133, 198, 234, 241, 243, 413, 414, 415
　activation and coupling protocol, 57
　hydrolysis rate, 60
　ligand coupling protocol, 60
　aqueous coupling, 63
　nonaqueous coupling, 62
　trans-esterification with methanol, 59
Nickel, 180
Nicolson, G.L., 325
Nilsson, K., 85, 394
Ninhydrin, 160, 282, 283, 284
　use in measuring activation level, 282
Nishikawa, A. H., 167
Nisonoff, A., 226, 373
Nitration
　of polystyrene, 43
Nitrene, 132
4-Nitrobenzoic acid hydrazide, 262
Nitrocellulose, 408, 409, 410
p-Nitrophenol, 57, 290
o-Nitrophenyl β-D-galactopyranoside (ONPG), 209
p-Nitrophenyl phosphate, 404
Nonenzymatic glycation, 390
Nonspecific binding, 5, 42, 65, 224, 225, 233, 244, 245, 246, 248, 294, 305, 306, 313, 348, 395, 403, 415
　blocking of, 233, 307
　by Fc, 373
　hydrophobic effects, 139

ionic effects, 139
  prevention by optimizing binding buffer, 307
  spacer arm contribution, 139
Nopper, B., 274, 340, 341
NP-40, 232
Nucleases, 174, 202
Nucleic acids
  coupling via CNBr, 258
  immobilization protocols, 255
Nucleotide homopolymers, 270
Nylon, 47, 257, 412

O'Carra, P., 139, 144
O'Shannessy, D. J., 110, 132
Octylamine, 184
Ohlson, S., 310, 397
Olfactory organelle, 412
Oligo(dT), 265
  coupling via carbodiimide, 263, 264
Oligonucleotides
  immobilization protocols, 255
Oligosaccharides
  bisoxirane coupling, 118
Olin, B., 179, 181
ONPG, 210
  measurement of galactosidase activity, 209
Organic sulfonyl chlorides. *See* Tosyl chloride or tresyl chloride
Orientation
  of ligand, 139
  site directed immobilization, 140
Oscilloscopes, 412
Oxazolone, 28, 90
Oxidants
  immobilized, 360
Oxirane, 37, 118, 153, 188

Pace, B,. 273
Palmer, J. L., 226
Papain, 206, 207, 372, 377, 379
  digestion of IgG, 377
  IgG cleavage, 218
  immobilized, 377, 379
Parikh, I., 190
Particle size, 400
Particle size distribution, 3

Passive adsorption, 42, 49, 232, 233, 236, 239, 408
  deficiencies of, 43
  to polystyrene, 231
Patchornik, A., 277, 382
Paul, R., 64
PBS
  formulation of, 306
Peak broadening, 397
Peak resolution, 300, 394
Pentaerythritol dimethacrylate, 31
Pepsin, 172, 205, 226, 275
  fragmentation of IgG, 218, 373
  fragmentation of IgM, 375
  immobilized, 372, 373, 375, 376
  preparation of immunoglobulin fragments, 205
Pepstatin A, 172, 173
Peptide, 210, 360
  coupling through amines or carboxylate functions, 211, 214
  coupling to DADPA spacer, 213
  coupling via active hydrogens, 216
  coupling via diazonium, 216
  coupling via sulfhydryls, 214
  digests, 204
  mapping, 207
  reduction of, 383
Perfusion chromatography, 39
Periodate, 11, 21, 33, 36, 70, 75, 76, 142, 151, 170, 196, 200, 204, 205, 207, 208, 209, 214, 223, 238, 245, 246, 247, 249, 252, 253, 254, 260, 270, 272, 354, 369, 395, 403
  oxidation of antibody carbohydrates, 228, 230
  oxidation of glycidol modified supports, 76
  oxidation of polysaccharide supports, 69, 110
  oxidation of RNA, 270
Peristaltic pumps, 300, 301
Persson, T., 258
pH
  affect on binding buffer, 304
Pharmacia BIAcore system, 415
Phenol red, 128, 176, 178, 292, 293
Phenol-sulfuric acid, 161
Phenols, 127
Phenyl azide, 50, 132, 134, 242
Phenylalanine, 150, 205, 369, 370, 372

Phosphoamino acids, 343
Phosphodiester, 264
Phosphoethanolamine
    immobilization of, 403
    immobilized, 403, 404
Phospholipids
    removal of, 351
O-Phospho-L-serine, 343
Phosphoproteins, 180
    purification of using iron chelate
        chromatography, 343
Photoactivatable cross-linking reagents, 132, 134
    heterobifunctional, 132
    ligand coupling protocol, 135
    use in coupling to polystyrene microplates, 241
Photochemical reactions, 49, 241
Photolysis, 132, 134
α-Picolines, 127
Pineapple stem, 379
Pivalic acid, 34
Plasma, 403, 404
Plasma kallikrein, 358
Plasmin, 358
Plasminogen activators, 167, 317, 359
Plastic, 399
Plaut, A.G., 364
Pneumococcal somatic C polysaccharide, 403
Poe, M., 186
Poly(A)
    coupling via CNBr, 259
Poly(dA)
    coupling via epoxy, 270
Poly(dC)
    coupling via epoxy, 270
Poly(dG)
    coupling via epoxy, 270
Poly(dT)
    coupling via epoxy, 270
Poly-L-lysine, 150, 151
Polyacrylamide beads, 16–19, 25, 27, 77, 148, 412
    advantages, 16
    aminoethylamide derivatives, 18, 116, 147, 149, 150
    boronate derivative, 273
    coupling of melibiose, 158
    disadvantages, 16
    entrapment of DNA, 257
    exclusion limits, 16

flow rates, 18
glutaraldehyde activation, 18
handling, 18
immobilization methods, 18
mechanical stability, 18
structure and properties, 16
Polyacrylonitrile, 47
Polyamide, 47
Polyclonal, 223, 332
Polyester, 47
Polyethylene, 19, 47
Polyethylene glycol, 31
Polyethyleneimine, 2, 41
Polymer
    coating of microplates, 239
Polymerases, 174
Polymerization
    free radical, 240
    using vinyl monomers, 242
Polymyxin B
    immobilization of, 354
Polypropylene, 47
    minicolumns, 295
Polysaccharides, 164, 253
Polystyrene, 5, 41–45, 135, 232, 251, 281, 307
    activation by photoreactive cross-linkers, 135
    activation of microtiter plates, 44
    activation using chlorosulfonic acid, 43
    alkylamine beads, coupling to using Sulfo-SMCC, 234
    antibody coupling, 231
    assay of coupled protein using BCA, 284
    balls or beads, 231, 232, 234, 399
    covalent attachment, 233
    structure and properties, 41
    carboxylated, 64
    chromatographic matrices, 39
    coupling to, 233
    coupling to hydrazide beads through
        oxidized carbohydrates, 237
    coupling to using photochemical reagents, 241
    covalent modification of, 43
    graft copolymer derivatives of, 44, 242
    hydrazide beads
        coupling to using glutaraldehyde, 236
    immobilization methods, 42
    microplates, 42
        covalent attachment, 239

minicolumns, 295
nitration, 43
nonspecific effects, 306
Poros matrices, 39–41
Polystyrene/divinylbenzene, 39, 41
Polysulfone, 47
Polyvinylalcohol, 412
Polyvinylchloride, 412
Polyvinylidene difluoride, 47
Polyvinylpyrrolidone, 327
Poonian, M. S., 258, 263
Porath, J., 88, 104, 105, 118, 164, 179, 180, 181, 274, 340, 343
Pores, 4
  diffusive, 39
  pore size distribution, 394
  throughpores, 39
Poros, 39–41
  chemical stability, 40
  flow rates, 311
  handling, 41
  immobilization methods, 41
  mechanical stability, 40
  particle size, 40
  structure and properties, 39
Potentiometric, 410
Potuzak, H., 255, 267
Pregnancy test devices, 406
Primary antibody, 399, 409
Process scale chromatography, 300, 311
Procion Blue MX-3G and MX-R, 174
Procion Brilliant Blue MX-R, 175
Procion Red HE-3B, 176
Proline, 375, 380
Protease inhibitors, 360
Proteases, 202, 302, 312, 361
  removal of, 355, 358, 360
α1-Proteinase inhibitor, 174
Protein
  assay of using BCA, 284
  assay of using Coomassie Blue, 290
Protein A, 41, 49, 91, 224, 225, 246, 247, 248, 305, 314, 331, 334, 335
  affinity matrix capacities, 245
  avidin conjugate, 319
  binding buffer pH, 304
  binding buffer salt strength, 305
  DMP immobilization of antibody, 224
  high salt binding buffer formula, 306
  immobilization of, 244

immobilized, 335, 374, 375, 379
  use in HPLAC, 396
  immunoglobulin binding affinities, 248
  interaction with immunoglobulin subclasses, 245, 334
  interaction with mouse and human immunoglobulins, 245
  recombinant, 245
Protein A/G, 248, 331
  binding buffer pH, 304
  coupling protocol, 249
  immunoglobulin binding affinities, 248
Protein B, 251
  coupling via reductive amination, 252
Protein denaturants, 308
Protein G, 247, 248, 331
  binding buffer pH, 304
  immobilization of, 246
  immunoglobulin binding affinities, 246, 248
  interaction with Fab, 246
  interaction with immunoglobulin subclasses, 246
  recombinant, 246
Proteinases, 329
Proteins
  immobilization of, 195
  reduction of, 383, 386
Pryridyl disulfide, 214
Psoralen, 44
  grafted to polystyrene plates, 243
Purification, 317
Pyridine, 265, 268
Pyridyl groups, 132
Pyridyl disulfide, 103
  activation protocol, 104
  ligand coupling protocol, 105
Pyrogens, 346
  removal of, 354
Pyruvate carboxylase, 169

Quarles, R. H., 132
Quinaldines, 127

Radioimmunoassays. *See* RIA
Radioimmunolocalization, 379
Radioisotope, 399

Radiolabel, 52, 75, 398
Rao, Y. S., 90
RCA I, 323, 325
RCA II, 323, 325
Receptors, 185, 410
Recycling
   of sample, 311
Red blood cells, 389, 392
Reducing agents, 69, 70, 112, 115, 377. *See also* Sodium borohydride and sodium cyanoborohydride
   activation of papain, 206
   immobilized, 360, 382, 384, 385, 386
Reducing sugar, 116
Reductive amination, 21, 33, 36, 41, 49, 69, 75, 112, 115, 142, 150, 164, 170, 245, 329, 395, 403, 413
   coupling of albumin, 196
   coupling of antibody, 229
   coupling of anti-human serum albumin, 223
   coupling of avidin, 200
   coupling of DADPA, 142
   coupling of melibiose, 158
   coupling of peptides, 214
   coupling of protein B, 252
   coupling of streptavidin, 200
   coupling reducing disaccharides, 116
   ligand coupling protocol, 76
   use of glutaraldehyde, 77
Refractive index, 410
Regnier, F. E., 12
Renin, 172
Renthal, R. P., 253
Resonance angle, 416
Restriction enzymes, 202
Reverse phase, 361, 370
RIA, 41, 399
Ribonuclease A, 275, 277, 368, 368
   oxidized, 362
      digestion of, 370
   RCM, 382
   reduction of, 386
Ricin, 323, 325
Ricinus communis agglutinin, 323, 325
Riddles, P. W., 293
Ring activators and deactivators, 128
RNA, 272
   coupling via diazonium, 259
   coupling via hydrazide, 272
   coupling via periodate oxidation/reductive amination, 270
   immobilization by irradiation, 257
   immobilization protocols, 255
   oxidation by periodate, 270
   single stranded
      coupling via CNBr, 259
Roitt, I., 218
Rosenberg, M., 272, 273
Rousseaux, J., 218
Rudders, R. A., 375
Ruhemann's purple, 282, 283

Sahni, G., 208
Sakakabara, S., 57
Salt gradient, 341
Salting out effect, 309
Sample, 312
   application of, 302
   clarification of, 312
   containing proteases, 312
   effect on capacity, 313
   equilibration with binding buffer, 312
   incubation on affinity support, 311
   removal of competing substances, 312
   removal of lipids, 312
Sandwich assay, 399
Sanger reagent, 43
SAP 381, 382
Scavenging, 346–360
   affinity supports useful in 346
Scheller, F., 410
Schiff's base, 69, 70, 115, 236, 240, 270, 272, 390, 413
Schmer, G., 146
Schneider, C., 224
Schott, H., 273
Schrock, A. K., 132
Schuster, G. B., 132
Schwartz, B. A., 115, 116
Secondary antibody, 400, 409
Seeds
   extraction of, 321
Sensor chip, 415, 416
Separon, 34
Sephacryl, 22–25
   chemical stability, 23
   exclusion limits, 23
   handling, 25

immobilization methods, 23
particle size distribution, 23
structure and properties, 22
Sephadex, 23, 263, 114
   hydroxyalkoxypropyl derivative. *See*
      Lipidex 1000
Sepharose, 7, 9, 55, 58, 79, 86, 87, 88, 97, 108, 118, 142, 143, 153, 154, 161, 163, 164, 166, 170, 176, 177, 181, 182, 184, 188, 194, 196, 199, 200, 201, 204, 205, 207, 208, 209, 249, 251, 252, 253, 254, 258, 259, 260, 274, 325, 329, 336, 338, 369. *See also* agarose
Seppala, I., 334
Sepsis, 403
   CRP concentration in, 405
L-Serine, 343
Serine proteases, 167, 203, 356
Serum, 331, 343, 397, 403, 404
   mouse
      fractionation on immobilized protein A, 335
      myeloma, 336
Shainoff, J. R., 69
Sharon, N., 253
Shaw, E., 167
Shepard, E. G., 132
Sica, V., 141, 150
Silane, 14, 394, 413
Silanol, 12
Silica, 5, 11–14, 119, 351, 394. *See also* Controlled pore glass
   coating with silane, 14, 307, 395
   nonspecific effects, 306
   pH stability, 395
Silicone, 239
Silicone-rubber, 399
Singhal, R. P., 273
Site directed immobilization, 140, 214, 216
   of antibody
      through carbohydrate, 228
      through sulfhydryls, 226
   of peptides, 211
SMCC, 100
Smith, P.K., 284
$SnCl_2$, 43
SNPA, 123, 124
Sodium borohydride, 69, 70, 78, 182, 197, 200, 204, 236, 239, 262, 268, 270, 272
Sodium cyanoborohydride, 69, 70, 77, 79, 115, 116, 143, 152, 197, 200, 204, 206, 207, 208, 209, 236, 239, 246, 247, 249, 252, 254, 255, 262, 270
Sodium deoxycholate, 355
Sodium dithionite, 120, 125, 240, 262
Sodium hydrosulfite, 43, 240, 262
Sodium nitrite, 120, 126, 216, 262
Sodium periodate. *See* periodate
Sodium sulfate, 93, 95, 309
Sorbitol, 390
Sorensen, K., 284
Sorensen, P., 132
Soybean trypsin inhibitor
   immobilized, 355, 358
Spacers, 4, 43, 58, 70, 82, 83, 91, 98, 101, 108, 110, 112, 146, 160, 167, 170, 214, 234, 287
   aminated epoxides, 153
   amino acids, chemical structures of, 151
   aminocaproic acid, 146
   bisoxirane, 118
   blocking or capping excess, 156, 157
   capping with acetic anhydride, 158
   chemical structures of, 140
   DADPA, 141
   detection with TNBS, 287
   diaminohexane, 144
   ethylene diamine, 147
   extension by anhydrides, 153
   hydrophilicity, 139
   ionic effects, 139
   site directed immobilization, 140
   succinic anhydride, 112, 153, 155
   types of, 140
   use of, 137
   use of amino acids, 150
   use with small peptides, 212
Specific activity
   measurement of, 315
Spheron, 34
ssDNA. *See* DNA, single stranded
Staphylococcus aureus, 244, 247
Staphylococcus aureus V8 Protease, 207
   immobilized, 381
Staros, J. V., 82
Statine, 172
Steers, E., 138
Stell, J. G. P., 167
Sterilization
   using NaOH, 21

Steroids, 188
  removal of, 351
Stewart, J. M., 81
Stich, T. M., 284
Streptavidin, 169, 171, 172, 197, 308, 320
  coupling via CNBr, 199
  coupling via reductive amination, 200
  immobilized
    assay of activity, 288
    purification, 319
Streptococcus sp, 247
Strick, N., 240
Styrene-divinylbenzene copolymer, 350
Subclasses, 341, 372
  isolation, 334
  of IgG, 334
Substrate, 400, 401, 402, 404, 405, 406, 407, 409, 410
Succinic anhydride, 112, 153, 155, 287. *See also* succinylation
Succinimidyl-4-(*N*-maleimidomethyl) cyclohexane-1-carboxylate. *See* SMCC
N-Succinimidyl-*p*-nitrophenylacetate. *See* SNPA
Succinylated DADPA, 112
Succinylation, 112
  monitoring with TNBS, 287
  of DADPA, 168
Sugar, 158, 252
  immobilized, 322
  oxidation of, 360
Sulfhydryl, 395
  immobilized, 382, 383, 384, 385, 386
    measurement by Ellman's reagent, 293, 383, 384
  reaction with TNBS, 287
  stabilization of, 306
  sulfhydryl reactive chemistries, 98
Sulfhydryl protease, 206, 377, 379
Sulfo-NHS, 82
Sulfo-NHS ester, 82
Sulfo-SANPAH, 241
Sulfo-SMCC, 44, 101, 103, 234, 244
  use in coupling to alkylamine polystyrene beads, 234
Sulfonates, 85, 395
Sulfonation
  of polystyrene, 43
Sulfone group, 274, 340
Sulfonyl chlorides. *See* tosyl chloride or tresyl chloride

Sulfosuccinimidyl-4-(*N*-maleimidomethyl) cyclohexane-1-carboxylate. *See* Sulfo-SMCC
Sulfosuccinimidyl-6-(4'-azido-2'-nitrophenylamino)hexanoate. *See* Sulfo-SANPAH
Sulkowski, E., 179
5-Sulphido-2-nitrobenzoate, 293
Sundberg, L., 118
Superose, 6, 339
Surface area, 4, 400
Surface plasmon resonance, 415
Synthetases, 174
Synthetic supports, 15–50
  definition, 15
  use in process scale separations, 15

TAME, 203
Target molecule, 3, 4, 132, 186, 281, 294, 304, 306, 313, 314
  purification of, 317
Taylor, R .F., 195, 410
TBS
  formulation of, 306
TED, 181
Tenner, A. J., 336
Ternynck, T., 77
Testosterone, 190
Tetrahydrocannabinol. *See* THC
THC, 193
Then, R. L., 186
Thermoreactive, 134
Thioctic acid, 279
  immobilized, 277
Thioether bonds, 98
Thioglycolic acid, 275
Thiol ester linkage, 81
Thiol-agarose, 109
5-Thio-2-nitrobenzoic acid. *See* TNB
Thiophilic adsorption, 340, 341
  preparation of thiophilic support, 274
Thomas, D,. 413
Thymidine-5'-phosphate, 264
Thymol blue, 176, 178
Tijssen, P., 231, 398
Titanium, 180
TLCK-α-chymotrypsin, 371
TNB, 293
TNB-Thiol, 107, 109, 214

activation and coupling chemistry, 107
ligand coupling protocol, 109
monitoring release of TNB groups, 110
TNBS, 113, 143, 144, 145, 146, 149, 150, 152, 153, 155, 158, 160, 344
qualitative test for amines, sulfhydryls, and hydrazides, 287
α-N-p-Toluene-sulfonyl-L-arginine methyl ester hydrochloride (TAME), 203
p-Toluenesulfonyl chloride. See tosyl chloride
Tomasi, Jr., T. B., 364
Toms, E. J., 197
Tosyl activation, 21, 23, 33, 36, 395
L-(Tosylamido-2-phenyl) ethyl chloromethyl ketone (TPCK), 203
Tosyl chloride, 85
 activation protocol, 86
 ligand coupling protocol, 86
Towbin, H., 233, 307
Toyopearl, 31–34, 76, 146, 148, 246, 247, 395
 chemical stability, 33
 glycidol modification, 33
 handling, 33
 immobilization methods, 33
 mechanical stability, 31
 nonspecific effects, 306
 particle sizes, 33
 structure and properties, 31
TPCK-trypsin, 361, 368
Transamidination, 18, 27, 147, 148, 149, 158
 protocol for modification of polyacrylamide usin,g 149
Transducer, 410, 415
Transfer cannula, 407
Transferases, 174
Traut, R. R., 133
Travis, J., 342
Tresyl chloride, 21, 23, 33, 36, 41, 85, 145, 245, 395
 activation protocol, 87
 coupling of diaminohexane, 144
 coupling of peptides, 214
 ligand coupling protocol, 87
Triazine ring, 97, 174
Trichloro-s-triazine. See cyanuric chloride
Trichlorophenol, 57
Triethanolamine, 226
Triethylamine, 59, 63, 79
2,2,2-Trifluoroethanesulfonyl chloride. See tresyl chloride

2,4,6-Trinitrobenzenesulfonate. See TNBS
Tris buffer, 156, 197, 200, 209, 254
 use in blocking excess amine reactive sites, 158
Tris(carboxylmethyl)ethylenediamine, 181
Tris-glycine, 306
Trisacryl, 5, 19–22, 76, 144, 145, 151, 247, 395
 chemical stability, 21
 exclusion limits, 20
 glycidol modified, 21, 69, 152
 handling, 22
 immobilization methods, 21
 mechanical stability, 20, 22
 structure and properites, 19
 varieties of, 19
Triton X-100, 232, 348, 350
Trypsin, 203, 204, 207, 275, 318, 319, 358, 361, 366, 369, 381
 activity, 204
 digestion of IgM, 363
 immobilized, 361, 363
 purification, 317
 removal of, 356
Tryptophan, 150, 369, 370
TSK-Gel, 5, 31, 395
Tumor imaging, 362, 363
Tween, 20 232, 233, 241, 244, 296, 307
Tyrosine, 129, 150, 180, 205, 206, 216, 369, 370

Ultrogel AcA, 25–28, 77
 chemical stability, 27
 exclusion limits, 25
 handling, 27
 immobilization methods, 27
 particle size distribution, 25
 structure and properties, 25
Uracil, 259
Urea, 309, 352
Urokinase, 167, 317, 359
 removal of, 356
Uy, R., 116, 161

V8 protease, 207, 381
Valeric acid, 169, 198
Valine, 372

Valkirs, G. E., 406
Van Regenmortel, M. H. V., 233, 307
Variable regions, 218
Veerkamp, J. H., 351
Veis, A., 172
Vicinal diols, 70, 76
Viljanen, M. K., 233
Villarejo, M. R., 150
Vinyl monomers, 257
  grafting to polystyrene, 242
  vinyl addition, 413
Vinyldimethyl azlactone, 28, 90
Viruses, 345
Vitamin H, 169, 197
VLDL, 352
Vogt, R. F., 233, 307
Vretblad, P., 163
Vunakis, H. V., 210

Watson, D. H., 176
Weak affinity chromatography, 310, 397
Weetall, H. H., 12
Weinstein, D. B., 351
Weiss, L. J., 167
Weissbach, A., 263
Weith, H. L., 272
Weston, P. D., 77

Wheat germ
  extraction of, 322
Wheat germ lectin, 321
  purification of, 322
Wheat-germ agglutinin, 163
White blood cells, 392
Whiteley, J. M., 186
Wilchek, M., 54, 58, 59, 60, 110, 169, 198, 288, 319
Wintersberger, U., 267
Wold, F., 116, 161

Yamamoto, N., 413
Yoshitake, S., 100
Young, J. D., 81

Z-Asn-Phe-OEt, 380
Zabin, I., 150
Zeolites, 11
Zero-length cross-linkers, 81
Zimmerman, D., 233, 307
Zinc, 180, 181, 309, 360
Zinc chelate affinity chromatography, 344
$ZnCl_2$, 181, 345